INSTRUMENTED IMPACT TESTING OF PLASTICS AND COMPOSITE MATERIALS

A symposium
sponsored by
ASTM Committee D-20
on Plastics
Houston, TX, 11–12 March 1985

ASTM SPECIAL TECHNICAL PUBLICATION 936
Sandra L. Kessler, PPG Industries, Inc.
G. C. Adams, E. I. du Pont de Nemours & Co.
Stephen Burke Driscoll, University of Lowell
Donald R. Ireland, Ireland and Associates
editors

ASTM Publication Code Number (PCN)
04-936000-19

ASTM 1916 Race Street, Philadelphia, PA 19103

Library of Congress Cataloging-in-Publication Data

Instrumented impact testing of plastics and composite
  materials.

  (ASTM special technical publication; 936)
  "ASTM publication code number (PCN) 04-936000-19."
  Includes bibliographies and index.
  1. Plastics—Impact testing—Congresses.   2. Composite
materials—Impact testing—Congresses.   I. Kessler,
Sandra L.; Adams, G. C.; Driscoll, Stephen Burke;
Ireland, Donald R.   II. ASTM Committee D-20 on Plastics.
III. Series.
TA455.P5I495   1987   620.1'9235   86-28692
ISBN 0-8031-0937-7

NOTE

The Society is not responsible, as a body,
for the statements and opinions
. advanced in this publication.

Printed in Baltimore, MD
January 1987

Foreword

The symposium on Instrumented Impact Testing of Plastics and Composite Materials was held 11–12 March 1985 in Houston, Texas. ASTM Committee D-20 on Plastics was sponsor of the event. The symposium chairman was Sandra L. Kessler, PPG Industries, Inc., who also served as editor of this publication. Also serving as editors were G. C. Adams, E. I. du Pont de Nemours and Co., Stephen Burke Driscoll, University of Lowell, and Donald R. Ireland, Ireland and Associates.

Page 119 table

Related
ASTM Publications

Quality Assurance of Polymer Materials and Products, STP 846 (1985), 04-846000-19

Physical Testing of Plastics, STP 736 (1981), 04-736000-19

Wear Tests for Plastics: Selection and Use, STP 701 (1980), 04-701000-19

A Note of Appreciation
to Reviewers

The quality of the papers that appear in this publication reflects not only the obvious efforts of the authors but also the unheralded, though essential, work of the reviewers. On behalf of ASTM we acknowledge with appreciation their dedication to high professional standards and their sacrifice of time and effort.

ASTM Committee on Publications

ASTM Editorial Staff

Contents

INDEXES

Overview

The last two decades have witnessed an explosion in the use of polymeric materials and their composites for structural applications. This has resulted in more demanding acceptance criteria with respect to all mechanical and chemical properties of these materials. In addition to the usual static properties obtained at very low testing speeds, it has become essential to know how polymeric systems behave under dynamic loading conditions, such as high-speed impact. Examples of such applications for these versatile materials are now commonplace: bumper backup beams, leaf springs, oil pan assemblies, car body panels, equipment container cases, gasoline tanks, helicopter blades, radiator shrouds, and other applications. In each instance the success of the material in the application requires knowledge of its mechanical and chemical property profile and, in particular, characterization of its impact performance. This Special Technical Publication has been published as a result of the 1985 symposium on Instrumented Impact Testing of Plastics and Composite Materials, held in Houston, Texas, in an effort to communicate state-of-the-art technology to those actively engaged in these studies. The symposium was the outgrowth of work within ASTM Subcommittee D20.10.02 on Impact and High-Speed Properties, a subcommittee of ASTM Committee D-20 on Plastics.

It is well known to those involved in impact testing that there frequently appear to be more variables than constants available to the engineer. The engineer is required to make the right choices to characterize the material properly for its ultimate use. It is generally true that an impact event can be thought of as the contact of a high-speed projectile [velocity ≥ 203 m (8000 in.)/min] of some specified geometry with a supported test specimen. During this impact event the specimen absorbs and transmits energy that can be expressed in simplest terms as the integral of the force-displacement curve. The shape of the curve provides information on the initiation, yielding, and propagation of energy during the event. The choices for the actual test configuration can be extremely diverse: the specimen may be notched or unnotched, may be supported as a cantilever or flexural beam, and may be a flat plaque, a rectangular bar, or the fabricated component. The striker (or tup) may be driven or free-falling, and the test data may be instrumented or noninstrumented. Consideration of the inertial effects of equipment may or may not be required. The engineer may be more interested in characterizing the impact fatigue life of the material, using low-blow impacts at energies less than that required for failure. Whether the material becomes work-toughened or fails catastrophically after a particular number of impacts can then be determined. This can then provide insight into the useful life of a material under certain expected conditions.

In order to define the proper test configuration for a material and application, the stress states and likely conditions need to be identified. What constitutes failure, the mode of failure, and its likely cause need to be established. This provides the guidance necessary to determine what combination of tests is required and what environmental conditions need to be considered. In the development of materials for new applications, instrumentation of the impact event is essential so that changes in the mechanism of failure can be quantified. Instrumentation provides data on the effective dynamic modulus of the material under the conditions of test, its strain and elongation capacities, and its energy storage capability. If no instrumentation is applied, a single number is obtained, which confounds these property values and does not provide the engineer with information for modifying the material and measuring exactly how that modification has affected the impact trace or force-displacement curve. In addition, instrumentation allows the engineer to define conditions of failure that do not require destruction of the component. For instance, a performance standard for a high-speed gear assembly could limit the strain to less than 0.5% at a ram energy of 11.3 N · m (100 in. · lb), although the part may actually yield or break only at a much higher energy level.

The collection of 19 papers published in this volume has been grouped into five major categories. Some papers could be placed in more than one category, and here an arbitrary selection has been made. These categories are methodology, impact testing for end-use applications, impact characterization of selected materials, partial impact testing and fatigue response of plastics, and fracture toughness.

Methodology for Impact Testing

The papers in the section on methodology are written with varying levels of technical depth, which provides those relatively new to the technology with specific guidelines for preparing the system for data collection, as well as evaluation of the data collected. In at least one paper the information is generic in nature and is independent of the specific type of data system, test geometry, and material tested. The differences between drop-weight machines and servohydraulic systems are also discussed. In this section the methodology for selecting impact tests applicable for automotive composites and interpreting their data is presented in detail. The approach that has been taken would be suitable for any application and consists of four basic steps: establishing the functional requirements for an application (a composite fender, for example), listing the stress states that could occur for the range of impact conditions for each functional requirement, determining the controlling variables for each stress state, and listing the failure limits for each functional requirement. It is often the case that several impact configurations and test conditions are required to test a material for a given end use. Also included in this section is an analysis of the stability and reproducibility of a driven ram impact tester and recommendations on improving the reliability of the data. The issue of changing velocity of the impact ram and its implications on the test data collected have been addressed. Digital filtering has also been reviewed in this section, with guidance provided as to its valid use and its misuse, which leads to anomalous data and incorrect analysis.

Impact Testing for End-Use Applications

The section on impact testing for end-use applications is, by its title, of a more applied nature and demonstrates the techniques used for various end uses. Those specifically reviewed in this section are impact measurements on low-pressure thermoplastic foam, material impact characteristics in the use of cushioning systems, a detailed survey of ten test methods for characterizing materials for automotive components, and impact testing for a variety of products such as tires, reinforced thermosetting pipe, boat hulls, and baseball helmets. In this section the effects of strain rate and temperature on the relative brittleness/ductility of materials is discussed, as well as the influence of the thickness and cross-sectional uniformity of the material. In the case of the survey of ten impact tests, conclusions regarding the relative discrimination powers of the tests, in comparison with each other, and their correlation or lack of correlation with each other are presented. The paper on cushioning systems presents a technique for quantifying the damping capacity of materials by the use of instrumentation of the impact event, which can discriminate between recoverable, elastic deformation and permanent, nonrecoverable deformation.

Impact Characterization of Selected Materials

The third section, on impact characterization of selected materials, covers more of the fundamental, research-oriented characterization of materials. As described in the paper on polyether sulfone, an effort was made to effect material failure by machining a central hole 1 mm in diameter in the flat plates, which were subsequently impacted by a falling dart. This work was done since the material would not fail under no-notch conditions. This preliminary work pointed to a brittle-ductile transition that was influenced by the presence of the machined hole, but only within a defined thickness range. Another interesting study was reported on the influence of test rate and temperature on the fracture behavior of rubber-modified acrylonitrile-butadiene-styrene (ABS) and polyvinyl chloride (PVC), with particular emphasis on the location of the

brittle-ductile transition. The crazing mechanism due to the presence of the
rubber modifiers was demonstrated to be responsible for the toughening capacity of these materials. Varying levels of rubber modifier were also considered.

The remaining papers address high-performance composites reinforced
with aramid and graphite fibers. In the paper on new composite materials for
aerospace applications, the influence of new thermosetting resins on the impact resistance of graphite composites was evaluated. The authors found that
the through-penetration or puncture test provided the majority of impact response data, but it was, by itself, insufficient to describe the conditions that
would be encountered in service. Studies using a number of impact energy
levels were recommended for better characterization of incipient damage,
augmented by the use of ultrasonics. The paper involving impact testing of
aramid composites compares the impact damage tolerance of these fibrous
composites to those reinforced with carbon and glass fibers. The fact that
aramid fibers are efficient energy absorbers with a level of recoverable deformation during impact was demonstrated by the use of instrumented impact
testing of flat plates, honeycomb aerospace panels, and filament-wound pressure vessels.

Partial Impact Testing and Fatigue Response of Plastics

There has been a growing interest in the behavior of materials under impact conditions that are within the initiation phase of the force-displacement
curve, prior to maximum load. In the section on partial impact testing and
impact fatigue, three papers address different aspects of this subject. The
paper on fatigue studies the use of low and constant impact energy as a
method for providing toughness measurements on polymers. The paper concludes that crystalline polymers appear to have better fatigue performance
than amorphous polymers, and that there seems to be a different energy
absorption process occurring in multiple impact tests than in single-blow impact tests. The point is made that the increasing use of polymeric materials in
hinges, gears, springs, and automated arms has made fatigue performance a
growing concern. The fundamental difference between the fatigue curves of
brittle and ductile materials is identified. A means for quantifying the absorbed energy as the area within the closed fatigue loop (force-displacement)
is described. Two papers on the subject of incipient crack formation and the
impact response at varying depths of penetration show that, at least for the
materials studied, the impact trace taken to less than maximum load mirrors
traces taken to complete failure. This provides technical justification for using this technique to identify the mechanism of incipient failure of materials.
By using shims on a falling dart tester the authors could raise the test specimen height in order to control the distance the tup travels into the specimen.
The impact characteristics at crack formation could then be analyzed. Com-

puter simulation of the impact event was explored to determine whether the use of wave mechanics was needed to model specimen deformation and failure.

Fracture Toughness

The last section, on fracture toughness, includes two papers. One of these uses laser-Doppler techniques for velocity measurement in order to characterize the impact behavior of materials. This eliminates the complications of the ringing of a transducer attached to the tup. The results of the paper indicate that this is technically feasible, particularly when the data are stored and analyzed via microcomputer. The other paper investigates the applicability of linear elastic fracture mechanics for treating the fracture of polymers under dynamic loading. Acetal and polymethyl methacrylate (PMMA) were the materials of choice, the first being crystalline and the second amorphous. This approach proved fruitful, with the plane-strain fracture toughness observed to be relatively constant with increasing crack-length-to-specimen-width ratios, except at the highest loading rate. In addition, it was found that fracture toughness was significantly influenced by loading rate, with transitions observed for both polymers. An explanation for these transitions was postulated.

The papers briefly outlined here should provide the reader with much of the very latest information in the area of instrumented impact testing. Virtually all possible combinations of test conditions have been addressed within this volume, as well as types of material and equipment. The symposium committee gratefully acknowledges the efforts of the authors and ASTM personnel that have made this publication possible.

Sandra L. Kessler

PPG Industries, Inc., Shelby, NC 28150; symposium chairman and editor.

Methodology for Impact Testing

Matthew C. Cheresh[1] *and Steven McMichael*[1]

Instrumented Impact Test Data Interpretation

REFERENCE: Cheresh, M. C. and McMichael, S., **"Instrumented Impact Test Data Interpretation,"** *Instrumented Impact Testing of Plastics and Composite Materials, ASTM STP 936,* S. L. Kessler, G. C. Adams, S. B. Driscoll, and D. R. Ireland, Eds., American Society for Testing and Materials, Philadelphia, 1987, pp. 9–23.

ABSTRACT: This paper provides a guide for those becoming acquainted with instrumented impact testing and faced with acquiring accurate data and subsequently evaluating it. The discussion deals with (1) preparation of the system for data collection and (2) evaluating the data collected. The system preparation discussion covers the determination of the major data system parameters and gives examples of the most commonly occurring data collection problems. Procedures are also presented for determining the acceptability of the data produced, and guidelines are proposed for utilizing data to answer experimenters' questions.

KEY WORDS: impact testing, data collection, data resolution, data filtering, inertial effects, harmonic oscillations, data utilization, energy absorption

The industrial use of instrumented impact testing is presently an "emerging" technology rather than a well-defined and understood test method. As such, instrumented impact testing equipment is often introduced into a company or laboratory where there are no personnel with experience in impact test data evaluation or equipment operation. The problem is compounded by a lack of reference materials available in the open literature.

This paper provides a guide for those becoming acquainted with instrumented impact testing and faced with acquiring accurate data and subsequently evaluating it. The discussion is separated into two parts: (1) the preparation of the system for data collection and (2) the evaluation of the data collected. The system preparation discussion covers the determination of the major data system parameters and gives examples of the most commonly occurring data collection problems. Procedures are also presented for determin-

---

[1]Members of the technical staff, Dynatup Products, General Research Corp., Santa Barbara, CA 93111.

ing the acceptability of the data produced, and guidelines are proposed for utilizing data to answer experimenters' questions.

The information presented here is generally independent of the type of data system, test geometry, and material being tested. Several sections, however, are specifically related to drop-weight testing machines rather than servohydraulic systems, because of differences in the methods used for calculating deflection and energy.

Determining Data Collection Parameters

The parameters to be determined and set prior to performing a test include the test time and load ranges, the method of triggering, the impactor weight, the amplifier gain, and the degree of filtering. The following sections describe each of these.

Time and Load Ranges

The time and load range settings act as "window" into which the data must fit (Fig. 1). This window must be large enough to capture all the data required, yet small enough to provide good resolution. In general, most modern instrumented impact data systems provide very high resolution in time, load,

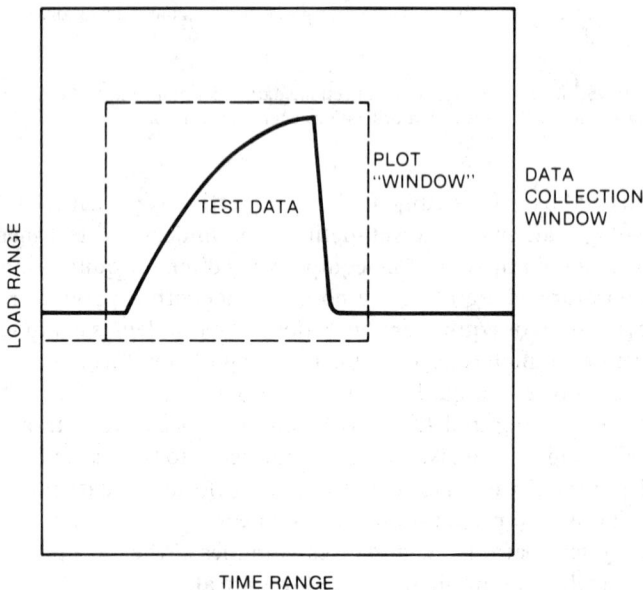

FIG. 1—Data collection and plot windows. Most digital data systems allow rescaling of data for presentation after collection and storage.

and deflection. These computerized systems also allow the test data to be easily rescaled. The following steps are a simple guide for determining a suitable time range setting for a specimen with unknown behavior:

1. Consider the specimen to be tested and estimate the deflection that will be required to obtain complete fracture.
2. Convert deflection to time by the following equation

$$t = \frac{d}{v}$$

where t = time, d = the expected deflection to complete failure, and v = the impact velocity. Be sure to use a consistent set of units.
3. Increase the time by a factor of two for safety. An even larger safety margin (such as four or five) may be appropriate if the test is a "low-blow" test, in which the impactor will bounce off the specimen.
4. Remember that the test time setting must be reevaluated each time the impact velocity or test specimen material thickness is changed.

To determine the load range for the test do the following:

1. Estimate the highest expected load. This can be done in a variety of ways, from pure experience and judgment to performing simple calculations or even static testing prior to the impact test series.
2. Multiply by two for safety.

An example of a test performed with incorrect test time and load range is shown in Fig. 2. In this case, the test time was too short and the load range too low. The only way to "save" data taken under these conditions is to extrapolate the missing portions of the curve manually, although extrapolated data must be identified as such when reported. Data from similar specimens will provide a guide for extrapolations.

Once a test has been performed and the entire test curve obtained, the data can then be evaluated to determine the resolution. The time resolution is simply the total time range selected divided by the number of points collected by the recorder.

$$r = \frac{t}{n}$$

where

r = recorder resolution with units of time per point,
t = test time, and
n = the number of points collected by the recorder.

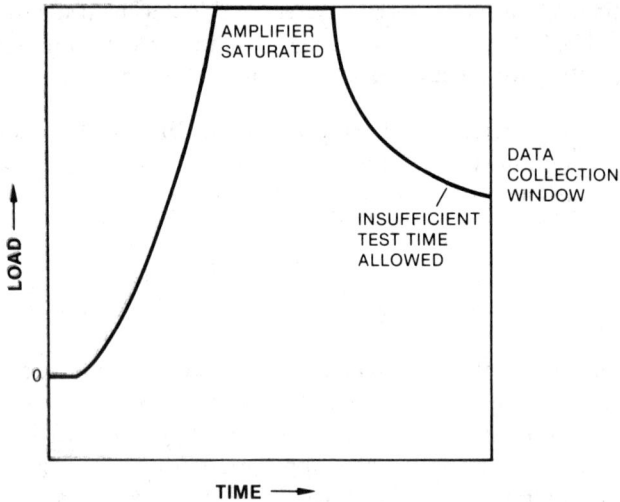

FIG. 2—*Examples of incorrect data collection parameters. The load range and test time are insufficient.*

Next, the number of points defining the test must be calculated by

$$p = \frac{t_t}{r}$$

where

p = the number of points actually defining the test event,
t_t = the total time required for the test, and
r = the recorder resolution.

While the acceptable number of points required to define a curve accurately depends largely on the type of behavior exhibited by the specimen and the detail required, a rule of thumb is to have at least 200 points defining a test plot.

Plots for specimens with ductile behavior and a relatively monotonic load increase and decrease (polycarbonates, for example) require fewer points than specimens that have sudden or discontinuous failures (such as graphite/epoxy composites). Often a data system will have a cursor that can be moved one point at a time along a data curve. This can be a useful diagnostic tool to graphically determine the resolution of test data.

Load resolution (the number of discrete values the converter can select) is determined by the analog/digital (A/D) converter and the range over which these "steps" are spread. For example, a typical system might have a twelve-bit A/D converter which yields 4096 discrete steps. These steps are usually spread from the positive full-scale load range setting to negative one eighth (or a similar percentage) of the full-scale load range (Fig. 3). With this resolution, a very small penalty is paid for a high load range value when testing unknown materials. If a 44 500-N (10 000-lb) load range is selected, the load resolution will be

$$R = \frac{44\ 500\ \text{N} + 1/8(44\ 500\ \text{N})}{4096} = 12.24\ \text{N}\ (2.75\ \text{lb})$$

Thus, if the specimen tested actually fractures at a load of only 445 N (100 lb), or 1% of the selected range, the resolution will still be better than 3%!

Triggering Method

Most instrumented impact test systems have two selectable methods of triggering data collection, which are analogous to internal and external triggers on a digital oscilloscope. The former uses a rise in the load signal above a preset threshold value to trigger data collection, and the latter generally uses the passage of a flag attached to the impactor through a light beam/photodetector to provide a trigger signal. Although both methods will provide identical data, and little reason exists for preferring one over the other, several precautions are in order when using each method.

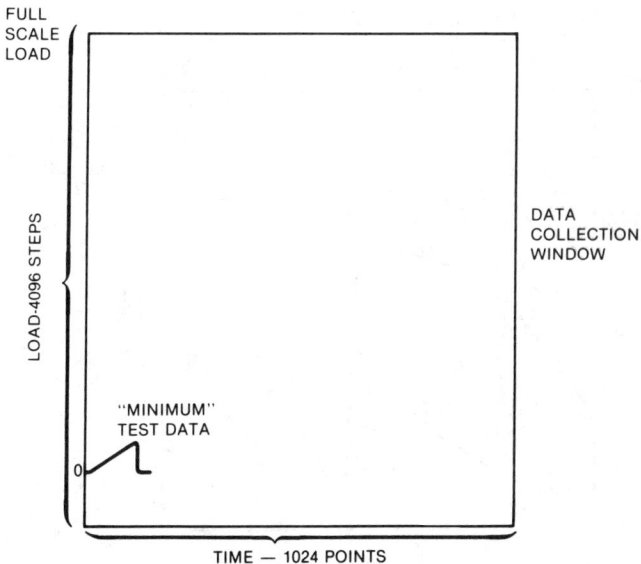

FIG. 3—*A too-large data collection window. Although the resolution for this test may be poor, these data indicate the correct settings for future tests.*

It is advisable, when using a signal (or internal) trigger, that the trigger level be set at roughly 10% of the expected maximum load to ensure that the trigger level is well above any background noise. It is often possible to set the trigger resolution lower than the A/D resolution described previously. Also, an adequate number of data points collected prior to the trigger signal must be saved. As a rule of thumb, save 10% of the points as "pretrigger" information.

If a load is already applied to the tup at the time data collection starts, automated scaling and data analysis routines can produce erroneous results. Computerized scaling routines often assume that the values of the first few data points collected are zero and will analyze the data based on that assumption. Photodetector triggers must, therefore, be set to allow an adequate "zero baseline" to be collected before the tup strikes the specimen. An example of a "late" trigger signal is shown in Fig. 4. The data can be corrected by offsetting each data point upward by a constant number of newtons (pounds) and recalculating the energy absorbed.

Filtering

Many data collection systems incorporate analog filters to reduce "noise" introduced by specimen and tup vibrations and to eliminate high-frequency noise from the computer and other external sources. Although these filters can almost always improve the "readability" of test data, their use should be restricted to situations in which the source of the removed noise is known and the effect on the data is understood.

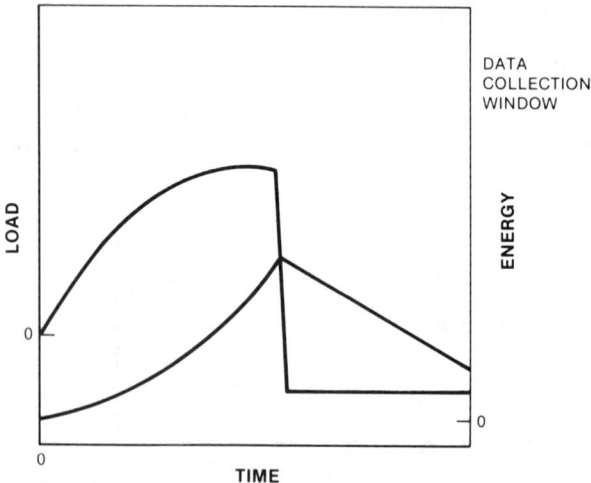

FIG. 4—*Example of a "late" trigger signal. The entire curve must be offset upward and the initial portion of the curve extrapolated to obtain accurate data.*

In laboratories where many different materials and specimen geometries are tested, it is advisable to use digital filtering methods (after saving the original, unfiltered data) rather than use analog front-end filters. In this way the filtered data can be compared with the original data, allowing the effect of the filtering to be easily seen. If the resultant data are judged to be overfiltered, the original can be recalled and a lesser degree of filtering used.

Data Evaluation

Above all, data evaluation depends on understanding the physical processes involved in impact testing. First, two extraneous physical phenomena that influence the data should be understood. These are inertial loads and harmonic oscillations. These dynamic effects are typically particular to the test setup and may obscure the actual material response. The goal, then, is to obtain data free of these effects and which reflect the material response alone.

Inertial Loads

The inertial load is simply the load required to accelerate the specimen from zero velocity up to the velocity of the tup. Inertial loads are most often characterized by a sharp spike (often followed by a decaying oscillation) at the beginning of the curve (Fig. 5). If this inertial load is high relative to the true mechanical load, inaccurate data may be collected or reported. An operator or automated data analysis routine may incorrectly select the maximum load (and possibly failure point values) if the inertial peak is the highest load value

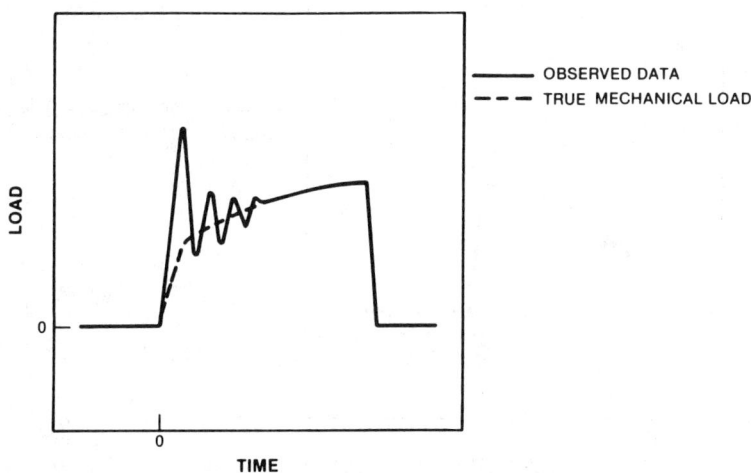

FIG. 5—*Example of high inertial load. The sharp spike at* t = 0 *is caused by load required to accelerate the specimen to tup velocity.*

recorded. In severe cases (high-impact velocity tests on heavy specimens with low mechanical strength), it is even possible for the inertial loads to obscure the desired data completely.

A simple diagnostic test to determine whether a given spike is caused by mechanical specimen response or by inertial loads is to repeat the suspect test using a different (usually lower) impact velocity. The magnitude of an inertial load is essentially proportional to the impact velocity [1] since $F = ma$, where F is the force applied by the specimen to the drop weight, m is the drop mass, and a is the acceleration of the drop mass. Therefore, if the data in question are caused by inertial loads, a lower impact velocity (and hence a lower acceleration) should reduce this load by a proportional amount. The mechanical response of materials is usually not nearly so strain-rate sensitive.

One classic test for inertial loads used in Charpy testing of steels with low ductility is to fracture the specimen, then tape the specimen back together and retest it. The mechanical loads should surely change between the tests, yet the inertial loads should not change.

Harmonic Oscillations

During an impact, the components involved (specimen and tup) react to the impact by oscillating at their natural frequencies. These oscillations are often detected and recorded by the instrumentation. If the oscillation amplitude is small relative to the signal produced by the material response, no problem is generally encountered (Fig. 6). If the amplitude of the oscillation is significant relative to the amplitude of the test loads (Fig. 7), then problems

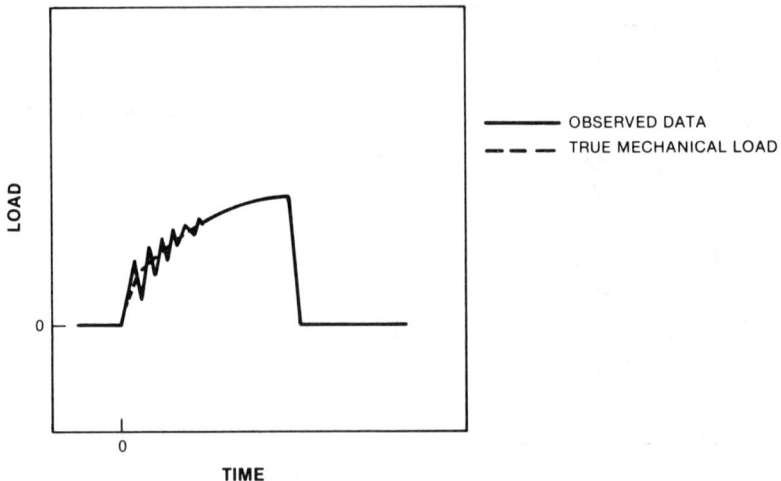

FIG. 6—Low-amplitude, high-frequency oscillations. No smoothing is required to obtain desired data.

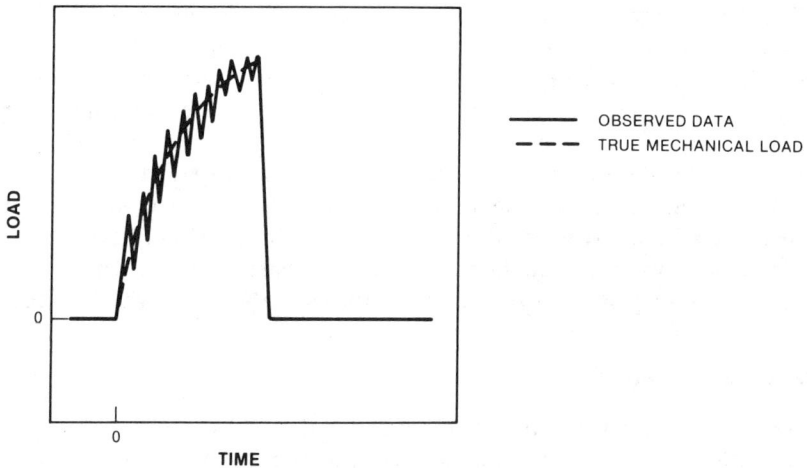

FIG. 7—*High-amplitude, high-frequency oscillations. Smoothing will produce accurate data because of good curve definition provided by the many cycles.*

in data analysis may result. The first step in separating the mechanical loads from these oscillations is to determine the source and frequency of the oscillations (often called "ringing").

Distinguishing between mechanical and oscillatory loads requires that the frequency of oscillation be compared to the natural frequencies of the tup. To determine the natural frequency of the tup, perform a test on a relatively strong but brittle specimen. The tup will continue to oscillate at its natural frequency after the specimen has fractured. Plotting the data collected after specimen fracture against time should allow the tup oscillation frequency to be determined. To determine directly if the oscillations are caused by the natural frequency of the specimen itself, perform several tests on specimens of the same material, varying only the thickness or other parameter (such as unsupported span) that is known to influence the natural frequency of the specimen.

Once the frequency and source of the oscillations are known, the effect on the signal can be estimated. Since the oscillations are harmonic about the mean (true signal) value, if sufficient cycles occur prior to the onset of fracture, then the energy values should be accurate, and only the maximum load data will be potentially incorrect. Previous literature [2] suggests procedures for determining the number of oscillation cycles "necessary" before the fracture to give confidence in the accuracy of the data.

Several techniques are available for reducing the effect of ringing. The first is, again, to reduce the impact velocity. The amplitude of the ringing is also proportional to the impact velocity, and a lower velocity will reduce the amplitude of the ringing (as well as lengthen the time of the test, allowing more time for the oscillations to decay). A layer of tape or other elastomer will also effectively decrease the ringing by providing a "dampener" between the tup and specimen. The energy absorption of the tape may, however, also have to be considered.

Filtering the data is another method for reducing the effect of ringing on the data. Digital filtering is again recommended, and the techniques for this have already been discussed. However, if the amplitude is significant and the period similar to that of the signal (Fig. 8), it becomes very difficult to separate material response information from the oscillation even with filtering.

Utilization of Test Data

The steps in using instrumented impact test data to solve real problems are essentially the same as with other mechanical tests. In general, instrumented impact test data are used in two ways: (1) parametric evaluation of a material for the purpose of establishing a generalized data base and (2) determination of material, design, or component suitability for a particular application.

Parametric Material Evaluation

Use of data for the parametric material evaluation is relatively straightforward, and the particular data points and features selected are generally understood. Instrumented impact test standards and industry material specifications based on instrumented impact test data are still few in number, and

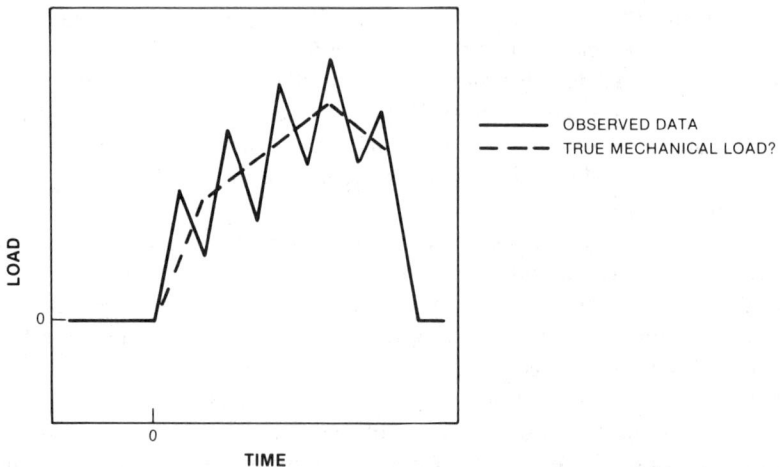

FIG. 8—*High-amplitude, low-frequency oscillations. An insufficient number of cycles are present prior to failure to define the curve features accurately even if they are smoothed.*

do not deal with data analysis and interpretation in specific terms. These standards, such as ISO Draft Specification 6603/2 and, the ASTM Test for High-Speed Puncture Properties of Rigid Plastics (D 3763-79), do, however, provide standard specimen sizes, test geometries, and, most important, test methods. Many other impact test standards, originally written as uninstrumented standards (ASTM Test for Impact Resistance of Plastics and Electrical Insulating Materials [D 256-81], for example), are commonly used as guides for instrumented testing and data collection.

The data generally recorded for homogeneous materials include maximum load, energy to maximum load, total energy, and energy to (the onset of) failure (if failure does not occur at maximum load). The data recorded for nonhomogeneous (composite) materials also often add load and energy at the first sign of damage to the list [3].

Suitability for an Application

The second use of instrumented impact data is more complicated and requires consideration of the relationship between the test and the actual service application. This relationship is normally established by comparing the service environment with the test environment and then equating the service failure criteria to the test failure criteria. The relationships that must be established include material, processing, specimen/test geometry, and impact velocity and energy. Several examples follow.

Applications such as automobile windshields have the primary requirement of energy absorption—for example, the important impact test criterion is that a foreign object cannot penetrate. In this case, a plaque specimen impacted by a simulated shape of foreign object (hemispherical or sharp, perhaps) would be a reasonable approximation of the actual service geometry. The total energy absorbed during a test up to the point of penetration is likely the most useful data.

In contrast, the windscreens used in jet fighter applications not only are required to resist penetration during a prescribed impact, they cannot deflect more than a certain distance during the event. Again, plaque specimens are often used, but equating the response of these to the highly complex shapes of actual windscreens may be difficult. In this case, the conventional instrumented impact test may be best used as a screening tool until a comparison is made with tests on specimens with a more accurate shape representation. In tests on the windscreen materials, energy to failure and dynamic compliance are critical data.

For liquid detergent bottles made of an acrylonitrile-butadiene-styrene (ABS) material, the presence of a leak may be considered failure. Flat plaques may again be good specimens, but actual bottles, sectioned for mounting, will include processing variables such as parting lines that may greatly affect impact performance. For this testing application, the load and energy at the first point of failure are used for material comparisons (Fig. 9). Also, since the actual damage sequence of concern is actually an energy-limited event (a full bottle falling from a given height), tests using a limited amount of energy, just enough to produce failure in all specimens, may be the best simulation.

Initially, subcritical impact damage to an aerospace structural composite material may propagate by fatigue and lead to failure. Test geometries can be selected to simulate the threats seen in service including rock damage, tools being dropped, or footsteps on "no-step" areas, and the first point of damage to the material (either matrix or fiber) may be the most relevant data point (Fig. 10).

The failure criteria for a composite battery case material may also be the first sign of cracking (Fig. 11). In this case, the load or energy at that point is most useful.

The failure of a drill motor dropped off a ladder is generally an energy-limited event, similar to the case of the liquid detergent bottle. The failure criterion is concerned with whether or not the motor case can withstand a given energy input. In this case again, flat plaque testing followed by sectioned molded component testing may be most useful, with the first point of fracture selected as the failure criterion.

An automotive component may experience only a limited load because of strength limitations in the path through which the load is transmitted—for example, the failure of another component. For these applications, the maximum load observed during the test may be the most valid parameter.

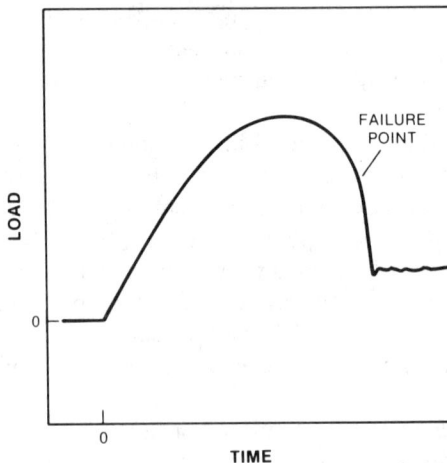

FIG. 9—Sample data plot for through-penetration test on an ABS disk specimen. The failure point indicated was determined by tests performed at energy levels above and below the suspected failure point.

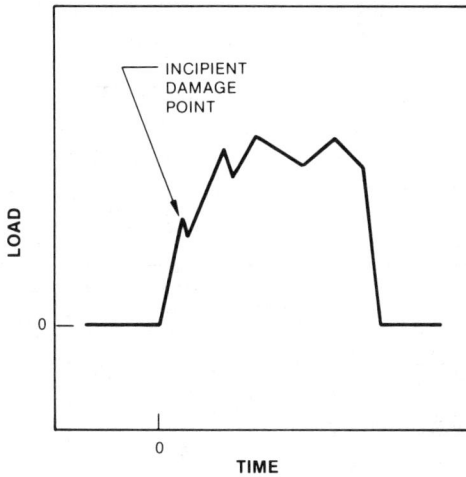

FIG. 10—*Example of a test record showing subcritical damage. Fatigue may cause extension of damage and ultimate component failure.*

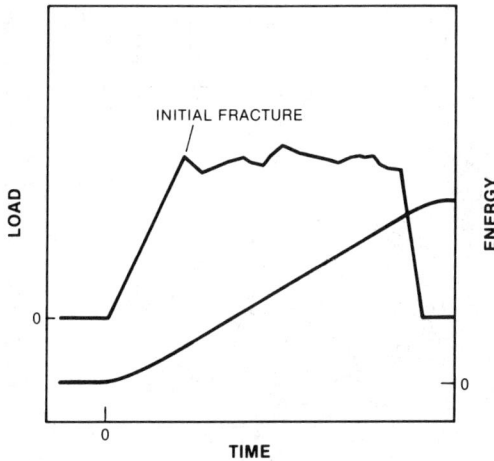

FIG. 11—*Test data from relatively brittle polycarbonate. The first indication of cracking may be the most accurate failure criterion for the intended leak-tight application.*

It should be noted that the relationship between the tests being performed and the application under consideration should be examined and understood *before* any tests are conducted. This process is often overlooked, resulting in wasted time and material, or even worse, important conclusions being made based on truly irrelevant data.

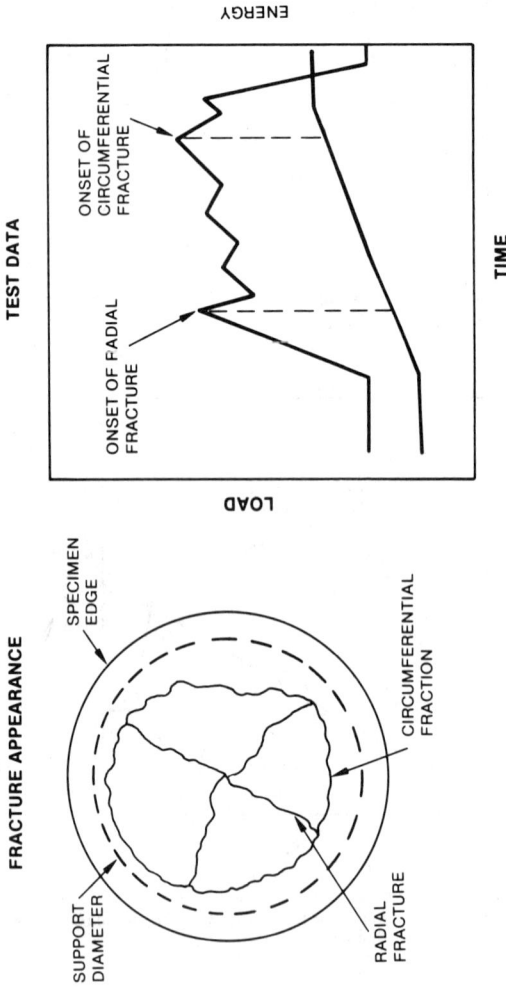

FIG. 12—A comparison of the failure progression in a brittle polycarbonate specimen compared with the data plot. The total energy is heavily dependent on the fixture size.

Test-Specific Effects

A final caution is that test-specific energy absorbers be recognized and "subtracted" when evaluating data. An example of these is fracture arrested by the fixture itself. This can distort valid energy absorption values. For example, larger fixtures will have higher energy absorption, as is shown by the fracture pattern in Fig. 12.

Conclusions

Instrumented impact testing provides both an opportunity to gain a great deal more information concerning impact performance than was previously available through uninstrumented tests. However, this extra information requires a more educated user to be used correctly. The examples and notes contained in this paper are based on the observances of newcomers to the world of instrumented impact testing. It is hoped that the explanations presented will help experimenters to obtain valid results quickly using less material, fewer specimens, less time, and lower cost than was possible with their uninstrumented impact test equipment.

References

[1] Saxton, H. J., Ireland, D. R., and Server, W. L., "Analysis and Control of Inertial Effects During Instrumented Impact Testing," in *Instrumented Impact Testing, ASTM STP 563,* American Society for Testing and Materials, Philadelphia, 1974, pp. 50–73.
[2] Ireland, D. R., "Procedures and Problems Associated with Reliable Control of the Instrumented Impact Test," Technical Report 73-25R, General Research Corp., Santa Barbara, CA, 1974.
[3] Herrick, J. W., "Multi-Directional Advanced Composites for Improved Damage Tolerance," *Proceedings,* Composites in Manufacturing Third Annual Conference, Society of Manufacturing Engineers, Los Angeles, CA, 1984, Paper EM84-104.

James L. Roche[1] and S. Norm Kakarala[1]

Methodology for Selecting Impact Tests of Composite Materials in Automotive Applications

REFERENCE: Roche, J. L. and Kakarala, S. N., **"Methodology for Selecting Impact Tests of Composite Materials in Automotive Applications,"** *Instrumented Impact Testing of Plastics and Composite Materials, ASTM STP 936,* S. L. Kessler, G. C. Adams, S. B. Driscoll, and D. R. Ireland, Eds., American Society for Testing and Materials, Philadelphia, 1987, pp. 24–43.

ABSTRACT: The concept of a "good impact test" has two aspects: the quality of the test results and the applicability of the test results. The quality of output from various impact tests is the subject of a companion paper. This paper presents a methodology for selection and interpretation of impact tests for automotive composites. Proper selection and interpretation can provide good correlation between test results and performance in application.

A necessary condition for obtaining correlation between test results and service performance is that the key impact parameters in an end-use impact event are identified and duplicated in the test method. The key impact parameters—stress state, controlling variable, and failure limit—are defined and developed from the special characteristics of composite materials.

The starting point of the methodology is a set of simple statements of the impact-related functional requirements of the application. A procedure is presented for systematically transforming these statements into a characterization of the application in terms of key impact parameters. Test methods that match this characterization are specified.

As an example, the methodology is applied to an automobile fender. A set of test methods is specified. Characterizations of six additional automotive applications are summarized, and implications for future test method development are discussed.

KEY WORDS: composite materials, reinforced plastics, impact behavior, impact functional requirements, impact controlling variables, key impact parameters, impact stress state, impact failure limit, impact performance criteria, impact testing

There are two aspects to the concept of a "good impact test." From the viewpoint of the laboratory technologist, a good impact test provides clear

[1]Senior project engineer and supervisor, respectively, Advanced Engineering Staff, General Motors Technical Center, Warren, MI 48090-9040.

differentiation in ratings for a wide variety of materials. From the viewpoint of the application engineer, a good impact test provides material ratings that correlate closely with in-service product performance. The first view is concerned with the quality of test the output [1]; the second with the applicability, or usability, of the test output.

This paper addresses the application engineer's need for a good impact test. A methodology is presented for selecting and interpreting impact tests for composites used in the automobile industry. Proper selection and interpretation will provide good correlation between test results and performance in application. Variation in material rating by different test methods indicates material performance variation with changes in application conditions.

There is no predictive model of failure that can be applied to composites in general. The application of fracture mechanics to composites has had very limited success. The models that have been successful apply only to narrow classes of materials in narrow classes of applications, and are generally appropriate only for aerospace components.

Therefore, with composites it is necessary that an impact test correspond more closely to the actual impact event. In fact, since any component will be exposed to several different types of impact events, a set of impact tests may be required to predict a composite material's performance in a particular application.

Methodology Overview

Scope

The methodology is completely general with respect to material. The test configurations and data interpretation procedures specified are appropriate for evaluation and quantitative comparison of any polymeric materials from high-glass thermosets to unreinforced thermoplastics. This is an important positive feature, because often materials from widely different classes are competing candidates for an application. The methodology even allows metals to be evaluated and compared directly with alternative composite materials.

Key Impact Parameters

Many parameters and variables are relevant to a complete description of an impact event. A practical impact test will, by design, neglect the majority of these variables. For example, an impact test configuration will consist of a specimen, a specimen support, and an impactor, all of simplified and idealized geometry. This idealization of the impact event does not necessarily make the test invalid as a predictor of impact behavior in application. The critical requirement is that the key impact parameters in an impact event be identified and duplicated in the impact test method.

Key impact parameters are those characteristics of an impact event that have the greatest influence on the response of the impacted object. The impact response of an object can abruptly change from strong to weak or from tough to brittle when a key impact parameter is varied. These are the features of the impact event which must be faithfully duplicated in the design of the impact test method. Parameters and variables that have no significant effect or that have a smooth, consistent, and predictable effect on the response of the impacted object need not be duplicated by the test method.

Impact test methods can also be characterized in terms of key impact parameters. The characterization of an application can then be matched to a master table of impact tests. Any automobile component can be characterized according to key impact parameters related to the material and the application. This characterization becomes the substance of an impact test method specification.

The pertinent key impact parameters are listed in Table 1. Also given are the major characteristics of composite materials to which they relate. Each will be discussed in detail later.

Procedure

The starting point of the methodology is a set of simple statements, in sentence form, of the impact-related functional requirements of the application. Information is extracted from these word statements and, together with considerations regarding the application and geometry, is systematically transformed into a characterization of the application in terms of key impact parameters.

Impact Stress States

The inherent structure of a composite material is highly anisotropic. As a result, localized stresses and strains and molecular energy absorption mechanism responses to impact loading will be highly sensitive to location and ori-

TABLE 1—*Application-determined key impact parameters.*

Key Impact Parameter	Related Characteristic of Composite Materials
Stress state	anisotropic structure
Controlling variable	large deflections
Failure limit	complex failure

entation. This dependence is of critical importance in determining the material's response to impact, but there is no quantitative model that describes this dependence. Therefore, the loading geometry, or stress state, of an impact application is a key impact parameter and must be identified and duplicated in the test method configuration.

Definitions of Possible Stress States

A total of 13 impact stress states have been identified which might occur in automotive applications. They are defined in Table 2.

Referring to Table 2, pure uniaxial tension does not occur, but the configuration shown for uniaxial tension (Stress State 1) comes close if the span is long enough. The Variation 2 of the uniaxial bending configuration (Stress State 2) might occur in some facia applications.

For the ratio stress states (Stress States 3, 4, and 6 in Table 2) material stiffness is very important. In biaxial bending/tension (Stress State 4) the stress state at failure for a fixed configuration can vary from pure plate bending for sheet molding compound (SMC) to pure tension in a puncture failure for unreinforced reaction injection molding (RIM).

Tearing (Stress State 13) is included because of its practical importance. For convenience, it is listed as an individual stress state.

Factors that Determine Stress State

Each statement of an impact functional requirement will specify one or several load distributions as well as the load direction. The functional requirement will also imply a range of impactor sizes.

Unsupported area and boundary conditions will be determined by taking into account details of the application geometry and the range of possible impact locations for each functional requirement.

The range of materials under consideration will determine the part stiffness.

To determine the stress states for a given application, first study the geometry and the boundary conditions of the component. Consider one functional requirement at a time, and list the stress states that can occur for the range of impactors and for the range of possible impact locations. Also consider the possibility and effects of changing boundary conditions as the impact event progresses.

Stress States in a Fender Example

Figure 1 illustrates the possible stress states in a fender example. An important concept is illustrated in Case 3. The cross-section view shows a steel rail behind the composite fender. If the impactor possesses sufficient energy, the

TABLE 2—*Impact stress states in automotive applications of composite materials.*

1. UNIAXIAL TENSION   • RELATIVELY LONG NARROW IMPACTOR   • PART CONSTRAINED IN PLANE, OUTSIDE OF THE IMPACTOR ZONE, BY ATTACHMENTS OR STIFFENERS INHERENT TO THE APPLICATION GEOMETRY   • RELATIVELY LONG SPAN

2. UNIAXIAL BENDING
 	
   1. VARIATION 1
     • RELATIVELY LONG NARROW IMPACTOR
     • PART UNCONSTRAINED IN PLANE, OUTSIDE OF THE IMPACT ZONE
     • RELATIVELY SHORT SPAN

2. VARIATION   2

3. UNIAXIAL BENDING/TENSION   • SAME IMPACTOR GEOMETRY AS 1 AND 2
       • LIMITED CONSTRAINT OF PART IN PLANE, OUTSIDE OF IMPACT ZONE
       • RATIO IS A FUNCTION OF
         • SPAN LENGTH
         • DEGREE OF CONSTRAINT
         • MATERIAL STIFFNESS

4. BIAXIAL BENDING/TENSION   • IMPACTOR ASPECT RATIO ≅ 1
       • IMPACTOR SMALL RELATIVE TO UNSUPPORTED AREA SURROUNDING IMPACT ZONE
       • RATIO IS A FUNCTION OF
         • SPAN RATIO
         • PART CONSTRAINT
         • MATERIAL STIFFNESS

fender will deflect far enough to come in contact with the rail. The steel rail then becomes a secondary support, and the relevant stress states are completely different. One important point is that both sets of stress states must be considered. Another major point is that the critical factor in this impact event may be the ability of the component to deform sufficiently to engage the secondary support. Thus, the controlling variable may be deflection rather than energy. This is a second key impact parameter, which will be addressed in the next section.

TABLE 2—*Continued.*

5. NORMAL SHEAR • POINT OF IMPACT IS VERY NEAR TO EDGE OF SUPPORT, EITHER AN EXTERNAL PART SUPPORT OR A STIFFENER WHICH IS A FEATURE OF THE PART GEOMETRY

6. NORMAL SHEAR/BENDING
 	
• SAME AS 5, EXCEPT THAT POINT OF IMPACT IS SOMEWHAT REMOTE FROM SUPPORT
• RATIO IS A FUNCTION OF
   • DISTANCE FROM SUPPORT
   • MATERIAL STIFFNESS

7. NORMAL COMPRESSION
8. IN-PLANE BENDING
9. IN-PLANE COMPRESSION

Impact Controlling Variables

Some composite materials are very flexible and will undergo large deflections with little or no damage during an impact event. It is possible that a composite component will simply deflect out of the path of the impactor. A component may deflect and come in contact with a secondary support, which will absorb the bulk of the impact energy. In such cases the important vari-

TABLE 2—*Continued.*

10. IN-PLANE SHEAR

* POINT OF IMPACT VERY NEAR TO EDGE OF
  SUPPORT, EITHER AN EXTERNAL PART SUPPORT OR
  A STIFFENER WHICH IS A FEATURE OF THE PART
  GEOMETRY

11. IN—PLANE SHEAR / CANTILEVER BENDING

* SAME AS 10 ABOVE, EXCEPT THAT POINT OF IMPACT
  IS SOMEWHAT REMOTE FROM SUPPORT
* RATIO IS A FUNCTION OF
  * DISTANCE FROM SUPPORT
  * MATERIAL STIFFNESS

**12. BUCKLING (~ UNIAXIAL
BENDING/COMPRESSION)**

* SAME GEOMETRY AS 9
* OCCURENCE IS DEPENDENT ON
  * PANEL UNSUPPORTED ASPECT RATIO
  * MATERIAL STIFFNESS

**13. TEARING (~ UNIAXIAL TENSION WITH
STRESS CONCENTRATION)**

**13. VARIATION
(~ IN-PLANE BENDING WITH
STRESS CONCENTRATION)**

able to monitor is the deflection that occurs during impact, not the energy absorbed.

In the case of a compliant support such as a bumper or a steel mounting rail in a collision, the bulk of the impact energy is absorbed by the support deflection. The requirements that must be met by the composite component are to match the deflection of the support and, in the case of direct loading, to bear the force of impact as intermediary between the impactor and the com-

CASE 1.
HANDLING AND ASSEMBLY

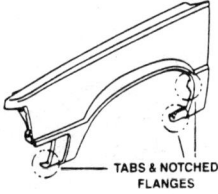

TABS & NOTCHED
FLANGES

2. UNIAXIAL BENDING

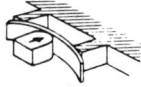

13. TEARING

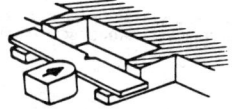

CASE 2. STONES AND ROAD DEBRIS

4. BIAXIAL BENDING/TENSION

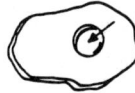

CASE 3.
SIDE IMPACT

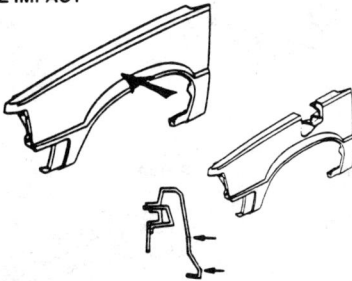

4. BIAXIAL BENDING/TENSION

3. UNIAXIAL BENDING/TENSION

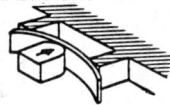

8. IN-PLANE BENDING

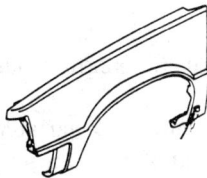

FIG. 1—*Stress states in the fender example.*

CASE 3.
SIDE IMPACT, CONT.

*CHANGE IN STRESS STATE IF DEFLECTION IS
SUFFICIENT TO ENGAGE SECONDARY SUPPORT

7. NORMAL COMPRESSION

5. NORMAL SHEAR

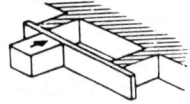

4. BIAXIAL BENDING/TENSION

CASE 4.
FRONT IMPACT

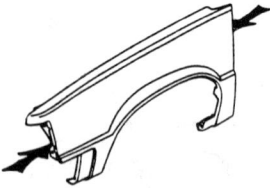

10. IN-PLANE SHEAR

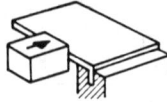

12. BUCKLING

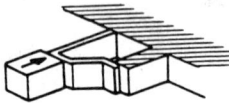

13. TEARING

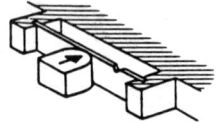

FIG. 1—*Continued.*

pliant support. Thus, the important variable to monitor would be the deflection, or the force, or both.

Application conditions determine the controlling variable for an impact event. It is easy to visualize that a material which ranks high in ability to absorb energy during an impact might rank low in ability to absorb deflection. For this reason the controlling variable of·an impact event is a key impact parameter and must be identified and monitored in the impact test.

Table 3 indicates how the controlling variables are determined from the application conditions.

Impact Failure Limit

The most commonly used impact tests, such as the Izod test, measure only one quantity—that is, the total energy absorbed by the specimen during the impact event. The failure of composites is complex and progressive. The point of interest during an impact event will depend on the performance criterion from the functional requirement statement. It may be different from the total—the end of the impact event. In the previous section, the idea was presented that the significant variable in an impact event, the controlling variable, may be deflection or force rather than energy. The third application-related key impact parameter, the failure limit, is the subject of this section.

Figure 2 shows three types of composite material response records from an instrumented impact test. The first material is brittle. It undergoes a linear elastic deformation and then shatters. The second material is ductile. After an elastic deformation, the material passes through a yield point and undergoes plastic deformation before it reaches the breaking point. Both of these material behaviors are analogous to metal behaviors. The term "yield" is

TABLE 3—*Impact-controlling variables as determined from application conditions.*

Application Conditions	Stress State/Support Stability	Impact Controlling Variable
Low energy	no change in stress state and no significant deflection of support	energy
Moderate energy— no secondary support	no change in stress state and no significant deflection of support	energy
Moderate energy— secondary support	change in stress state and no significant deflection of support	energy and/or deflection
Moderate energy— compliant support	significant deflection of support	force and/or deflection
Collision—direct loading	significant deflection of support	force and/or deflection
Collision—indirect loading	significant deflection of support	deflection
Collision—toward passenger compartment	. . .	energy
Deflection out of path of impactor	. . .	deflection

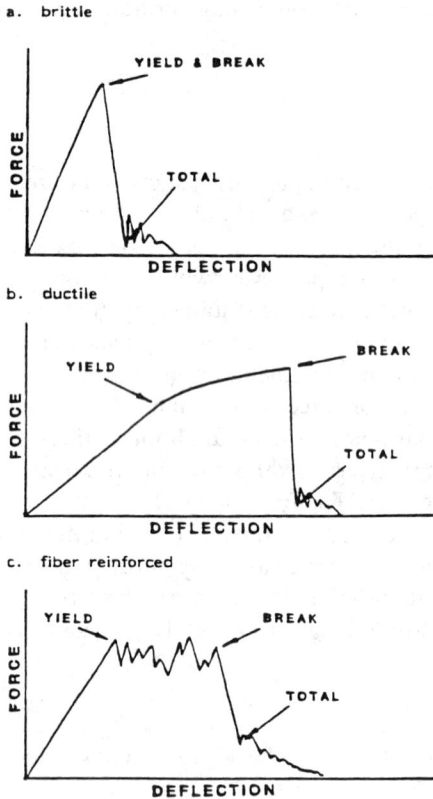

FIG. 2—*Examples of impact response records for various types of composite materials. (The term "yield" refers to the material characteristic known as proportional limit.)*

used in the paper to represent the material characteristic commonly known as "proportional limit."

The third material in Fig. 2 is reinforced with fiber. Again, an elastic portion can be seen. The start of material failure is abrupt, with no plastic deformation. Instead, there is a region of fiber breakage and pullout ending with an abrupt break. The fiber breakage region is characterized by oscillations in load capacity. In this region, load may remain essentially flat, drift up, or drift down.

In all three cases, the load may not drop to zero after the break point. The examples in Fig. 2 are from a dart test. After the break, some force may be measured as result of bending and friction as the probe travels through the broken plaque. The bending is included when determining the total point, but the final friction load is neglected.

A consistent labeling scheme has been developed for the major features of these response records. Referred to as "failure limits," they relate directly back to the performance criteria from the impact functional requirements. The relationship is as shown in Table 4.

Incipiency of failure is only occasionally observed in high-glass thermoset structural composites. As a sharp oscillation in the elastic slope portion of the curve, it indicates microscopic matrix cracking and a measurable degradation in static properties, such as stiffness.

It is easy to visualize that a material may rank low at the yield point but high at the break or total point. For this reason the failure limit is a key impact parameter and must be defined and used for data interpretation.

Procedure—Analysis of the Application

Impact Application Characterization

An impact application characterization can be compiled conveniently in a table format. Each column would be generated as one step in a four-step procedure. The four steps, which have been discussed separately in the preceding sections of this paper, are summarized in the following section. A complete impact characterization for the fender example is given in Table 5.

Format for Requirements

Figure 3 lists five impact-related functional requirements for a fender example. Each is described in a simple verb phrase, although the list is somewhat structured to fit a standard format. Each requirement must contain two elements: (1) the description of a type of impact event that might occur and (2) the amount of damage that would be acceptable.

These impact-related functional requirements can be summarized in four steps:

Step 1—Write a set of simple statements, in sentence form, of the impact-related *functional requirements* for the application.

TABLE 4—*Impact failure limits as determined from performance criteria.*

Functional Requirement Performance Criterion	Failure Limit
No property degradation	incipiency of failure
No visible damage	yield
Minimum damage	yield and break
Maximum energy absorption	total

TABLE 5—*Impact application characterization for the fender example.*

Functional Requirement	Stress State	Controlling Variable	Failure Limit
Handling and assembly (A1)	uniaxial bending	deflection	yield
		energy	yield
No visible damage	tearing	deflection	yield
		energy	yield
Stones and road debris (A2)	biaxial bending/tension	energy	yield
No visible damage	normal shear	energy	yield
	normal compression	energy	yield
Low-energy object (B1)	uniaxial bending/tension	energy	yield
No visible damage	biaxial bending/tension	energy	yield
	normal shear	energy	yield
	normal compression	energy	yield
	in-plane bending	energy	yield
Collision normal to panel surface (C1)	uniaxial bending/tension	deflection	yield
			break
Minimum damage	normal shear	deflection	yield
			break
	normal compression	force	yield
			break
	in-plane bending	deflection	yield
			break
	tearing	deflection	yield
			break
Collision in plane of panel surface (C2)	in-plane compression	deflection	yield
			break
		energy	total
Minimum damage	in-plane shear	deflection	yield
Maximum energy			break
absorption		energy	total
	buckling	deflection	yield
			break
		energy	total
	tearing	deflection	yield
			break
		energy	total

Step 2—For each functional requirement, list the *stress states* that can occur for the range of impactors and for the range of possible impact locations.

Step 3—For each stress state, consider the application conditions to determine the *controlling variables*.

Step 4—For each functional requirement, list the *failure limits* corresponding to the functional requirement performance criterion.

Table Format

The functional requirements can be put into tabular form, as shown in Table 6. The table columns list six impact classes and four performance criteria,

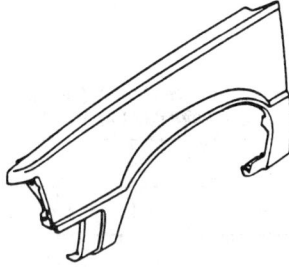

1. WITHSTAND CHANCE IMPACTS DURING HANDLING AND ASSEMBLY
   WITH NO VISIBLE DAMAGE.

2. WITHSTAND REPEATED SIDE IMPACTS AND GLANCING FRONT
   IMPACTS FROM ROCKS AND ROAD DEBRIS WITH NO VISIBLE
   DAMAGE.

3. WITHSTAND REPEATED SIDE IMPACTS FROM DOOR EDGES,
   SHOPPING CARTS, AND VARIOUS HAND HELD OBJECTS WITH
   NO VISIBLE DAMAGE.

4. WITHSTAND A SIDE COLLISION WITH A CAR, BARRIER, OR
   POLE WITH MINIMUM DAMAGE.

5. WITHSTAND A FRONT COLLISION LOAD WITH MINIMUM DAMAGE
   AND/OR MAXIMUM ENERGY ABSORPTION.

FIG. 3—*Impact-related functional requirements for a fender.*

which cover the full range of automotive applications. Note that two columns do not apply to the fender example.

Additional details regarding the impact classes are given in Table 7. The six impact classes can be assigned design priorities, as shown in Table 8. These priorities carry through from the functional requirements to the impact test specification and become useful for establishing test program priorities.

Specification of Impact Test Methods

Impact Test Method Characterizations

Impact test methods can be characterized according to the same three key impact parameters that have been defined for characterizing impact applications. Such a characterization of ten commonly used test methods is given in Table 9.

TABLE 6—*Impact-related functional requirements for a fender example.*

Impact Class and Performance Criterion	Fender Requirement[a]				
	1	2	3	4	5
Impact class					
A1—handling and assembly	X				
A2—stones and road debris		X			
B1—low-energy object			X		
B2—moderate-energy object					
C1—Collision normal to panel surface				X	
C2—Collision in plane of panel surface					X
Performance criterion					
No property degradation					
No visible damage	X	X	X		
Minimum damage				X	
Maximum energy absorption				X	X

[a]The five fender requirements correspond to those shown in Fig. 3 and Table 5.

TABLE 7—*Data on the impact classes tested.*

Impact Class	Impactor			Number of Impacts during Product Life Cycle
	Mass	Velocity	Energy	
A1—handling and assembly	NA[a]	low to moderate	low to moderate	1 to 10
A2—stones and road debris	low	high	low	100 to 1000
B1—low-energy object	low to moderate	low to moderate	low	10 to 100
B2—moderate-energy object	low to moderate	moderate to high	moderate	1 to 10
C1—collision normal to panel surface	high	moderate to high	high	1
C2—collision in plane of panel surface	high	moderate to high	high	1

[a]NA = not applicable.

TABLE 8—*Design priority guidelines of impact classes.*

Impact Class	Guidelines
A1, A2	minimum requirements
B1, B2	working objectives for optimization
C1, C2	desirables for secondary consideration

TABLE 9—*Impact test method characterizations.*

		Measured Characteristic	
Stress State	Impact Test	Controlling Variable	Failure Limit
Uniaxial tension	tensile, instrumented	deflection force energy	yield break total
Uniaxial bending	flex, instrumented	deflection force energy	yield break total
Biaxial bending/tension	dart, instrumented	deflection force energy	yield break total
	falling weight	energy	yield total
	Gardner, 31.75-mm (1.25-in.) ring	energy	yield total
Normal shear	Gardner, standard	energy	yield total
Normal compression	Gardner, anvil	energy	yield total
In-plane bending	Izod, unnotched	energy	total
Tearing (in-plane bending with stress concentration)	Izod, notched Charpy	energy energy	total total

It should be noted that test methods exist for only 7 out of 13 stress states. Some test methods are redundant.

The standard Gardner test only approximates a normal shear stress state. The actual stress state will be a ratio of biaxial bending/normal shear depending on material stiffness.

The first three test methods listed are instrumented and provide a full set of measured characteristics. Impact test methods that are not instrumented provide a single number, an energy, by which to characterize the material behavior. The Izod and Charpy tests measure total energy. The falling weight and Gardner tests measure something between yield energy and total energy, depending on the material toughness and the operator's definition of failure.

Impact Test Method Specification for the Fender Example

An impact test method specification is generated simply by matching up the test method characterization table with the application characterization table. The test methods can be assigned priorities that trace back directly to the first statement of the impact application functional requirements.

The impact test method specification for the fender example is given in Table 10. The terms in parentheses indicate test methods that are not available or response features that are not measured because of the lack of instrumentation.

TABLE 10—*Impact test method specification for the fender example.*

Impact Test	Measured Characteristic[a]	
	Controlling Variable	Failure Limit
PRIORITY A—MINIMUM REQUIREMENTS		
Flex, instrumented	deflection	yield
	energy	yield
Notched Izod or Charpy	(deflection)	(yield)
	energy	(yield)
Dart, instrumented	energy	yield
Gardner, standard	energy	(yield)
Gardner, anvil	energy	(yield)
PRIORITY B—WORKING OBJECTIVES FOR OPTIMIZATION		
(Uniaxial bending/tension)	(energy)	(yield)
Unnotched Izod	energy	(yield)
PRIORITY C—DESIRABLES FOR SECONDARY CONSIDERATION		
(In-plane compression)	(deflection)	(yield)
		(break)
	(energy)	(total)
(In-plane shear)	(deflection)	(yield)
		(break)
	(energy)	(total)
(Buckling)	(deflection)	(yield)
		(break)
	(energy)	(total)

[a]The terms in parentheses indicate test methods that are not available or response features that are not measured because of the lack of instrumentation.

Summary of Six Example Applications

A total of six additional example applications were characterized according to the procedures presented in this paper. This was done to develop a perspective and direction for the future development of impact tests. The six applications, which are listed here, were selected to represent a broad range of conditions.

(a) fender,
(b) door outer panel,
(c) front facia,
(d) hood outer panel,

a. for all design priorities

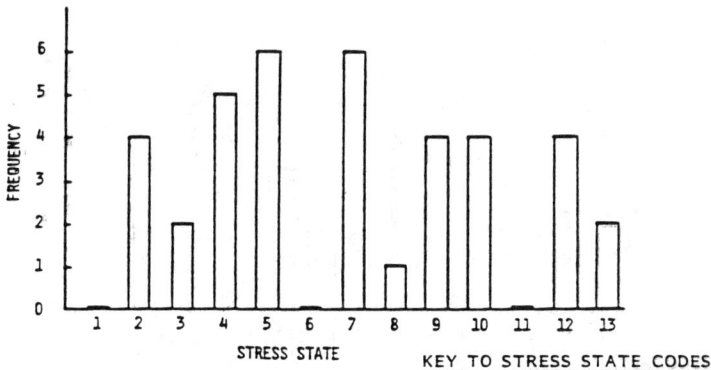

KEY TO STRESS STATE CODES

1. uniaxial tension
2. uniaxial bending
3. uniaxial bending/tension
4. biaxial bending/tension
5. normal shear
6. normal shear/bending
7. normal compression
8. in-plane bending
9. in-plane compression
10. in-plane shear
11. in-plane shear/compression
12. buckling
13. tearing

b. for design priorities A and B only

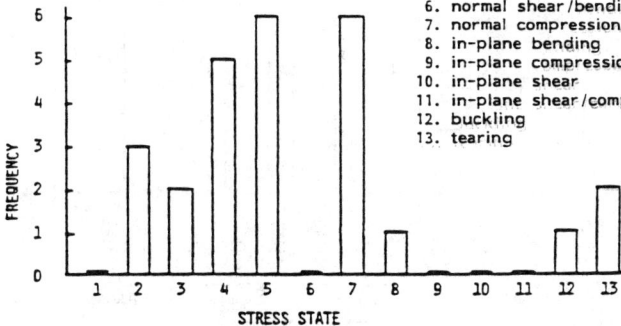

FIG. 4—*Frequency of occurrence of stress states in characterization of six example applications.*

(e) engine side panel, and

(f) grill.

Figure 4 shows that 3 of the 13 possible stress states that were defined in the Impact Controlling Variables section did not occur at all in the characterization of the six applications. Also, if Design Priority C in Fig. 4 is neglected, the number of significant stress states is reduced to only six.

The frequency of occurrence of controlling variables and failure limits are shown combined as response features in Fig. 5. It can be seen that deflection is as important as energy as an impact controlling variable. The yield and break responses are more frequently of interest than the "total" response as a failure limit. Total energy is the response feature of interest in only 13% of impact application situations. It has previously been noted that impact test methods that are not instrumented provide a single number, an energy, by which to characterize the material behavior.

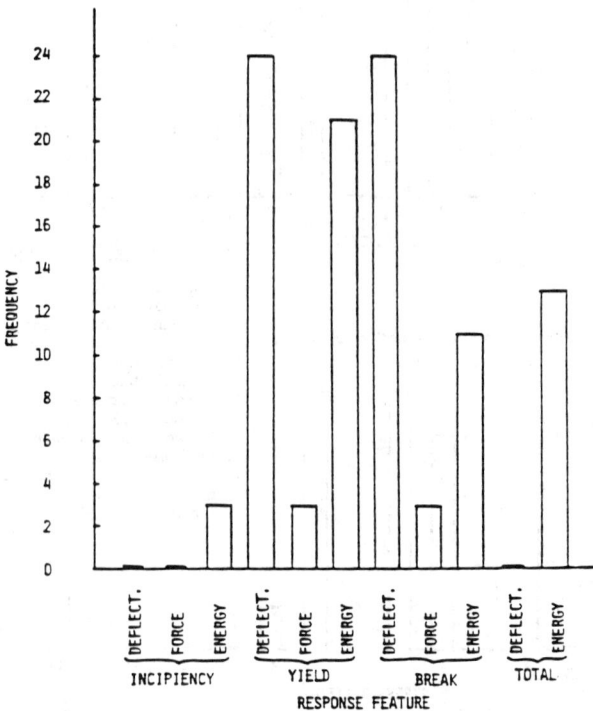

FIG. 5—*Frequency of occurrence of* response features *in characterization of six example applications.*

Conclusions

The concepts and observations presented in the development of this methodology and illustrated by the fender example lead to some general conclusions relative to impact characterization. The summary of six example applications supports these conclusions regarding impact test development.

Impact Characterization

1. The methodology presented in this paper provides a means, based on application considerations, for specification of an appropriate set of impact tests.
2. A set of three or more impact test methods will probably be required to provide a material characterization for any reasonably complicated application.
3. Material impact performance ratings vary between different test methods. Therefore, the test methods must be assigned a weighting factor based on design priorities.

Impact Test Development

1. Development priorities can be assigned to impact test methods as follows:

(*a*) First priority—biaxial bending/tension, normal shear, and normal compression;
(*b*) Second priority—uniaxial tension, uniaxial bending, and tearing.

2. Impact tests must be instrumented to provide the response data necessary to characterize the material behavior.

Reference

[*1*] Kakarala, S. N. and Roche, J. L., "Experimental Comparison of Several Impact Test Methods," this publication, pp. 144–162.

Charles W. Knakal[1] and Donald R. Ireland[2]

Instrumented Dart Impact Evaluation of Some Automotive Plastics and Composites

REFERENCE: Knakal, C. W. and Ireland, D. R., **"Instrumented Dart Impact Evaluation of Some Automotive Plastics and Composites,"** *Instrumented Impact Testing of Plastics and Composite Materials, ASTM STP 936,* S. L. Kessler, G. C. Adams, S. B. Driscoll, and D. R. Ireland, Eds., American Society for Testing and Materials, Philadelphia, 1987, pp. 44–57.

ABSTRACT: This paper presents a guide for evaluation and utilization of data provided by the instrumented dart impact test. This guide was developed after evaluating seven different plastics and composites, different probe diameters, impact velocities, and different test temperatures. These were instrumented tests that yielded complete load and absorbed energy profiles as a function of probe penetration distance. A special high-speed video technique was used to develop an understanding of and confidence in interpretation of the major features of the load-deflection profile.

The concepts and comments are applicable to other techniques of dart impact testing. An important element of this paper is its application to widely differing data records and the recognition that the data analysis should not be forced for certain types of data records.

KEY WORDS: impact testing, instrumented impact test, plastics, composites, puncture, automotive plastics, automotive composites

The instrumented impact test is potentially a very useful tool for evaluating the dynamic response of materials subjected to a specific set of geometric loading conditions. The biaxial loading imposed by a hemispherical dart on a flat plate specimen rigidly supported over a circular annulus is a typical load-

[1]Supervisor—Development Engineering, Advanced Engineering Staff, General Motors Corp., Warren, MI 48090.
[2]Consultant, Ireland and Associates, Santa Barbara, CA 93101.

ing condition. Work is in progress by ASTM Committee D-20 on plastics to develop a practical test method for instrumented dart impact testing.

The two most common methods for producing the dart impact condition are use of a free-falling weight and a servo-controlled hydraulic ram. The latter has a distinct advantage because of its ability to vary the impact velocity over a wider range (lower speeds). Limitations are imposed by the data analysis of the load signal, which sets similar upper limits on the impact velocity for both methods. These limits are strongly dependent on the fracture mode of the material, which is influenced by both test temperature and dart impact velocity. The hydraulic ram's lower controlled velocities are not limited by minimal deliverable energy/velocity constraints.

To date, there has been no definitive guide for the analysis of impact records from instrumented impact tests on the types of plastics and composites intended for automotive applications. This paper represents a portion of the results of an extensive program that was designed to yield an understanding of various impact tests and their relationship to evaluating end-use conditions of widely different plastic and composite materials. The program examined material responses at different temperatures and impact velocities. The apparatus variables of load sensor position and dart diameter were also evaluated with respect to the materials' responses.

This paper will not be comparing these materials and variables. It is intended as a guide to impact record analysis. Special high-speed video techniques were used to compare visual observations of the fracture process with the interpretations used for data selection from the load signal. This paper presents the results in terms of comparing real records to an idealized load deflection. The use of the suggested procedure is presented with examples of selected responses from actual data records.

The limitations of the suggested analysis procedure are discussed. The apparent relationships of typical impact data values to specimen thickness variations and end-use structural relationships are suggested.

Apparatus

The instrumented impact test equipment used in this study was a Rheometrics high-speed RIT-8000 purchased in 1980. The general features of the testing apparatus are shown in Fig. 1. The material specimen is a flat plate, approximately 127-mm (5-in.) square, which is securely clamped over a 76-mm (3-in.)-diameter annular anvil. The test consists of complete penetration of the specimen by a 25.4-mm (1-in.)-diameter hemispherical probe, which is guided through the center of the annulus under conditions of near constant velocity. It is possible, on tough materials, to have the undesirable condition of large velocity variations of the probe during penetration. Proper use of the dart test should be limited to materials that are within the capabilities of the test equipment.

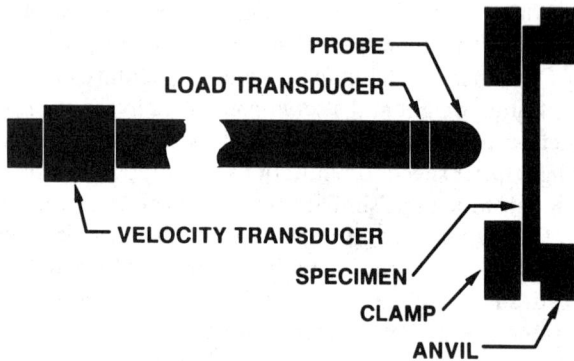

FIG. 1—*General features of the test apparatus.*

The probe is equipped with transducers for measurement of velocity and the load interaction between the probe and specimen. Two optoelectric sensors are used to start and stop data acquisition. Three thousand load data points are collected during the 76 mm (3 in.) of probe travel that is considered the impact event.

The velocity and load transducers provide complete profiles of the deformation response of the specimen from initial impact through final penetration. This measured information is sufficient to provide the additional parameters of deflection and energy.

Velocity

A linear variable displacement transducer (LVDT) type of transducer provides continuous monitoring of velocity. This velocity signal is used for (1) feedback to control probe speed and (2) driving a voltage-controlled oscillator (VCO), which determines load data acquisition timing and deflection calculations.

Load

The load is the force required to drive the probe through the specimen. Discrete values provided by the load transducer are sampled and stored at the end of each VCO cycle [each 0.025 mm (0.001 in.) of probe travel].

Deflection

Deflection is defined as the linear motion of the probe in the direction of penetration. The VCO generates a signal at a frequency proportional to the measured velocity. Each VCO cycle is equal to 0.025 mm (0.001 in.) of probe travel.

Energy

Energy is a two-dimensional parameter defined as the sum of the incremental products of load and deflection. It is the area under the load-deflection curve. Energy is automatically computed from measured values of load and velocity.

Data Acquisition

The stored load points are presented on a cathode-ray tube (CRT) display as a continuous load-deflection curve. The operator uses the CRT display to adjust for optimum viewing of the curve and to select specific points of interest. Vertical "tick" marks are shown on the curve as reminders of the positions selected. The computer will then produce a hard-copy plot of the load-deflection curve and tabulated results of the selected points of interest (load, deflection, and energy). Load is shown on the vertical axis and displacement on the horizontal. Included with this record are the test parameters of impact velocity, specimen thickness and full-scale values of the curves. A computer-calculated value of specimen stiffness, based on operator point selection, is also given. Typical hard-copy graphical and tabulated records from the equipment are shown in Fig. 2.

Idealized Dart Deformation

The relative shape of the load-deflection record is indicative of the deformation and fracture history of the specimen. It is convenient to interpret the shape of the load record by concepts that are similar to those employed for the conventional tensile test. It should be remembered, however, that the dart test imposes a biaxial flexural deformation to the material.

The load record can be subdivided into deformation stages, which are connected by transition points (see Fig. 3). For any test, the load-deflection record will be composed of all or portions of the idealized stages indicated in this figure. The stages and transition points identified in Fig. 3 are defined as follows:

Stage A—dynamic offset. These expected initial smooth-loading responses may be complicated by several dynamic factors of the apparatus and specimen. The load values are *not* representative of those required for the indicated deflection by static mechanics relationships. The perturbations are generally small and insignificant.

Transition A—start of the linear load-deflection deformation.

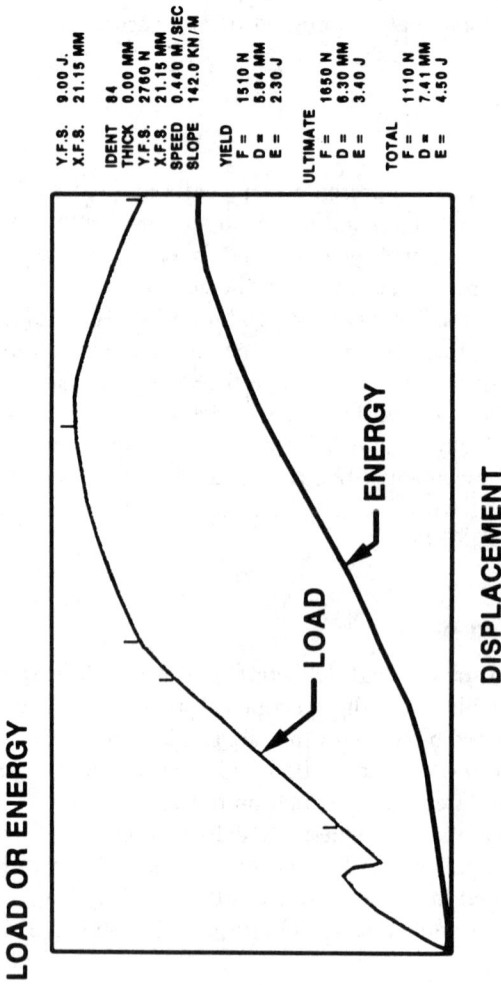

FIG. 2—*Graphical and tabulated data report.*

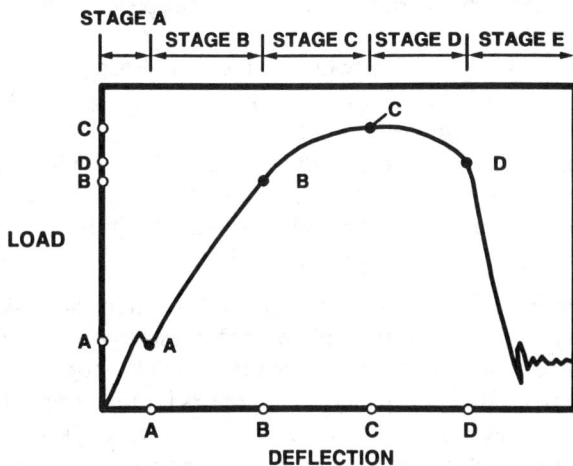

FIG. 3—*Idealized deformation stages and transitions for puncture testing of flat-plate specimens by a hemispherical probe.*

Stage B—linear load-deflection deformation. Although a localized plastic indentation may occur at the load point, the specimen is essentially reacting as an elastically loaded structure. The removal of the applied load would result in an essentially complete recovery to the original specimen position.

Transition B—yield. This implies the onset of plastic or permanent deformation and is not necessarily an indication of cracking. For some fiber-filled materials, the yield point is characterized by a sudden decrease in load, followed by an apparent second stage of linear load-deflection deformation but at a *reduced* slope from Stage B.

Stage C—first major permanent deformation. The damage is generally distributed over a relatively large volume, so that a decrease in load is not observed. For fiber-filled materials, this can be associated with the development of extensive microcracking or interlaminar shearing of the matrix. With unfilled plastics, this stage can be an extension of the plastic deformation initiated at Transition B.

Transition C—maximum load. This is defined as the onset of deformation, which does *not* result in an increase in load. This point has also been identified as the "ultimate," or "peak" load. This transition is usually associated with the first appearance of visible cracks on the tension surface of the specimen.

Stage D—stable or slow-rate deformation after Transition C. For unfilled plastics, this deformation can be identified with localized thinning of the specimen around the circumference of the dart. For the filled materials, this stage is usually associated with the extension of the cracking initiated at Transition C. During the deformation in this stage, the specimen still retains significant structural integrity.

Transition D—end of the test, which is the end of Stage D deformation, that is, the onset of unstable macrocracking or fracture, for which the specimen does *not* have structural integrity.

Stage E—a nondescriptive portion of the load-deflection record essentially showing the probe sliding through the puncture. Load, deflection, or energy data for this stage *do not* have any utility for describing the impact fracture resistance of the material.

These definitions of deformation stages and transitions are based on previous experience with other instrumented impact tests and an extensive study of the high-rate servo-controlled dart puncture test. The latter included high-speed video studies of the dart penetration process for the seven different materials used in the study (see Table 1). The testing utilized variations in temperature and impact velocity to produce major changes in the impact fracture resistance of the materials.

The idealized record shown in Fig. 3 can be complicated by the deformation mode of the material and the dynamics of loading at a specific velocity. These complications are revealed as oscillations or perturbations of the load signal. Application of the idealized curve analysis to actual load records is discussed in the next section.

TABLE 1—*Characteristics of materials used in this study.*[a]

Material Type	Resin	Reinforcement	Potential
SMC-R28	polyester	28% chopped glass fiber	front-end panels, body panels
SMC-R65	vinyl ester	65% chopped glass fibers	beams, structural reinforcements
RIM	low-modulus polyurethane	none	fascias
RRIM	high-modulus polyurethane	20% glass flakes	fenders, door skins
Nylon RRIM	nylon with 20% polyol	20% milled glass fibers	fenders, door skins
ABS	acrylonitrile-butadiene-styrene	none	interior panels
PP	polypropylene	none	interior trim

[a]Thickness of specimens: SMC-R28 = 0.33 cm; SMC-R65 = 0.264 to 0.292 cm; RIM = 0.356 cm; RRIM = 0.30 cm; nylon RRIM = 0.269 cm; ABS = 0.254 cm; PP = 0.33 cm.

Data Selection

The data selection requirements for dart impact tests should be based on general usage and not necessarily on end-use conditions. The end-use conditions become important later when the impact calculations from the data points selected are evaluated. A consistent rationale for selecting and presenting data will enhance the ability to collect and compare data.

The relative stiffness of the material is determined by the slope of the initial portion of the curve. It is computed from two cursor positions selected within deformation Stage B.

This value is a relative measure of the elastic response of the specimen before any significant damage has occurred. Selected cursor positions for computation of the slope do not have to define the boundaries of Stage B. Two typical examples of cursor positions for slope computation are shown in Fig. 4. The general shape of the load record is not a good indicator of relative slope, since the electronic scaling frequently changes. The slope value is a measurement of specimen stiffness, which is related to the specimen thickness.

Problems can occur as a result of abnormal oscillations of the load signal (Fig. 5a) or when there is no distinct linear portion (Fig. 5b) of the curve. It is

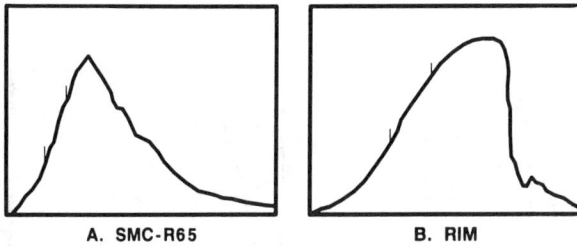

A. SMC-R65   B. RIM

FIG. 4—*Typical load-deflection records. The cursor position for slope computation is indicated by vertical tick marks.*

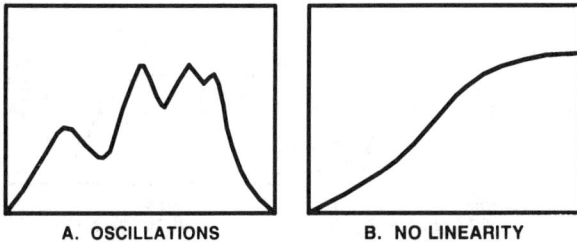

A. OSCILLATIONS   B. NO LINEARITY

FIG. 5—*Slope evaluation and possible problems.*

strongly recommended that the operator not force the data selection when presented with these problems. When a problem occurs with respect to selecting data coordinates, the uncertainty should be clearly identified and reported, not a "guessed" value. It may be necessary to change the test parameters and perform other tests.

The yield position is a visual estimate of Transition Point B. It is the first major deviation from linear load-deflection behavior. The determination of an exact yield point is not a precise procedure. The exact concept of yielding as applied to a standard uniaxial tensile test does not generally apply to the bending of a flat-plate specimen. It can be useful when used to define the onset of plastic damage. This damage could be a shear-induced localized plastic delamination, localized plastic instability (denting or thinning), or some combination of these events. The yield point for a puncture test is not always associated with the first appearance of cracks or tears in the specimen.

The yield characteristic of a load-deflection record can take many forms. Four typical examples are shown in Fig. 6. The specimen temperature and probe velocity at impact are shown for each record. The cursor position selected for each yield point is indicated by a small vertical "tick" mark on the load record. These marks are identified by an arrow Fig. 6a and d. The yield shown for nylon reinforced reaction injection molding (RRIM) in Fig. 6a is probably indicative of the first appearance of cracks or tears on the tension surface of the specimen. The record shown in Fig. 6c for reaction injection

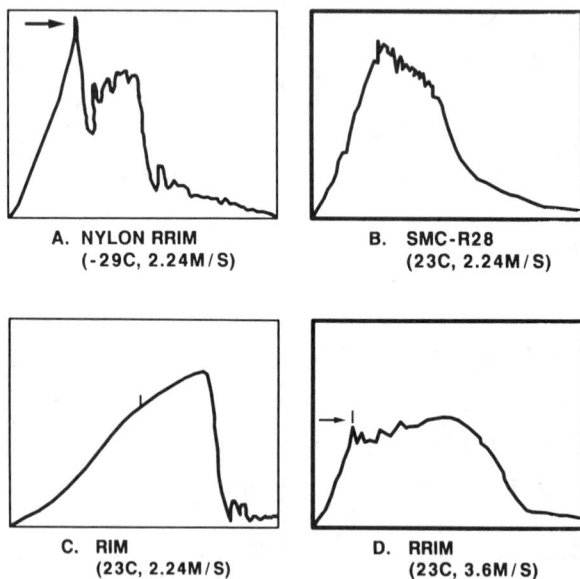

A. NYLON RRIM
(-29C, 2.24M/S)

B. SMC-R28
(23C, 2.24M/S)

C. RIM
(23C, 2.24M/S)

D. RRIM
(23C, 3.6M/S)

FIG. 6—*Typical examples of yield point selection.*

molding (RIM) selects a yield at a linear slope change when "denting/thinning" occurs. The other two illustrations (Fig. 6*b* and *d*) for sheet molding-compound–random 28% fiber (SMC-R28) and RRIM probably reflect a mixture of those two types of damage after yield. Problems with the selection of the yield point are similar to those for stiffness.

The maximum load point (Transition C) is arbitrarily defined as when the load first reaches a maximum value before a significant load loss. The cursor position for determination of ultimate parameters is illustrated for four different load-deflection records (Fig. 7). The specimen temperatures and probe velocities at impact are indicated. The cursor position for the ultimate point are indicated by a vertical tick mark. The records for SMC-R65 and RIM, shown in Fig. 7*a* and *c*, indicate a sharp decrease in load after the ultimate point, which can be readily associated with the occurrence of major cracking or tearing leading to failure. The records for RRIM and RIM in Fig. 7*b* and *d* do not give the same confidence for major crack initiation occurring at the ultimate point. The gradual decrease in load after the maximum load could be associated with localized cracking, leading to a more gradual failure.

Two problems that can occur for the operator in determining the ultimate point are illustrated in Fig. 8*a* and *b*. Figure 8*a* shows two small oscillations reaching the same value of maximum load (indicated by the dashed line) but at drastically different deflections, and hence absorbed energy values. These small oscillations can be caused by several factors (for example, electrical or

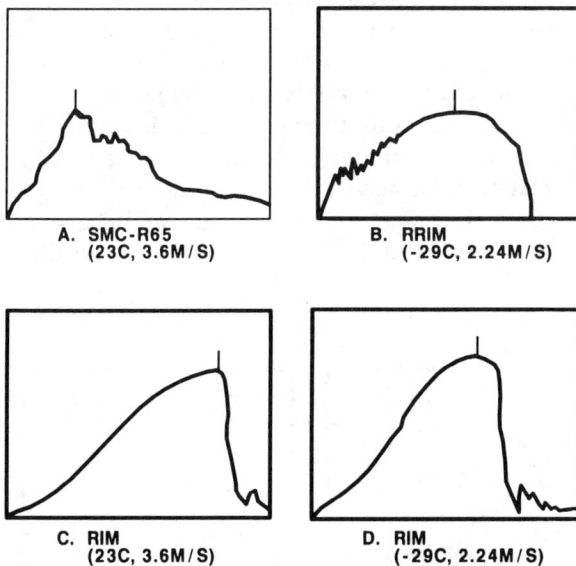

A. SMC-R65
(23C, 3.6M/S)

B. RRIM
(-29C, 2.24M/S)

C. RIM
(23C, 3.6M/S)

D. RIM
(-29C, 2.24M/S)

FIG. 7—*Typical examples of ultimate point selection.*

mechanical responses on the apparatus), which are not reflections of the specimen's material deformation characteristics. Curve fitting, either automated or visual, should result in the selection of a maximum load point at the second position in Fig. 8a. Figure 8b shows a curve with an unexpected flat top in the load signal. This could be caused by the load cell/amplifier saturating or by coarse digitizing by the transient recorder. The test should be rerun with appropriate adjustments to the apparatus.

The cursor position that defines the coordinates for the total point should be at the end of the useful data obtainable from the puncture test. This position is usually selected after maximum load and at a point where the load signal has decreased by 10 to 15%. Typical examples of the cursor position for total data coordinates are shown in Fig. 9. The cursor coordinates of load, deflection, and energy listed under the heading of Total in the tabulated report (Fig. 2) have no practical value. The original purpose of the puncture test was to define the impact conditions required to *initiate* cracking or tearing. The limits for *useful* data in a puncture test are between initial impact and maximum load. Other test configurations (such as Charpy and Izod) can provide useful applications for the coordinates associated with the total point.

Video Observations

High-speed observations of the dart penetration of the specimens were obtained through the use of a unique video system provided by Spin Physics, Inc. The high-speed video provided 2000 full frames per second (or 12 000 partial frames per second) for instant replay and analysis. The pictures were stored on magnetic tape packaged in a cassette, which can store 1 h of real-time recordings. The advantage of this recording technique is the ability to play back at very slow rates or for frame-by-frame viewing with no damage to the recording medium. For example, at a dart impact speed of 2.24 m/s, the specimen deformation could be viewed at a rate of 0.0373 cm per frame, and for a typical 0.25-cm-thick SMC specimen, there would be approximately 12 to 14 frames available between the start of impact and Transition B. The ob-

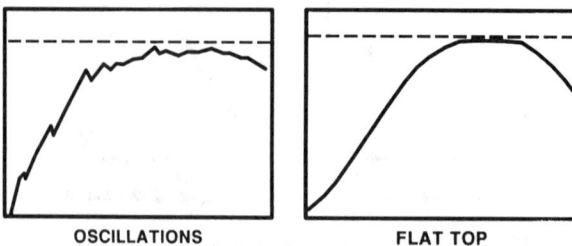

OSCILLATIONS   FLAT TOP

FIG. 8—*Ultimate evaluation and possible problems.*

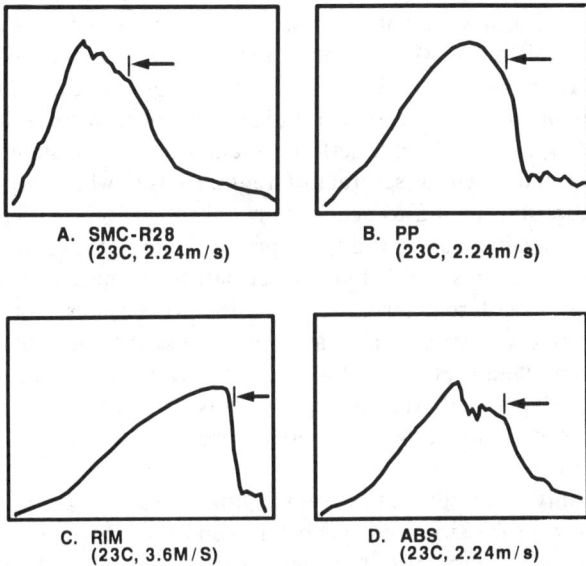

A. SMC-R28
(23C, 2.24m/s)

B. PP
(23C, 2.24m/s)

C. RIM
(23C, 3.6M/S)

D. ABS
(23C, 2.24m/s)

FIG. 9—*Typical examples of cursor position for the total point.*

servations from these high-speed video tests confirmed and revealed the analysis made for the idealized load-deflection record (see Fig. 3).

Data Evaluation Considerations

Since the usual objective of the high-speed puncture test is to evaluate material performance, the selected data values should be normalized for differences in specimen thickness, t. For example, the RIM and SMC-R65 materials had nominal thicknesses of 3.56 mm (0.14 in.) and 2.79 mm (0.11 in.), respectively. The authors suggest that the slope values be normalized for a hypothetical end-use thickness of 2.54 mm (0.10 in.) as follows

$$slope_{(normalized)} = \left(\frac{2.54}{t}\right)^3 slope_{(measured)}$$

with the following results:

		Slope, kN/m	
Material	Thickness, mm (in.)	Measured	Normalized for $t = 2.54$ mm (0.10 in.)
SMC-R65	2.79 (0.11)	1100	830
RIM	3.56 (0.14)	500	180

The importance of comparing material response values for equivalent specimen thicknesses is quite evident. At first the apparent differences in stiffness between SMC-R65 and RIM are approximately 2 to 1; however, normalizing for thickness reveals that the difference is actually greater than 4 to 1.

The yield point could be used as a design point (with a suitable factor of safety). The coordinates of load, deflection, and energy corresponding to the yield position can be used for several different purposes, which depend on the intended relationship to end-use conditions.

The deflection values (up to yield) are proportional to the strain; however, the computation requires input of an appropriate mechanical elastic modulus value, which is difficult to determine. This difficulty is centered on the uncertainty of the effects of stress-state variations across the specimen thickness and the inherent dynamics of the Rheometrics dart puncture test. Work is in progress to develop a method for using the measured stiffness (that is, slope) to estimate this modulus value and, thereby, provide a technique for estimating critical strain values up to yield.

The real utility of energy values is doubtful. Unless the test conditions of probe geometry, probe size, and specimen support anvil size are directly relatable to the intended end use, the energy values determined by the laboratory test have little value. It is better to evaluate impact through independent consideration of the coordinates of load and deflection.

The coordinates of load, deflection, and energy corresponding to the ultimate position also have limited utility in real life. They are useful for comparisons of different materials evaluated by the same conditions exactly. It is convenient to regard the ultimate load as a definition of the onset of significant cracking or tearing. Again, the energy value can be useful for comparisons of different materials tested under the same conditions, but the load and deflection values have more dimensional significance.

Conclusions and Observations

The following is a summary of statements regarding use of the instrumented dart impact test for evaluations of automotive plastic and composite materials. Although this work was based on the use of a servo-controlled hydraulic impact machine, the statements regarding analysis and utilization of data apply equally to work generated by instrumented drop-weight devices.

1. The load-deflection record is an excellent indication of the specimen deformation process. Key features of that record can be used as distinct characteristics of the material response to impact loading.

2. A single idealized load-deflection record, divided into different deformation stages, which are connected by transition points, can be used as a guide for data selection analysis of any record.

3. The data to be derived from this test consist of the following:

   (a) stiffness, which is the slope of the linear elastic load-deflection portion (Stage B) of the curve, and
   (b) load and deflection values at the following transition points:

   (1) Transition B, yield (end of linear elastic region), and
   (2) Transition C, ultimate point (when the curve first reaches a significant maximum load).

4. The load, deflection, and energy values for Transitions C and D should be used only for comparisons between specimens tested by *exactly* the same procedure. Published values should clearly define the test conditions.

5. Under no circumstances should any data obtained after Transition D be used to represent material impact performance. The dart sliding through the fracture does not reveal useful data. The concept of propagation energy has no meaning for the dart impact test.

6. Care must always be taken to avoid forcing any data analysis. When there are inherent oscillations of the load record, caused by specimen fracture mode and dart velocity, it should be identified as "uncertain for data analysis." A narrative description of the fracture appearance can be more reliable and useful than any "forced" data guesses of load, deflection, and energy.

7. Additional work is required to determine a clear method for relating dart impact parameters to product end use. The end-use conditions should be understood before selection of an impact test is attempted.

8. Specimen thickness and end-use thickness should be accounted for in any evaluation of the instrumented dart impact test.

M. R. Kamal,[1] Q. Samak,[1] J. Provan,[2] and Vagar Ahmad[2]

Evaluation of a Variable-Speed Impact Tester for Analysis of Impact Behavior of Plastics and Composites

REFERENCE: Kamal, M. R., Samak, Q., Provan, J., and Ahmad, V., "**Evaluation of a Variable-Speed Impact Tester for Analysis of Impact Behavior of Plastics and Composites,**" *Instrumented Impact Testing of Plastics and Composite Materials, ASTM STP 936,* S. L. Kessler, G. C. Adams, S. B. Driscoll, and D. R. Ireland, Eds., American Society for Testing and Materials, Philadelphia, 1987, pp. 58–80.

ABSTRACT: The Rheometrics variable-speed impact tester (RVSIT) was evaluated with regard to the stability and reproducibility of the velocity and load signals. A drop in the speed of the probe was observed upon impact. The extent of the drop depends on the impact speed and the material tested. Mounting the load cell ahead of the impact rod was found to improve the reproducibility of load-deflection signals and to alleviate spurious postpuncture noise. A new rod was designed to apply circumferential loads on notched specimens. Load-deflection data obtained with this type of loading were used to calculate the dynamic critical stress intensity factors (SIF) for sheet molding compound (SMC) and reinforced reaction injection-molded (RRIM) polyurethane panels in the opening and the sliding modes.

KEY WORDS: fracture toughness, impact testing, notching, reaction injection-molded (RIM) polyurethane, sheet molding compound (SMC)

Since the early days of the plastics industry, there has been an interest in developing structural and engineering materials that would combine strength, toughness, lower specific gravity, and good processability. Moreover, in recent years, emphasis on energy conservation has led to greater interest in the expanded utilization of plastics in automotive applications. In many of the aforementioned markets and applications, a major consideration has been the performance of plastics and polymeric materials under condi-

[1]Chairman and graduate student, respectively, Department of Chemical Engineering, McGill University, Montreal, Canada H3A 2A7.
[2]Associate professor and graduate student, respectively, Department of Mechanical Engineering, McGill University, Montreal, Canada H3A 2A7.

tions of dynamic loading, in general, and mechanical impact, in particular. A fundamental requirement for dealing with these issues relates to the development of appropriate impact tests, which could be useful in satisfying the needs of engineering design and product quality control [1].

The impact tests most commonly used in the plastic industry, such as the Izod and the Charpy tests, are adaptations of impact testing techniques developed for metallurgical applications. These tests have recently been instrumented to generate data describing load-time or load-deflection relationships in the course of an impact event [2-4]. Instrumented tests such as the Charpy and Izod are still of limited use in testing sheets or films. There are also limits on their capacity to assess the performance of tough materials having low moduli.

In order to overcome these shortcomings and to simulate dynamic field loading conditions better, a variety of other instrumented impact tests have been developed. These include the drop-weight test [5-8] and tests based on ballistic or variable-speed puncturing [1, 9-11]. The latest development in the latter category is the instrumented Rheometrics variable-speed impact tester (RVSIT).

In the present study, the performance of the RVSIT was evaluated as a constant-velocity, variable-rate impact tester. The evaluation was conducted by studying the impact behavior of reinforced reaction injection-molded polyurethane panels (RRIM), reinforced sheet molding compound panels (RSMC), and two thermoplastic polymers, polycarbonate and low-density polyethylene (LDPE).

Experimental Procedure

The Impact Tester

The RVSIT is built around a linear-displacement, velocity-controlled, hydraulically driven mechanism. The system drives a penetrating rod, the shape and size of which may be varied, to impact, short of or up to puncturing, a flat specimen or a formed part. The tester is instrumented and fitted with a data acquisition system, which presents the information on the impact event in the form of load-deflection signals. A detailed description of the RVSIT can be found elsewhere [10-12].

Materials

The materials tested in this study include reinforced reaction injection-molded polyurethane (RRIM) and polyester sheet molding compound (SMC) panels (supplied by the Advanced Processing and Materials Engineering Staff, General Motors Corp.). The SMC and RRIM panels were 3.2 and 2.3 mm thick, respectively. From these, 12.7 by 12.7-mm-square specimens were cut for testing. Polycarbonate and high-density polyethylene (HDPE) sheets were also tested. Commercial polycarbonate (Lexan) sheets were available in the form of 3.175-cm-thick sheets, from which 12.7 by 12.7-mm-square specimens were cut for testing.

HDPE specimens, 3.0 mm thick, were produced from sheets by injection molding at this laboratory.

Evaluation of the Impact Tester

The velocity development and variations of the RVSIT at different set speeds were evaluated with the help of an external oscilloscope. This was necessary, since the apparatus did not have the facility to record or exhibit velocity as a function of probe movement. Therefore, the appropriate electrical signals, corresponding to the probe velocity, were tapped and used to calculate the actual velocities and to compare these velocities with those commanded by the user through the tester controls.

Velocity verification tests were carried out, initially, in the absence of test specimens (dry firing).

Specimens of the four polymeric materials were then tested, under impact at different commanded speeds, and the variation of the velocity of the tester probe was determined with the help of the external oscilloscope.

The characteristics of the load-deflection signals obtained by the tester were also examined, and modifications to the probe load cell assembly were considered and implemented, in cooperation with the manufacturer.

The possibility of upgrading the use of the RVSIT, in order to elucidate the mechanism of fracture and related phenomena in polymeric materials, was also examined. In this regard, a new impact probe was designed and used, together with the proper modifications to the tester, to calculate the critical stress intensity factors (SIFs) (fracture toughness) for two of the tested materials in the opening and sliding (shearing) modes.

Results and Discussion

Evaluation and Modification of the RVSIT

Velocity Measurements—Voltage measurements of the velocity control circuit showed a linear relationship between the commanded velocity and the measured potential, covering all the operational velocity range of the machine (0.0127 to 12.7 m/s or 30 to 30 000 in./min). This relationship is shown in Fig. 1. An example of the velocity profile obtained with an external oscilloscope is shown in Fig. 2 for a commanded speed of 4.23 m/s (10 000 in./min) in dry firing (without the test specimen).

The actual velocities at dry firing were calculated from the potential signals registered and compared with the commanded velocity in every run.

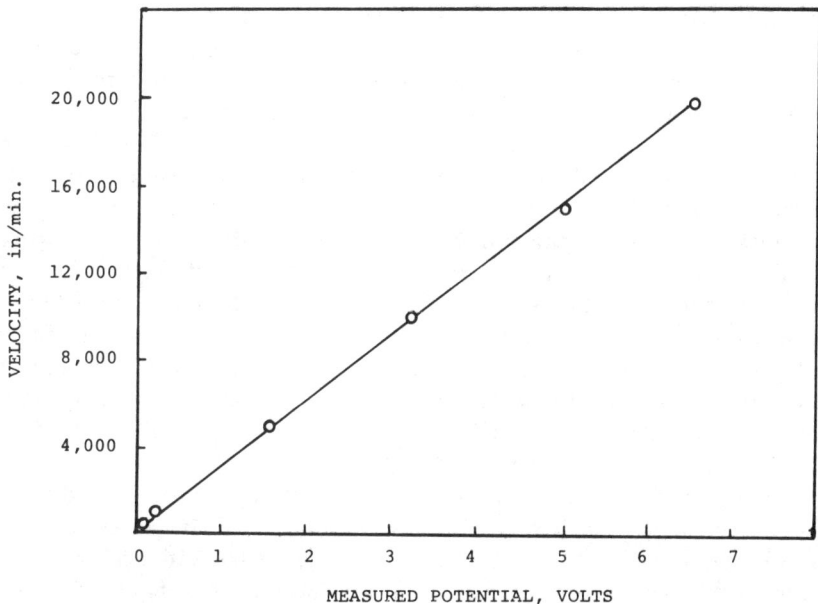

FIG. 1—*Characteristics of the velocity signal.*

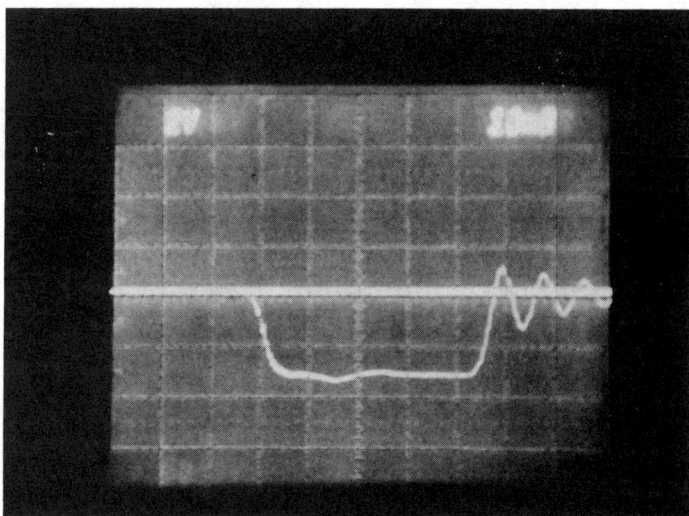

FIG. 2—*Velocity profile registered by an external scope on a dry firing at a commanded speed of 4.23 m/s (10 000 in./min).*

The comparison between the commanded velocities and the actual velocities registered under normal operating conditions shows that some differences may exist in certain ranges of velocities. This suggests that the tester velocity control circuitry should be checked and tuned regularly (the manufacturer specifies some simple tests and procedures to be performed for this purpose).

It has been observed that, while the probe rod travels with a constant speed, it experiences an instantaneous drop in velocity upon impact with a specimen. The extent of this drop was found to vary according to the impact speed (the inertia of the moving impact sled), the degree of toughness of the specimen, and its thickness. Figure 3 shows the extent of drop in speed obtained from testing a polycarbonate specimen at 6.345 m/s (15 000 in./min).

The velocity signals obtained by an external oscilloscope during the impact testing of the four materials were analyzed to establish the velocity drop upon impact in each case. The velocity drops calculated in this way for the four tested materials are plotted in Fig. 4.

The degree of resistance against the movement of the probe varied from material to material. At low speeds, polycarbonate and SMC polyester specimens brought the penetrating probe to a halt, before the feedback hydraulic control system could correct for the drop in velocity in order to complete the puncturing of the specimen.

Resistance of the materials to puncturing is also evident at the high end of the impact speed range, where polycarbonate specimens still caused a drop in velocity of close to 40%.

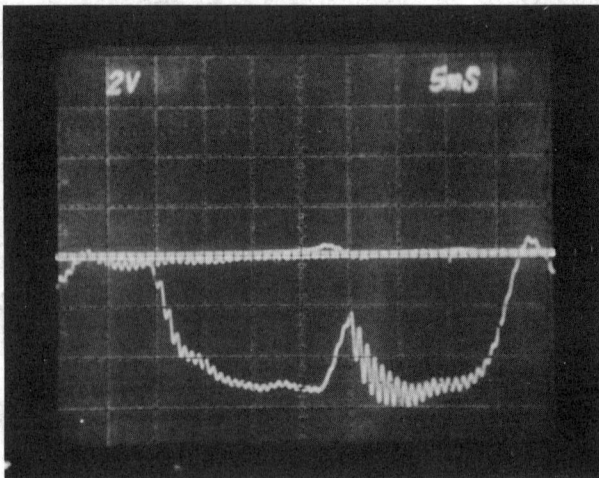

FIG. 3—*Velocity profile for testing a polycarbonate specimen at 6.345 m/s (15 000 in./min).*

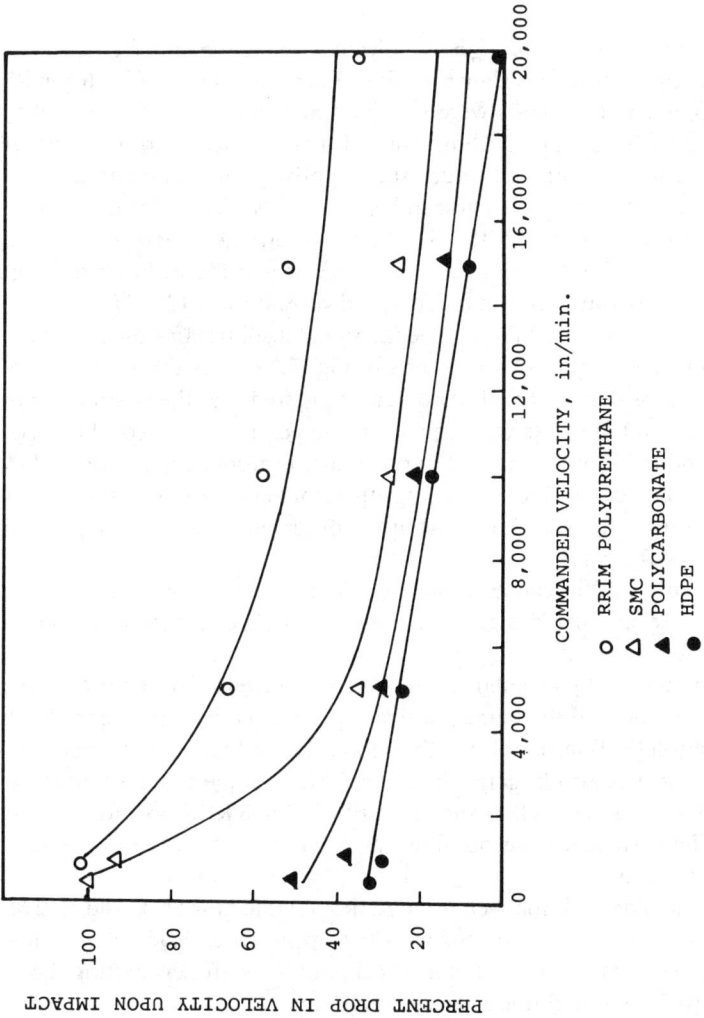

FIG. 4—*Percentage drop in velocity upon impact.*

Specimens of HDPE and RRIM polyurethane cause much less reduction in impact speed, which becomes insignificant at higher velocities.

It should be noted that these tests were carried out on a tester of 4448-N (1000-lb) nominal power. Some RVSIT models are now driven by a 221 240-N (5000-lb) power system. These models should produce a more stable velocity-displacement profile, in comparison with profiles observed in the present study.

Impact Load-Deflection Signals—Typical load-deflection signals obtained from the impact tests are shown in Fig. 5 for SMC and in Fig. 6 for RRIM polyurethane. Both signals were obtained at a commanded speed of 0.423 m/s (1000 in./min). The authors noticed that the clear and noise-free load-deflection signals, obtained under such relatively moderate speeds, become rather noisy and nonreproducible at higher speeds. For example, when signal reproducibility was tested for RRIM polyurethane specimens at a commanded speed of 0.212 m/s (500 in./min), it was found to be quite good. Conducting tests at a commanded speed of 8.46 m/s (20 000 in./min) resulted in poor reproducibility, while the signal itself became quite noisy in the postperforation stage, as can be seen in Fig. 7.

Further consideration of the problem suggested that the positioning of the load cell behind the penetrating probe is the possible source of the noise and poor reproducibility obtained at high speeds. Positioned in the rear, the load cell would be susceptible to picking up resonance and postpuncture back-traveling waves. This would result in the observed noise and the poor reproducibility of the results.

As a result, and in cooperation with the manufacturer, a new probe-load cell assembly has been designed, in which the load cell is mounted on the tip of the probe.

As expected, the placement of the load cell at the tip of the probe (that is, between the rod and the hemispherical tip) resulted in better reproducibility of the load-deflection signal. Furthermore, the problem of postpuncture spurious signals was also largely eliminated. An example of data obtained with the tip-mounted load cell is shown in Fig. 8 for RRIM polyurethane specimens. These signals were obtained at a commanded speed of 8.466 m/s (20 000 in./min).

A summary of the impact characteristics obtained for SMC and RRIM materials, at three different speeds with the tip-mounted load cell, is shown in Table 1. The data also include statistical parameters for evaluating the reproducibility of the signal under these conditions.

RRIM polyurethane, SMC polyester, and polycarbonate specimens were tested at room temperature at six different speeds, ranging from 0.212 m/s (500 in./min) to 8.466 m/s (20 000 in./min). A summary of the results obtained is shown in Table 2. In these data, the term *elastic energy* refers to the energy absorbed under impact up to the end of the elastic deformation range. The term *yield energy* refers to the total energy absorbed up to total failure. It

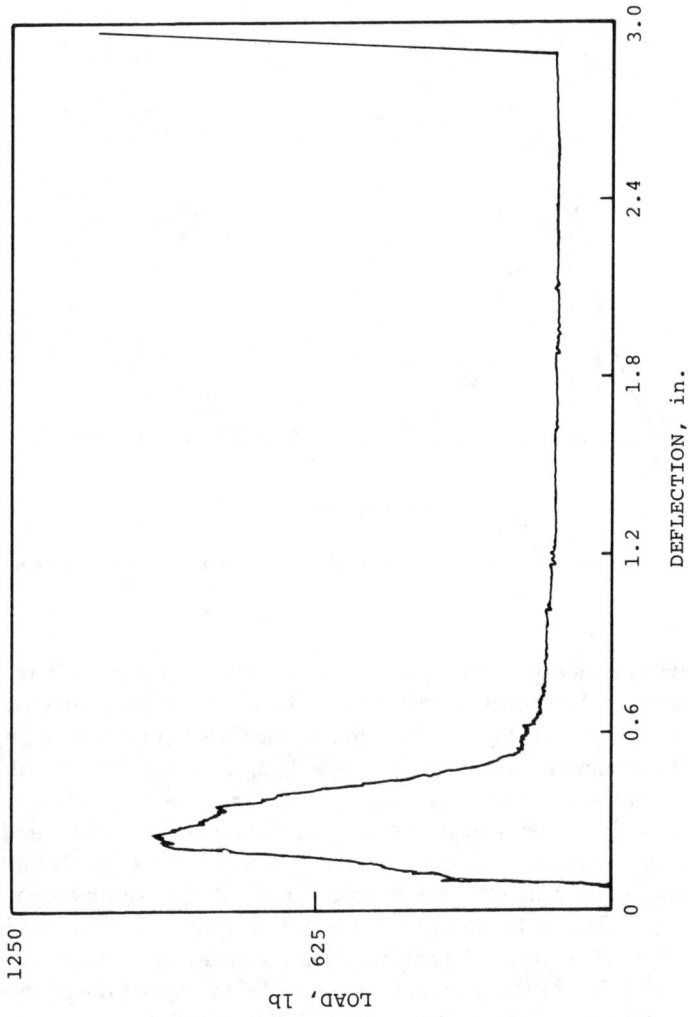

FIG. 5—*Load-deflection for SMC at a commanded impact speed of 0.423 m/s (1000 in./min).*

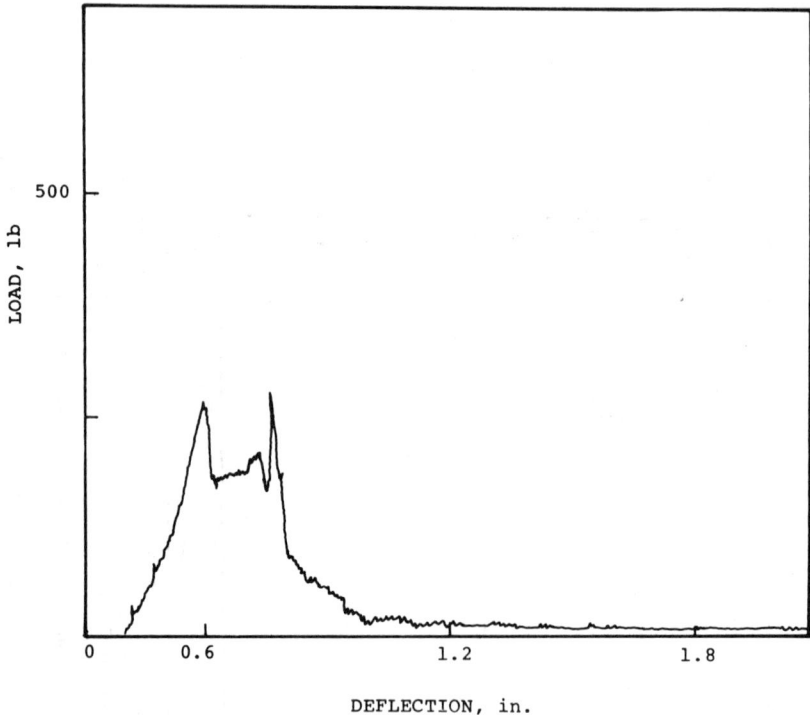

FIG. 6—*Load-deflection for RRIM polyurethane at a commanded speed of 0.423 m/s (1000 in./min).*

thus includes the energy elastically absorbed as well as that used up in permanently deforming the specimen prior to total failure when yield obtains.

The data represent averages of three to four measurements for each impact velocity. The absence of ultimate values reflects the absence of an identifiable yield section in the load-deflection signal.

A secant modulus, defined as the ratio between maximum load and the corresponding deflection, was calculated at different impact speeds for both RRIM polyurethane and SMC specimens. The results for both materials are shown in Fig. 9. The plots show that the secant modulus is more or less independent of impact speed for the polyester composite, while it shows a drop of more than 30% for RRIM polyurethane in the impact speed range between 0.423 m/s (1000 in./min) and 8.466 m/s (20 000 in./min).

Fracture Toughness

An attempt was made to evaluate the use of the RVSIT to measure material resistance to fracture. The approach was based on concepts related to the theory of linear elastic fracture mechanics (LEFM) [13].

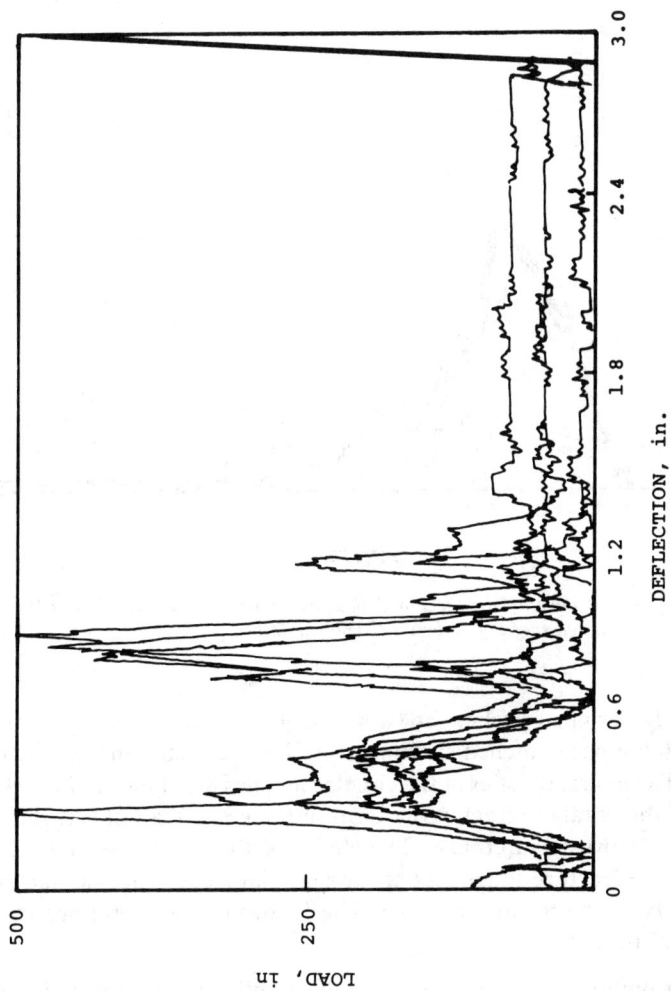

FIG. 7—Reproducibility of the load-deflection signal for RIM polyurethane at a commanded speed of 8.46 m/s (20 000 in./min).

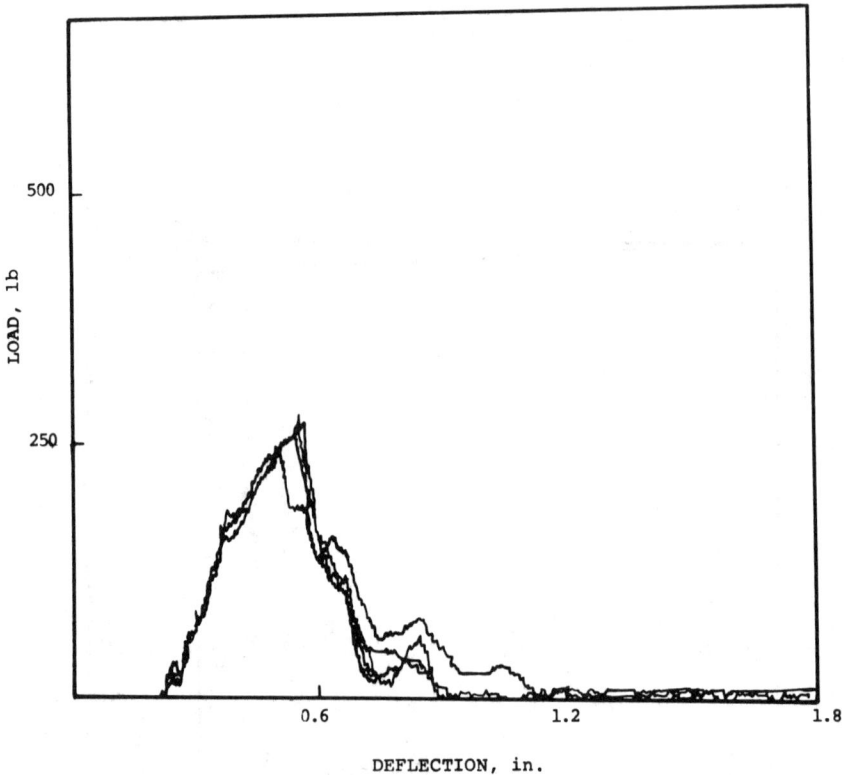

FIG. 8—*Reproducibility of load-deflection data using a tip-mounted load cell for RRIM poly-urethane at a commanded speed of 8.46 m/s (20 000 in./min).*

A specially designed probe, shown in Fig. 10, was used to impact rigidly supported (clamped) notched specimens of SMC polyester and RRIM poly-urethane. The characteristics of the circular notches are shown in Fig. 11. An example of the obtained fracture patterns of these two materials is shown in Fig. 12 for RRIM polyurethane. The choice of these notch geometries de-pended on the fracture toughness of the material tested and the maximum load capacity of the testing machine. The following considerations favored the choice of these geometries.

1. The geometries are axisymmetric, which both considerably reduces the complexity of the finite element analysis and ensures that plane strain condi-tions are maintained throughout the fracture process.

2. The choice of these geometries required minimum mechanical change in the RVSIT used and no electronic modifications in the circuitry.

3. The notch depths were dictated by the desire to have a notch size that would preclude extensive weakening of the specimens. The upper limit on the

TABLE 1—Reproducibility of selected impact parameters with the tip-mounted load cell.[a]

Material	Impact Parameter	Impact Speed					
		0.423 m/s (1000 in./min)		4.23 m/s (10 000 in./min)		8.46 m/s (20 000 in./min)	
		Average	Coefficient of Variation	Average	Coefficient of Variation	Average	Coefficient of Variation
RSMC	Elastic slope, lb/in.	4174 ± 350	0.119	4500 ± 445	0.098	5216 ± 320	0.098
	Energy at maximum load, lb · in.	138 ± 17	0.073	155 ± 15	0.11	161 ± 23	0.126
RRIM	Elastic slope, lb/in.	1744 ± 100	0.096	1610 ± 230	0.1	1520 ± 420	0.11
	Energy at maximum load, lb · in.	62 ± 4	0.078	58 ± 7	0.09	54 ± 10	0.159

[a]Metric conversion factors:
1 in. = 0.0254 m = 1000 mil.
1 lb/in. = 175.1 N/m.
1 lb · in. = 0.113 N · m.

TABLE 2—Impact characteristics as obtained from the RVSIT.ᵃ

Speed, in./min (m/s)	Elastic Slope, lb/in.	Yield Load, lb	Yield Deflection, mil	Elastic Energy, lb·in.	Ultimate Load, lb	Ultimate Deflection, mil	Yield Energy, lb·in.
RIM POLYURETHANE							
500 (0.211)	920	260	295	20	220	475	60
1 000 (0.423)	806	261	314	23	285	482	55
5 000 (2.115)	965	160	294	17	246	400	49
10 000 (4.23)	1135	179	295	14	250	395	46
15 000 (6.345)	1373	214	296	18	171	490	53
20 000 (8.46)	1328	199	298	14	160	477	41
SMC							
500 (0.211)	7890	900	274	50	...	...	...
1 000 (0.423)	5270	910	279	65	...	...	...
5 000 (2.115)	6250	920	266	68	...	...	...
10 000 (4.23)	7890	900	274	50	...	...	...
15 000 (6.345)	7930	1000	285	70	...	...	...
20 000 (8.46)	6320	940	307	90	...	...	...
POLYCARBONATE							
500 (0.211)	2620	1520	732	480	...	...	...
1 000 (0.433)	2880	1520	693	460	...	...	...
5 000 (2.115)	3990	1540	700	500	...	...	...
10 000 (4.23)	3240	1580	730	520	...	...	...
15 000 (6.345)	1940	1540	777	610	...	...	...
20 000 (8.46)	3830	1070	510	210	1460	708	470

ᵃMetric conversion factors:
1 mil = 0.0254 mm.
1 lb = 0.4536 kg.
1 lb/in. = 175.1 N/m.
1 lb · in. = 0.113 N · m.

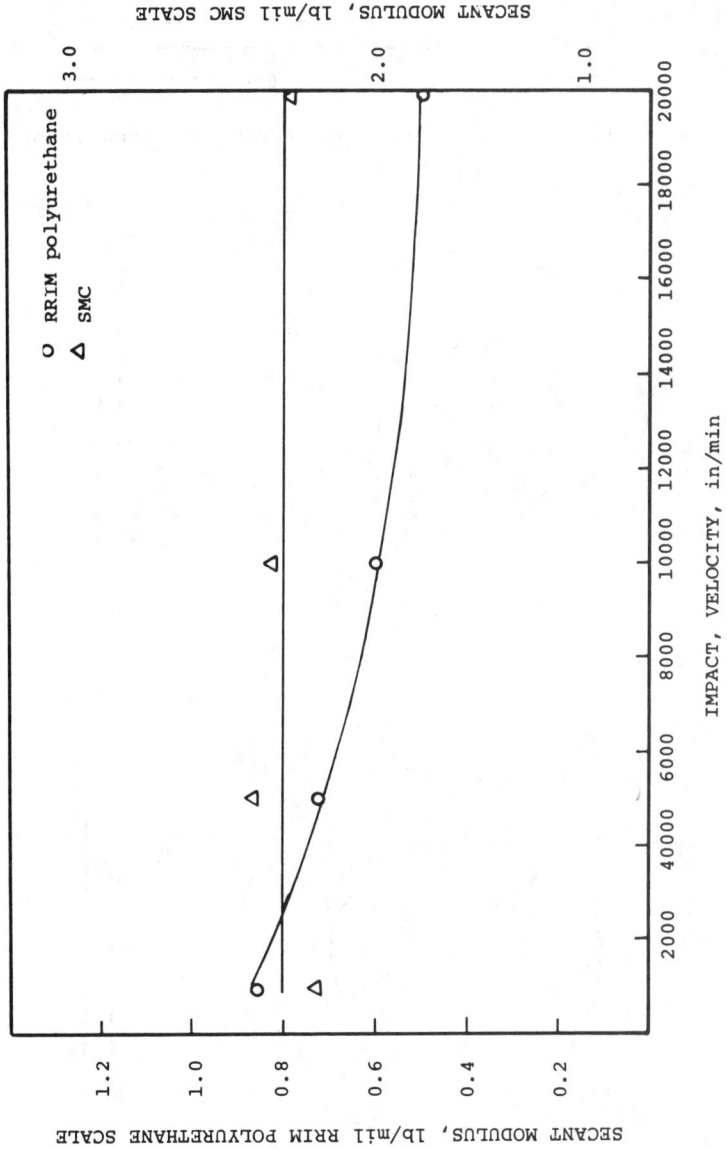

FIG. 9—*Variation of secant modulus as a function of impact velocity.*

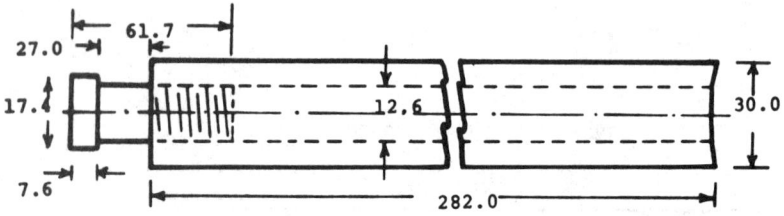

FIG. 10—*New probe for circumferential loading with a circularly notched specimen.*

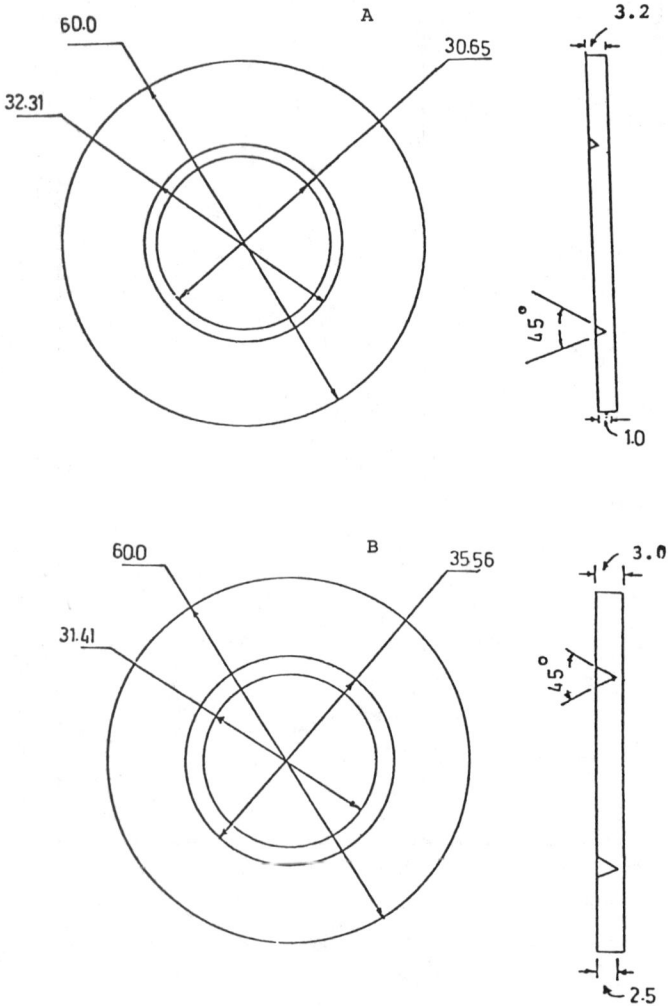

FIG. 11—*Notch and specimen geometry for circumferential loading testing with the new probe:* (a) *RRIM polyurethane;* (b) *SMC.*

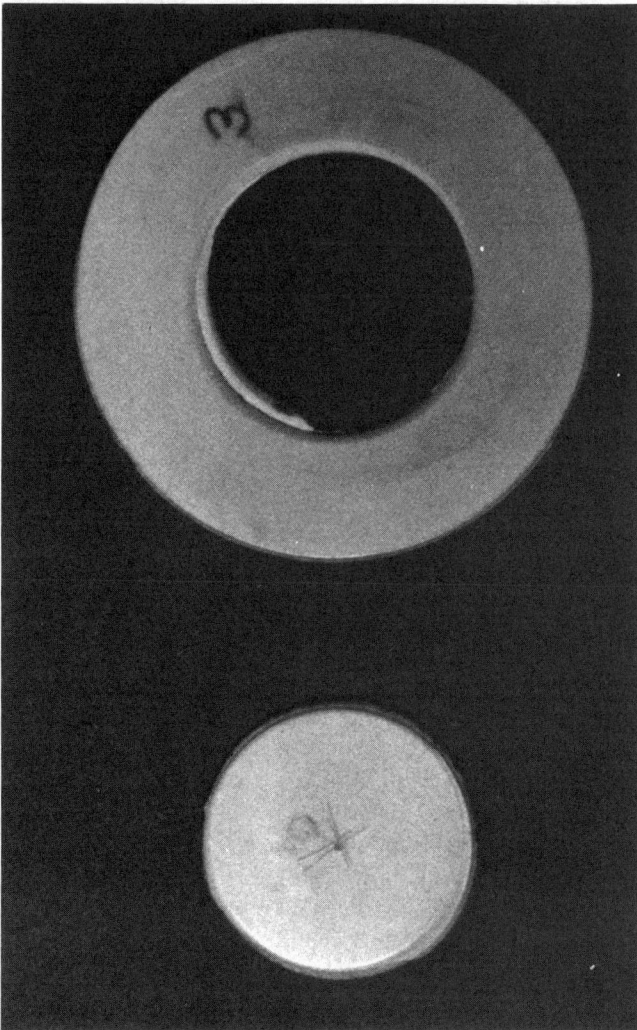

FIG. 12—*RRIM polyurethane notched specimen after impact by the ring load probe.*

strength of the specimens was the maximum hydraulic power of the tester. The chosen geometries ensured that both limitations were respected for both tested materials.

A typical load-deflection signal, obtained with the ring probe, is shown in Fig. 13 for RRIM polyurethane, at an impact speed of 4.23 m/s (10 000 in./min).

Finite Element Analysis (FE)—The FE approach used in this study is

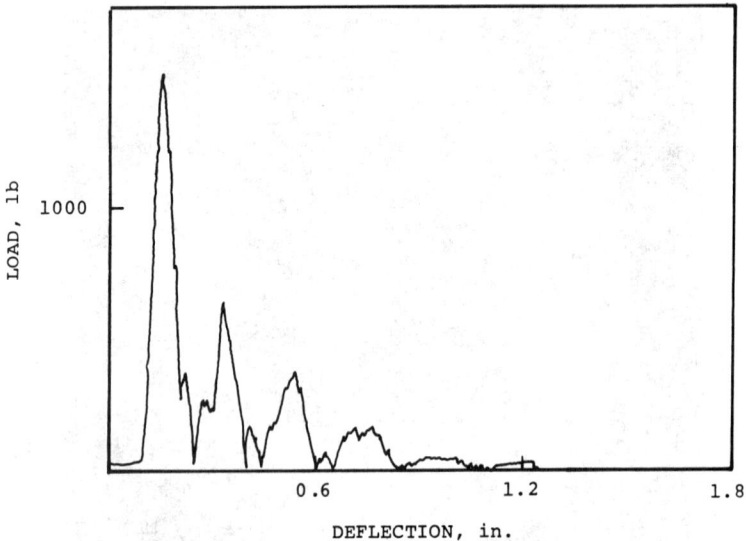

FIG. 13—*Load-deflection from a notched RRIM polyurethane specimen tested under circumferential loading conditions at a commanded speed of 4.23 m/s (10 000 in./min).*

based on two-dimensional axisymmetric techniques justified by the loading patterns and the geometry of the specimens tested. In this approach, the continuous stress-strain fields were approximated by isoparametric elements with the number of nodes defining the element shape corresponding to the number of those used in determining the interpolation polynomial. The analysis takes into account the singularity developing in the stress field around the crack tip (the notch) by choosing the isoparametric elements in a quadrilateral form with appropriately positioned side nodes, as proposed by Lynn and Ingraffea [14].

The Computation of SIF from Nodal Displacements—On the theoretical side, expressions for the two-dimensional displacements in the vicinity of a crack tip have been developed by combining the series solutions developed by Williams [15] with the SIF definitions. A convenient form of writing this displacement field is given, with reference to Fig. 14, in Ref 16.

$$u = \frac{K_{\mathrm{I}}}{4\mu} \left(\frac{r}{2\pi}\right)^{1/2} \left[(2k + 1) \cos \frac{\psi'}{2} - \cos \frac{3\psi'}{2}\right]$$

$$+ \frac{K_{\mathrm{II}}}{4\mu} \left(\frac{r}{2\pi}\right)^{1/2} \left[(2k + 3) \sin \frac{\psi'}{2} + \sin \frac{3\psi'}{2}\right]$$

$$(1)$$

$$v = \frac{K_I}{4\mu} \left(\frac{r}{2\pi}\right)^{1/2} \left[(2k + 1) \sin \frac{\psi'}{2}\right]$$

$$- \frac{K_{II}}{4\mu} \left(\frac{r}{2\pi}\right)^{1/2} \left[(2k - 3) \cos \frac{\psi'}{2} + \cos \frac{3\psi'}{2}\right]$$

where

$k = (3 - 4\nu)$ for plain strain ($\nu =$ Poisson's ratio),
$\mu =$ the shear modulus,
$u =$ displacement in the opening mode, and
$v =$ displacement in the sliding mode.

All other terms are defined in Fig. 14.

On the FE side, analyses leading to SIF from quarter-point crack-tip element nodal displacements are documented by Ingraffea [17] and Ingraffea and Cormeilliu [18]. With reference to Fig. 14, the displacement field along any edge of a quarter-point element (represented by the edge ABC), in terms of the nodal displacements is given by

$$u = u_A + [4(u_B - u_A) - (u_C - u_A)] \frac{r}{L}$$

$$+ [2(u_C - u_A) - 4(u_B - u_A)] \frac{r}{L}$$

$$v = v_A + [4(v_B - v_A) - (v_C - v_A)] \frac{r}{L}$$

$$(2)$$

$$+ [2(v_C - v_A) - 4(v_B - v_A)] \frac{r}{L}$$

For any specific value of ψ' that corresponds to the edge of an element, K_I and K_{II} can be found from the two simultaneous equations formed by equating the $r^{1/2}$ coefficients of Eqs 1 and 2.

Finite Element Grids—The FE grid used in the analysis of Geometry 1 (Fig. 11a) is shown in Figs. 15 and 17. The grid consists of 127 nodes and 76 axisymmetric standard transition and quarter-point elements. Elements 40 through 47 are crack-tip elements with their midside nodes, Nodes 66 through 70 and 98 through 101, shifted to their quarter-point locations. Elements 36 through 39 and 48 through 51 are transition elements with their isoparametric midside nodes, Nodes 52 through 56 and 84 through 87, shifted to their 0.46 L location [14]. The FE grid used in the analysis of Geometry 2

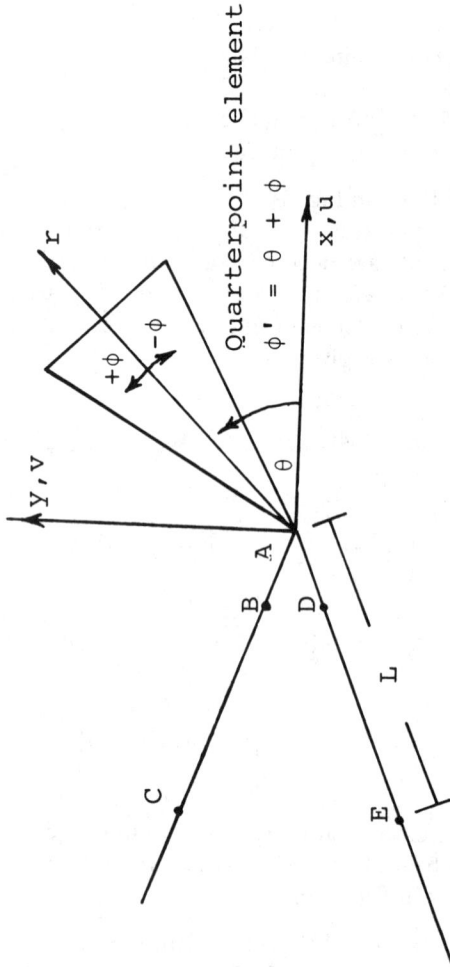

FIG. 14—*Crack-tip elements and coordinates.*

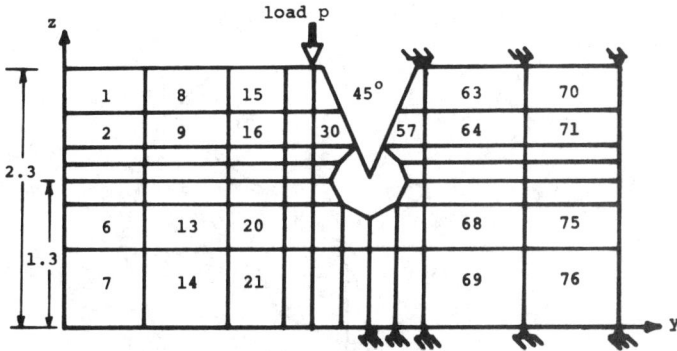

FIG. 15—*Final FE grid for Geometry I.*

(Fig. 11*b*) is indicated in Figs. 16 and 17. Apart from the overall specimen thickness and the notch depth, Geometry 2 corresponds as closely as possible to Geometry 1.

Calculation of SIFs—Load-deflection data were obtained for both geometries (corresponding to the two tested materials) at different speeds using the RVSIT. Since the analysis is linear elastic with zero initial displacement, any one circumferential load, P, could allow the determination of the linear relationship between the load and the stress-intensity factors for each geometry.

A finite-element program based on the NONSAP structural analysis program [19] was used according to the approach described earlier to produce displacements in the opening and sliding modes. The program calculated the displacements corresponding to the nodes used to establish the grids in both

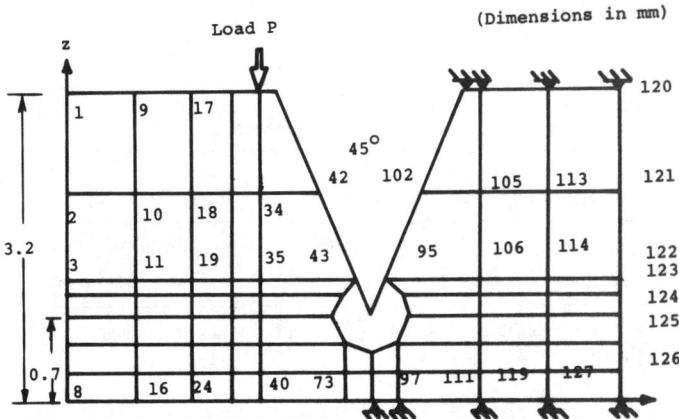

FIG. 16—*Final FE grid for Geometry II.*

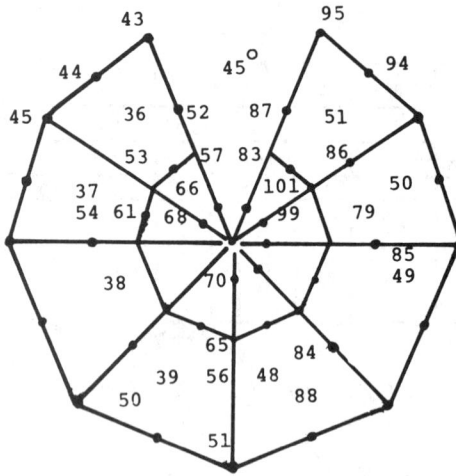

FIG. 17—*The quarter-point and transition elements of Geometries I and II.*

geometries for an arbitrary load, P, of 1 kN/rad (225 lb/rad). From the listing produced, appropriate nodal displacements were extracted and used for the calculation of K_I and K_{II} for each geometry. The results are listed in Table 3.

Using the failure dynamic loads obtained from the RVSIT and the relations shown in Table 3, K_{IdQ} and K_{IIdQ} were calculated at different speeds for both materials (and their corresponding geometries).[3] The results are shown in Table 4. The results also include estimates of the critical energy release rate, G_{dQ}, which is calculated from the obtained dynamic fracture toughness for each material.

TABLE 3—*The SIFs for both specimen geometries.*

Geometries	K_I, MPa m$^{1/2}$ (ksi in.$^{1/2}$)	K_{II}, MPa m$^{1/2}$ (ksi in.$^{1/2}$)
I	1.24 P^a (0.005 P)	0.95 P (0.0038 P)
II	0.79 P (0.0032 P)	1.1 P (0.0044 P)

$^a P$ is in kilonewtons per radian; pounds per radian are indicated in parentheses.

[3]The subscript d relates to the fact that the calculations are based on dynamic measurements, while the subscript Q indicates that the parameter has not yet been accepted by ASTM as a fracture toughness property.

TABLE 4—*Estimated values of* K_{IdQ}, K_{IIdQ}, *and* G_{dQ} *for RIM and SMC.*

Impact Speed, in./min (m/s)	RRIM			SMC		
	K_{IdQ}, MPa m$^{1/2}$	K_{IIdQ}, MPa m$^{1/2}$	G_{dQ}, kJ/m^2	K_{IdQ}, MPa m$^{1/2}$	K_{IIdQ}, MPa m$^{1/2}$	G_{dQ}, kJ/m^2
1 000 (0.423)	1.24	0.96	1.05	. . .	. . .	. . .
5 000 (2.115)	1.19	0.92	0.97	1.6	2.29	2.54
10 000 (4.23)	1.33	1.03	1.11	1.23	1.76	1.49
15 000 (6.345)	0.93	0.72	0.59	1.53	2.19	2.32
20 000 (8.46)	0.97	0.75	0.64			

Conclusions

The performance of the 4448-N (1000-lb) nominal power Rheometrics variable-speed impact tester was evaluated. The constant-velocity characteristics of the tester were examined under dry firing and specimen testing conditions.

Slight differences were found to exist between the commanded impact and the actual speeds with which the impact probe travels. Regular checking and tuning of the velocity control circuitry was found adequate to reduce these differences.

A drop in speed was observed upon impact of a specimen. The extent of this drop varied according to both the impact speed and the toughness of the tested material. While the tester control system corrects for any drop in speed, complete information on speed variations during puncturing of a specimen would be necessary, together with the load-deflection data, to form a complete characterization of the tested material. Information on velocity changes is also important, in view of the fact that the deflection data generated by the machine are not measured but calculated from averaged velocities.

Recent generations of the Rheometrics impact tester have a nominal power of 22 240 N (5000 lb), which would largely reduce the extent of drop in speed upon impact. These models are also equipped to generate simultaneous load-deflection and velocity-time data.

The load cell-probe assembly was found to generate data of acceptable reproducibility at low impact speeds. Reproducibility was, however, poor at high speeds. The placement of the load cell ahead of the impact probe substantially eliminated this problem.

The tester was modified, and a new impact probe was designed to apply circumferential loads to flat specimens with circular notches. The load-deflection signals thus generated were used to calculate dynamic critical intensity factors for SMC and RRIM polyurethane panels, in both the opening, K_{IdQ}, and the sliding, K_{IIdQ}, modes.

The RVSIT provides a flexible impact testing system. Its capacity to test both flat specimens and formed parts is a useful feature in impact testing. The tester can also be fitted with a clamping mechanism to carry out tension impact tests, within the operational speed range.

Acknowledgments

The authors would like to thank the Advanced Product and Manufacturing Engineering staff, General Motors Technical Center, Warren, Michigan, for financial support and for supplying some of the materials tested in this work. Also, they would like to acknowledge the cooperation of Rheometrics Inc. in implementing design changes in RVSIT. Financial support was also received from the Natural Sciences and Engineering Research Council of Canada and the Ministère de l'Education of the Government of Quebec.

References

[1] Cessna, L. G., Lehave, J. P., Ralston, R. H., and Prindle, T., *Polymer Engineering and Science,* Vol. 16, No. 6, 1976, pp. 419-425.

[2] Radon, J. C., *Journal of Applied Polymer Science,* Vol. 22, 1978, pp. 1569-1581.

[3] Gonzalez, H., *Journal of Applied Polymer Science,* Vol. 19, 1975, pp. 2717-2730.

[4] Yeung, P. and Broutman, L. J., *Polymer Engineering and Science,* Vol. 18, No. 2, 1978, pp. 62-72.

[5] Turner, S., Reed, P. E., and Money, M., *Plastics and Rubber Processing Applications,* Vol. 4, No. 4, 1984, pp. 369-378.

[6] Caprino, G., Crivelli Visconti, I., and Di Ilio, A., *Composite Structures,* Vol. 2, 1984, pp. 261-271.

[7] Wnuck, A. J., Ward, T. C., and McGrath, J. E., *Polymer Engineering and Science,* Vol. 21, No. 6, 1981, pp. 313-324.

[8] Delpy, U., *Kunststoffe,* No. 72, 1982, pp. 476-478.

[9] Klunder, J., *Kunststoffe,* No. 72, 1982, pp. 548-553.

[10] Owen, G. E., *Polymer Engineering and Science,* Vol. 21, No. 8, 1981, pp. 467-473.

[11] Nimmer, R. P., *Polymer Engineering and Science,* Vol. 23, No. 3, 1983, pp. 155-164.

[12] "High Speed Impact Tester: A Preliminary Manual," Rheometrics Inc., Union, NJ, 1978.

[13] Williams, J. G. in *Failure in Polymers, Advances in Polymer Science,* Vol. 27, Springer-Verlag, Berlin, 1978, Chapter 2, pp. 67-120.

[14] Lynn, P. P. and Ingraffea, A. R., *International Journal for Numerical Methods in Engineering,* Vol. 12, 1978, pp. 1031-1036.

[15] Williams, M. L., *Journal of Applied Mechanics,* Vol. 24, 1957, pp. 109-114.

[16] Rice, J. R., *Journal of Applied Mechanics,* Vol. 35, 1968, pp. 379-386.

[17] Ingraffea, A. R., *Proceedings,* Numerical Methods in Fracture Mechanics, Swansea, Wales, January 1978, pp. 235-238.

[18] Ingraffea, A. R. and Cormeilliu, M., *International Journal for Numerical Methods in Engineering,* Vol. 15, 1980, pp. 1427-1445.

[19] Bathe, K.-J., Wilson, E. L., and Idin, R. H., "NONSAP, A Structural Analysis Program for Static and Dynamic Response of Nonlinear Systems," Structural Engineering Laboratory, University of California, Berkeley, CA, February 1974.

Patrick J. Cain[1]

Digital Filtering of Impact Data

REFERENCE: Cain, P. J., **"Digital Filtering of Impact Data,"** *Instrumented Impact Testing of Plastics and Composite Materials, ASTM STP 936,* S. L. Kessler, G. C. Adams, S. B. Driscoll, and D. R. Ireland, Eds., American Society for Testing and Materials, Philadelphia, 1987, pp. 81–102.

ABSTRACT: Idealized load signals were created to represent the failure of various types of materials under impact loading. The frequency content was investigated using Fourier transform methods. Several types of filtering approaches were applied to these signals. A numerical model of the impact process was used to generate data to study sources of noise in the impact process. Based on these investigations, a filter design was proposed and evaluated on examples of data from experiments. The conventional approaches to filter techniques may not be suitable for high rate data.

KEY WORDS: instrumented impact testing, frequency analysis, Fourier transform, filters, impact testing

High-speed impact testing of materials, whether they are in a compressive tensile, fracture, or puncture mode, may lead to the problem of separating the significant signal from background noise. The information, such as peak load, final fracture point, or cracking prior to final fracture, may be masked by noise. Through proper design of fixtures and instrumentation this noise can be minimized but will be present because of the fundamental nature of the impact process itself.

Filtering may be desirable to clarify the data. If required, digital techniques offer the most promise. Analog filtering increases the risk that information will be lost or masked.

The approach taken was to investigate, first, the nature of the signals derived from impact phenomena by examining the characteristics of ideal data using a frequency domain or Fourier transform approach. The effect of adding noise and applying some simple forms of filters will then be discussed. Next, the sources of the noise in the impact process itself will be modeled and characterized. This model will be used to examine the effect of processing or

[1]Senior application engineer, MTS Systems Corp., Minneapolis, MN 55424.

filtering the data. The goal is to define filter types that would be useful in removing unwanted noise while preserving the important characteristics.

Sources of noise in impact problems of various geometries have been discussed in the past. Turner [1] refers to a model for noise generation that shows the oscillation as being triggered by the impact process itself. Ireland [2] discusses the effects of instrumentation response in the excitations of oscillations by the initial impact. Also, in Ref 3, Saxton et al. have discussed the effects of dynamics and instrumentation in the impact process.

The approach used here is based on frequency domain methods using the fast Fourier transform technique. The Fourier transform approach is interesting for several reasons. It can be applied to analyzing actual data since, in modern impact testing systems, data are collected by some type of digital storage device and can then be transferred to a computer for processing. The fast Fourier transform routines make the process of converting the data to a frequency representation and then modifying the spectrum a practical approach. Also, the frequency domain approach ties together the issues of noise, data content, and filter behavior since all can be analyzed using available methods in the frequency domain.

The results of this approach can then be implemented in a practical manner on a computer-equipped test system. The use of a general-purpose minicomputer allows flexibility in the handling of data and the ease of implementation of digital methods for processing the data.

Idealized Impact Histories

To understand the possible problems in filtering data, it is of interest to look at the Fourier representation of some idealized impact load histories. This will demonstrate that many shapes of technical interest have very wide frequency spectra and that key parameters, such as peak strength or location of fracture points, may be heavily dependent on the higher-frequency terms.

The Fourier series is often used in mathematics to solve problems involving functions that do not have a simple representation. These functions are then represented (Eq 1) as a series of sines and cosines of frequencies, f_i, and amplitudes, A_i and B_i.

$$F(t) = \sum_{i=0} (A_i \sin 2\pi f_i t + B_i \cos 2\pi f_i t) \qquad (1)$$

The same concept can be extended to represent sampled data [4], as might be taken from a high-speed data acquisition device in an impact test, in terms of a series of sinusoidal functions at discrete frequencies. The discrete Fourier transform representation of sampled data has a transform given by the following

$$x(n) = \frac{1}{N} \sum_{k=0}^{N-1} X(k) e^{j(2\pi/N)nk} \tag{2}$$

where N is the number of data samples in the record with time between samples. Each frequency coefficient, $X(\ )$, is a complex number representing the magnitude and phase of that component. An inverse transform is given by the following

$$X(k) = \sum_{n=0}^{N-1} x(n) e^{-j(2\pi/N)nk} \tag{3}$$

This transform pair allows one to represent sampled data in terms of its frequency components and then, after modifying the frequency components (which is the filtering operation), use the inverse transform to return to the time domain and examine the modified time history. An extension of this has been developed for implementation on computers. It is called the fast Fourier transform. Imposing certain constraints on the relationship between the frequencies and time intervals allows the transform and inverse transform to be computed very quickly. This has made frequency domain processing on actual data very practical for routine use.

In addition to the load–displacement history, the calculation of energy is an important part of the analysis of impact data. The filtering process, which can be expressed in an equation as

$$y(f) = H(f) X(f) \tag{4}$$

is the modification of the function $X(f)$, which is the transform of a time history, $x(t)$, by $H(f)$, which represents the filter process to produce $Y(f)$, the transform of the output, $y(t)$.

The effect of filtering on the energy calculation can be found using the series representation, Eq 1, of the load history. The energy, E, is given by

$$E = \int_{0}^{T} PV \, dt \tag{5}$$

where P is the load signal and V the velocity. Assuming the velocity is a constant V_0 and the load signal can be represented by

$$p(t) = \sum_{i=0}^{\infty} [A_i \sin 2\pi f_i t + B_i \cos 2\pi f_i t] \qquad (6)$$

then integrating

$$E = V_0 \sum_{i=0}^{\infty} \left[A_i \left(\frac{1 - \cos 2\pi f_i T}{2\pi f_i} \right) + \frac{B_i \sin 2\pi f_i T}{2\pi f_i} \right] \qquad (7)$$

the dependence of the energy on the higher-order terms will be less than the load history, since the denominator will be an increasingly large number.

One potential problem in any use of sampled data is the possibility of aliasing. This occurs when there is a frequency component higher than the sampling rate. For accurate definition of the data, the Nyquist criteria should be applied, and the sampling rate should be twice the highest frequency present in the data. As will be seen, many real signals may contain very high frequency components, so this criterion cannot be absolutely adhered to. However, in practical terms, it means that there should not be any significant harmonic content above one half the sample rate. Normally, this will be limited by the frequency response or bandwidth of the instrumentation itself, and it is also possible to apply antialiasing filters to limit these higher frequencies. If they are not limited, they will appear as content in the lower-frequency range and will result in the distortion of the signal.

However, as discussed in Ref 5, the filtering techniques themselves, even antialiasing filtering techniques, do have potential problems in that they can induce ringing in the sample time history. This is due to the shape of the cutoff of the antialiasing filter and distortion caused by phase shifting of the various frequency components. Fundamentally, a signal that does not contain ringing may have ringing after the signal has been passed through such a filter. This is due to filter characteristics and is not a true characteristic of the data.

Examples of some idealized load histories that might be expected from impact tests are shown in Figs. 1, 2, and 3. With each load history is presented the frequency spectra or amplitude of frequency components for that load history.

They are intended to approximate the behavior of materials with brittle, ductile-brittle, and ductile failure characteristics. The important feature to note is that for each shape, even that approximated by a single sinusoidal cycle, there is a wide range of frequency content. These higher frequencies come about from the fact that each of these load histories is really a combination of a simple geometrical form such as a ramp or sine wave and a step or boxcar function. Also, it is important to note that these higher frequencies are present even though the data are free of noise.

The boxcar function has the value of zero everywhere except over a certain

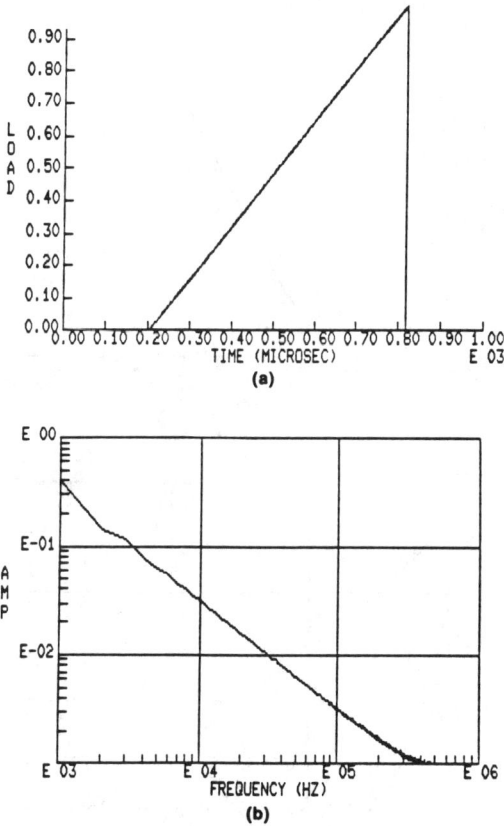

FIG. 1—*Time history* (a) *and frequency spectrum* (b) *for a material with brittle failure characteristics.*

range that has a value of 1. An example is the frequency spectra of the pulse represented in single sine cycle (Fig. 3). This is not only the frequency content associated with the sine wave but also that associated with the boxcar function, which reduces it to a single cycle. The basic boxcar function and its transform are shown in Fig. 4. This function requires higher harmonics in its Fourier representation. These higher harmonics are associated with the step function at either edge of the boxcar. In impact data phenomena related to sudden fracture or brittle behavior, resolution of the sharp edges and peaks requires the high frequencies of the wave shape be preserved. The symmetry of the transform pairs in Eqs 2 and 3 points to another phenomenon that can occur when filtering. If sharp discontinuities are made in the transform, oscillations will appear in the data. This is the complement to the case in which a sharp pulse in a time history excites many frequency components.

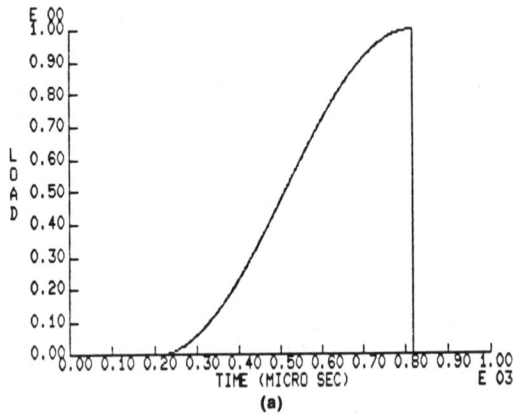

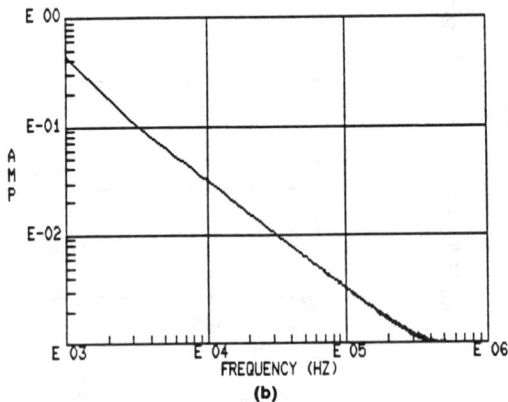

FIG. 2—*Time history* (a) *and frequency spectra* (b) *for a material with ductile-brittle failure characteristics.*

A highly idealized example of combining noise with the load history spectrum is shown in Fig. 5. The simple sinewave has been imposed on the complete time history for a ductile-brittle material (Fig. 2) and appears in the spectrum as an increased amplitude at a single frequency. In reality, the addition of noise, even of a single frequency, is not so straightforward because it may occur in only part of the time history. When that is the case, sinusoid function is combined with a boxcar, which results in additional frequency content and spreading of the peak.

One approach to filtering the data is shown in Fig. 6. In the first case, a very simple notch filter was constructed by removing the frequency content associated with the superimposed noise. However, complete removal of these

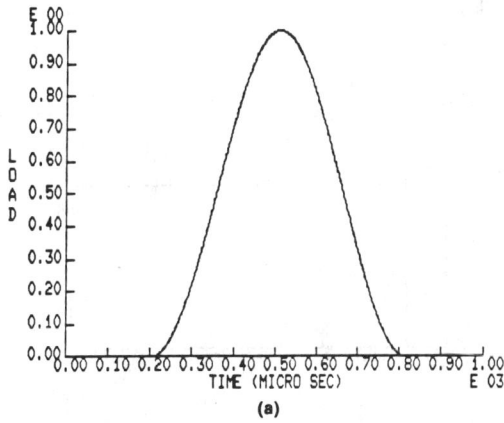

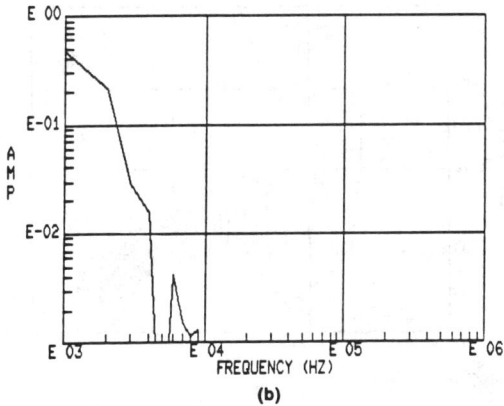

FIG. 3— *Time history* (a) *and frequency spectra* (b) *for a material with ductile failure charac-teristics.*

frequencies results in distortion of the reconstructed signal, since a certain portion of the harmonic content is required to reconstruct the actual load history.

In the example in Fig. 7, the spectrum is filtered by using a two-pole filter with characteristics similar to many instrumentation amplifiers. This also results in ringing because part of the frequency content removed was, in fact, necessary to accurately describe the shape of the original load history. The noise is still present, though at a reduced amplitude. This illustrates the basic problem in trying to design a filter; the objective is to remove the unwanted noise while maintaining the portions of the frequency spectra or harmonic content that describe the original history. The finite impulse response ap-

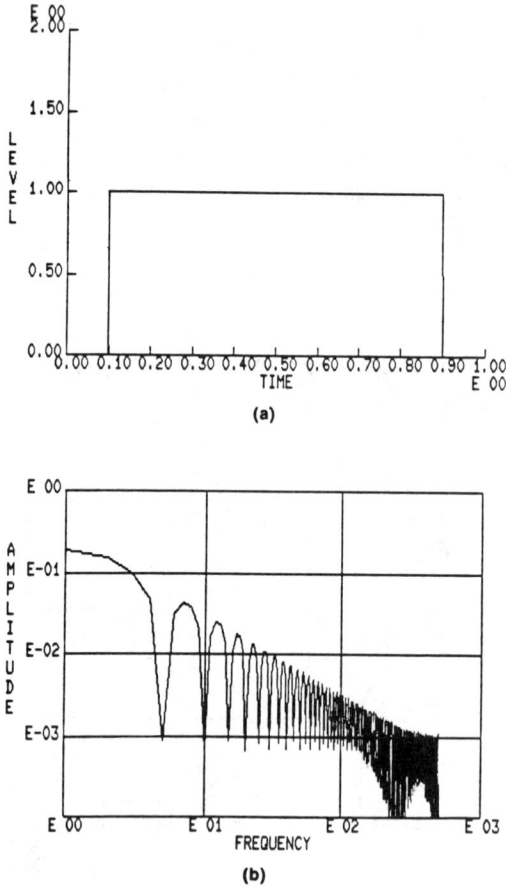

FIG. 4—*Time history of the boxcar function* (a) *and the frequency spectrum* (b).

proach to filter design [6] allows great flexibility in setting the shape. Since impact data can be processed in the frequency domain, it is not necessary to develop the time domain equivalent.

The problems with trying to implement an ideal filter with sharp edges are discussed in Ref 6. To obtain the best result, it is desirable to have a smooth transition at the edge. Following this concept, the use of a low-pass filter with the following characteristics was investigated.

$$|H(f)| = 0 \qquad \text{when } f < f_0 - Cf_0$$

$$|H(f)| = \frac{1}{2} + \frac{1}{2} \cos\left[\left(\frac{f + Cf_0 - f_0}{2Cf_0}\right)\pi\right]$$

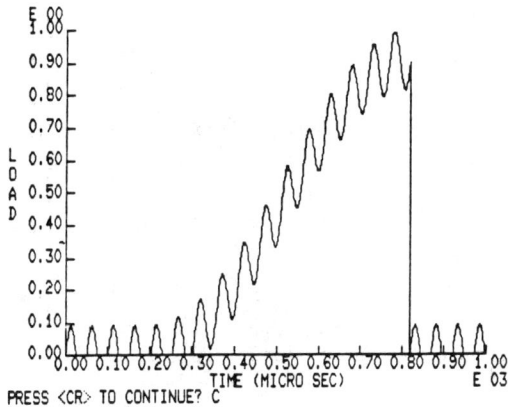

PRESS <CR> TO CONTINUE? C

(a)

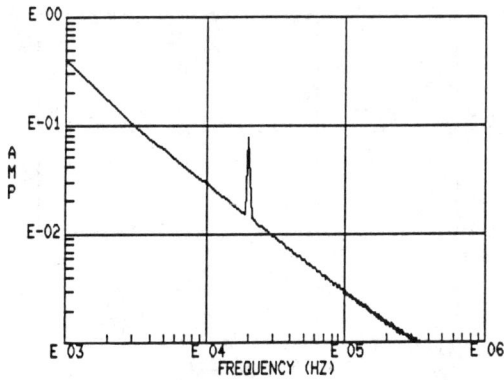

(b)

FIG. 5—*Load history with a single frequency noise added* (a) *and its frequency spectrum* (b).

$$\text{when } f_0 - Cf_0 < f < f_0 + Cf_0 \qquad (8)$$

$$|H(f)| = 0 \qquad \text{when } f > f_0 + Cf_0$$

The variable f_0 defines the corner frequency, and C defines the width of the transition. The filter was applied to the time history for a ductile-brittle material, and the result is shown in Fig. 8. The noise is removed, but there is some loss of definition of the breakpoint. The next step is to investigate more realistic noise conditions.

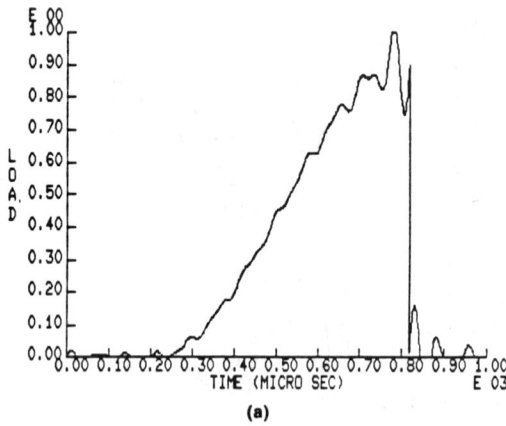

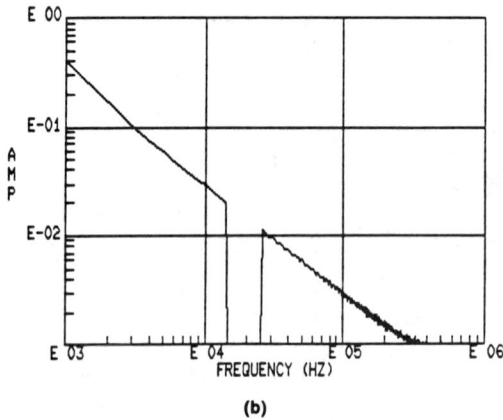

FIG. 6—*Time history* (a) *and frequency spectrum* (b) *after application of an ideal notic filter to the load history of a ductile-brittle material.*

Model of Impact Problem

The dynamic model for the impact test system in terms of mass and stiffness elements is shown in Fig. 9. This basic model is intended to represent a dart or puncture impact type of event. With minor changes in initial conditions, it can also represent other types of events such as those involving a moving specimen and a stationary probe, or a high-rate tensile test. In any of these cases, the results are very similar. The one that will be discussed is the dart impact problem.

The mass of the piston rod or other element driving the probe and the stiffness of its supporting structure are represented by M_R and K_R. The stiffness

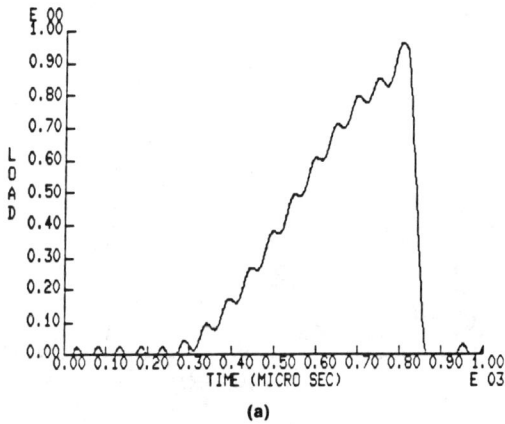

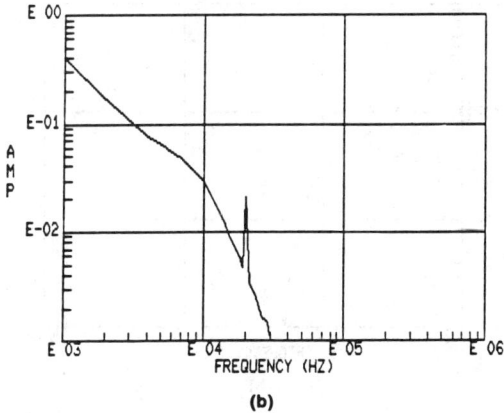

FIG. 7—*Time history* (a) *and frequency* (b) *after application of a two-pole filter to the load history of a ductile-brittle material.*

of the load cell and the mass of the probe are represented by K_{LC} and M_{LP}. The force, F, is the contact force between the probe and specimen and is the objective of the measurement. However, in reality, what the load cell reads is the force transmitted across the spring element, K_{LC}. The specimen is represented by a mass of M_S, and the stiffness of its supporting structure by K_S. The mass of the supporting load frame is represented by M_F, and the stiffness of the load frame relative to an inertial reference is represented by K_F. The equations of the mechanical system are as follows

$$M_R \frac{d^2x_1}{dt^2} = K_{LC}(x_2 - x_1) + K_R(V_0 t - x_1) \qquad (9)$$

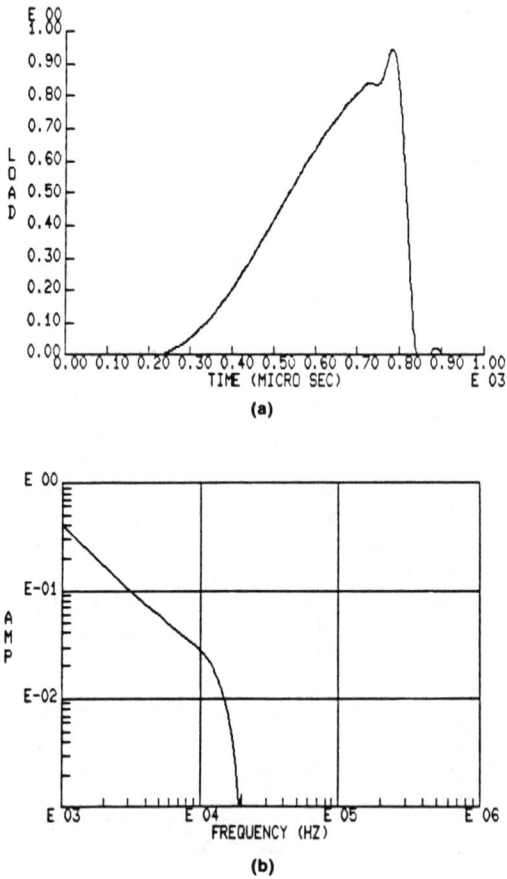

FIG. 8—*Time history* (a) *and frequency spectrum* (b) *after application of a low-pass filter to the load history of a ductile-brittle material.*

$$M_P \frac{d^2 x_2}{dt^2} = K_{LC}(x_1 - x_2) - F \tag{10}$$

$$M_S \frac{d^2 x_3}{dt^2} = K_S(x_4 - x_3) - F \tag{11}$$

$$M_F \frac{d^2 x_4}{dt^2} = K_F x_4 - K_S(x_3 - x_4) \tag{12}$$

And the initial conditions are as follows

$$x_1(0) = x_2(0) = x_3(0) = x_4(0) = 0$$

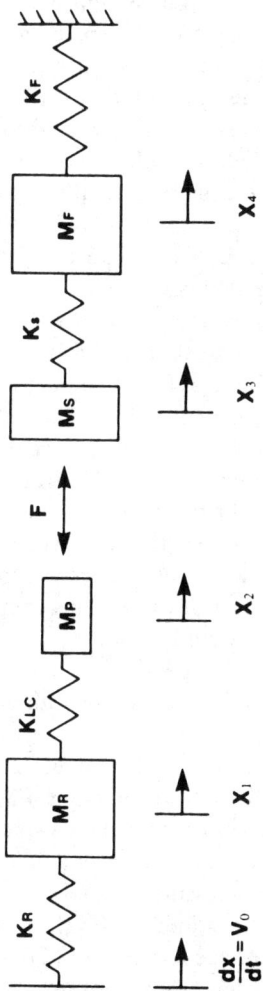

FIG. 9—Model of impact dynamics.

$$\frac{dx_1}{dt}(0) = \frac{dx_2}{dt}(0) = V_0$$

$$\frac{dx_3}{dt}(0) = \frac{dx_4}{dt}(0) = 0$$

This representation of the dynamic system will accommodate the major sources of noise. These equations are then solved numerically by integrating over time. This approach allows one to look at the state of the system at any point in time. The output of this computation is the time history for the contact force between the specimen and the probe and the time history of the force read by the load cell. The failure characteristics of the specimen can be varied in the calculation. The case investigated was a specimen exhibiting essentially ductile behavior with a rounding off of the force-time curve near the peak force, and a rapid dropoff of the force following the peak.

The output of the impact model is shown in Fig. 10. The first curve shows the time history of the actual contact force at the interface, and the second shows the time history of the load that would be read by the load cell. The primary difference between the two, of course, is the ringing that appears in the load cell signal. This ringing is initially excited by the contact and dies out slightly during the loading process. Following specimen failure, the release of the energy stored in the probe causes larger vibrations to be excited. In this process, specimens with more brittle behavior will release more energy into the probe vibrations, so that failure will be followed by larger oscillations. The frequency and amplitude of the vibrations are controlled by the mechanical design and the stiffness and mass of the various components. Such things as probe weight, load, cell stiffness, and sensitivity must be carefully matched to the application. However, the desire here is to explore noise effects, so those parameters have been chosen that resulted in a high level of noise being generated.

The frequency spectrum representation of both the force at the contact point and the force in the load cell are shown in Fig. 11. The probe ringing appears as a spike in the spectrum centered on a single frequency. Compared to the previous situation, in which noise was added to the basic signal, the spike related to the noise is much wider. This is because the sinusoidal noise signal is basically two separate sections within the test history. One is of lower amplitude from the initial contact excitation, and the other is of higher amplitude following the release of energy in the failure of the specimen. Thus, the addition to the spectrum is not as simple as that of sinusoid plus boxcar functions representing the leading and trailing edges of the sinusoid.

The noise content in the signal can be reduced by minimizing the amount of data reported following specimen failure in order to eliminate the content of the postfailure ringing. However, in a real situation, in which the exact failure point may be obscured by noise, there would be a risk of arbitrarily

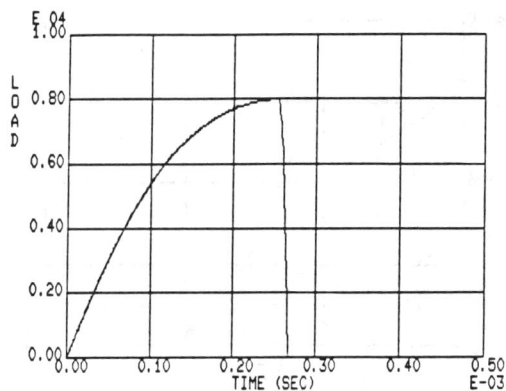

(a)

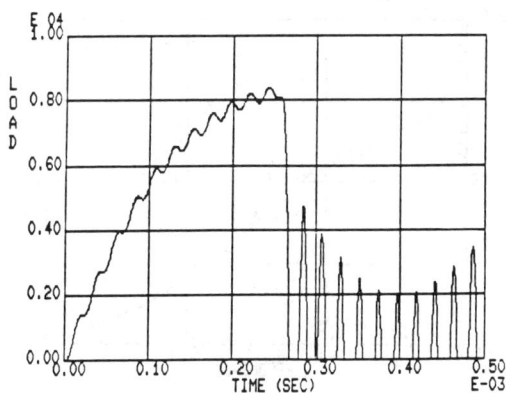

(b)

FIG. 10—*Time history* (a) *at the contact point* (b) *at the load cell.*

cutting off the specimen. Another consideration is that by including a portion of the time history following a failure that contains the noise, the frequencies associated with the noise may become more clearly defined.

Noise Characterization

The nature of the noise can be characterized by looking at the difference between the actual force signal at the contact point and the measured force at the load cell. Let $F(t)$ and $L(t)$ be the actual force and the measured force at the load cell, respectively. Since the response is that of a linear mechanical system, the noise, $N(t)$, is equal to the difference between the measured force and the actual force as expressed in the following equation. Then, taking the

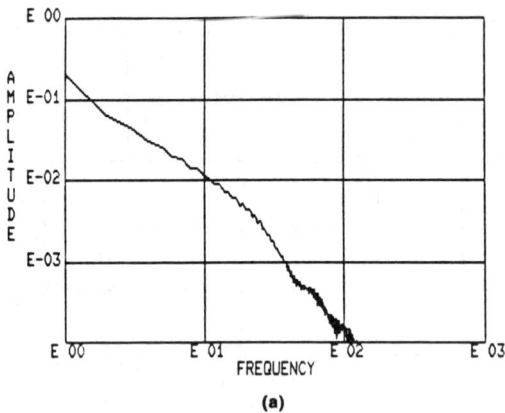

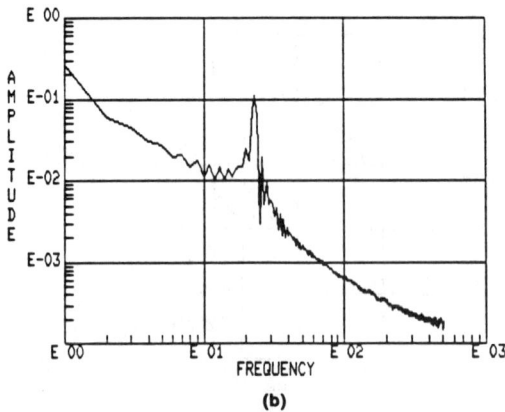

FIG. 11—*Frequency spectra for the force* (a) *at the contact point* (b) *at the load cell.*

Fourier transforms and forming the equivalent equation in the frequency domain, we have the following, where the time functions are replaced by their transforms.

$$L(t) = F(t) + N(t) \qquad (13)$$

$$L(f) = F(f) + N(f) \qquad (14)$$

If we let $H(f)$ be a frequency representation of the filter process, we would like to have its features in frequency domain such as in the following equation

$$H(f)\,L(f) = F(f) \qquad (15)$$

$$H(f) = \frac{F(f)}{L(f)} \tag{16}$$

Then the desired filter shape wil be expressed as a function of frequency using the spectra for the load cell reading and the contact force.

Using the data in the previous example, we can numerically determine the ratio of the spectra for the actual contact force and the load cell reading. This is shown in Fig. 12.

The ratio indicates that perhaps the best type of filter would be a combination of a notch filter with an additional filter which rolls off at about 20 dB per decade. The notch filter should not reduce the amplitude to zero.

The low-pass filter described in Eq 8 was applied to the model using a corner frequency matching the noise peak. The result is shown in Fig. 13. The noise is greatly reduced, but there is overshoot at the edges.

Application to Experimental Data

The low-pass filter was evaluated on data taken from tests of two materials—one, a polystyrene that had a brittle failure characteristic and the other a high-density polyethylene that had a relatively ductile failure characteristic. The initial data for the high-density polyethylene and its frequency spectrum are shown in Fig. 14. Data were obtained on an MTS hydraulic impact system using a probe equipped with a quartz load sensor and an LVDT to measure displacement.

The low-pass filter, with a corner frequency of approximately 2000 Hz and a C value of 0.4, was then applied to the data. The results are shown in Fig. 15. A very clean representation of the load form was obtained with a fair

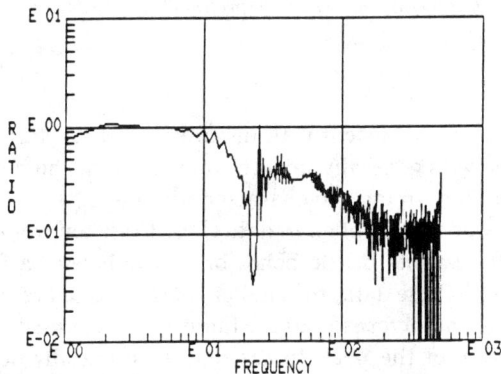

FIG. 12—*Ratio of the frequency spectrum amplitude for the contact force to the amplitude for the load cell reading.*

(a)

(b)

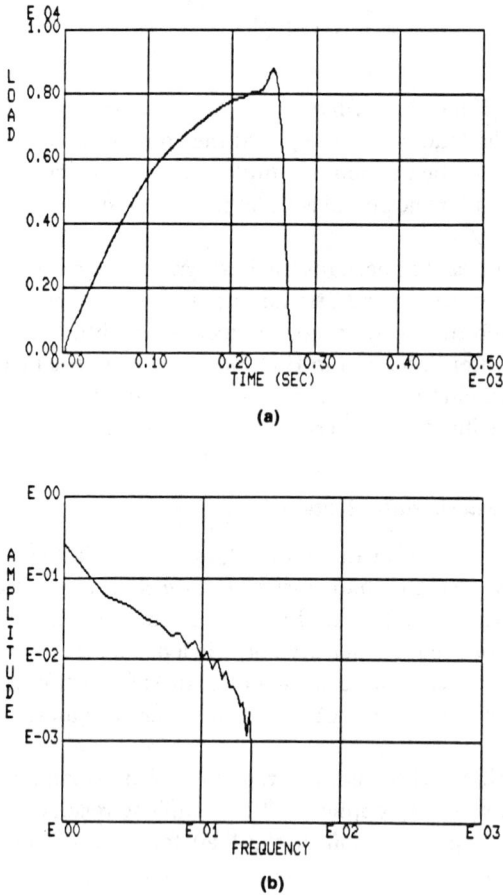

FIG. 13— *The time history* (a) *and frequency spectrum* (b) *following the application of a low-pass filter.*

amount of detail to be consistent with the previous characterization. Since it is a ductile process with generally smooth transitions, the high-frequency content is not critical in defining the characteristic shape.

The technique was then applied to data (Fig. 16) taken from a specimen of polystyrene which exhibits brittle behavior and fails with a fair amount of noise imposed. With these data, the challenge, then, is to preserve as much of the sharp breaking characteristics of the failure zone as possible while obtaining better definition of the peak forces that were initially obscured by the noise. After application of the low-pass filter, the noise level on the load history is considerably reduced (Fig. 17) and there is better definition of the be-

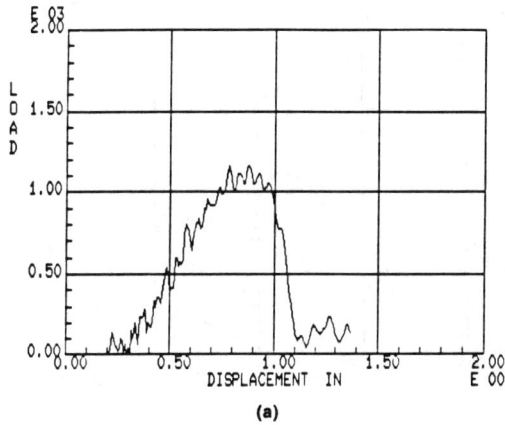

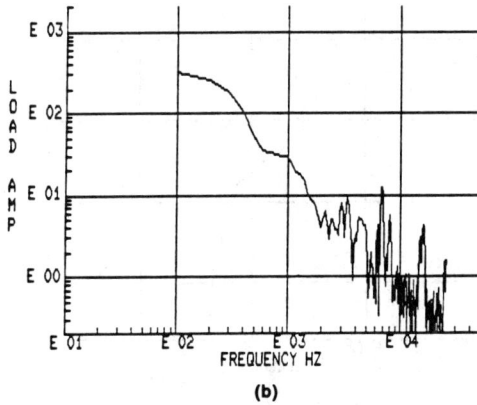

FIG. 14—*Time history* (a) *and frequency spectrum* (b) *for an impact test of high-density polyethylene.*

havior. It is interesting to observe that the discontinuity in the load spectrum is not necessarily the sign of an imposed noise signal but may well be a significant part of the frequency content. The steepness of the failure portion of the load curve is reduced because of the truncation of the higher-frequency components. Also, the deflection at failure is altered.

Conclusions

The first conclusion that can be drawn is that approaching the filtering from the viewpoint of working with the signals in the frequency domain pro-

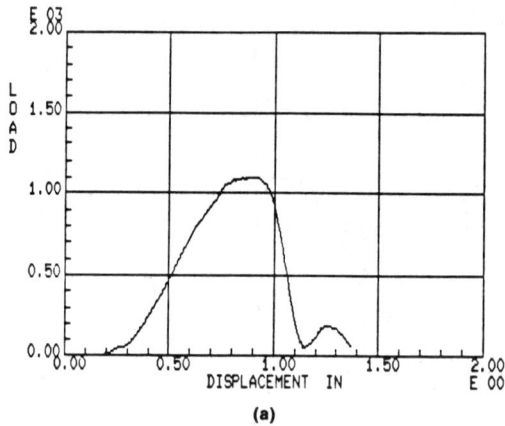

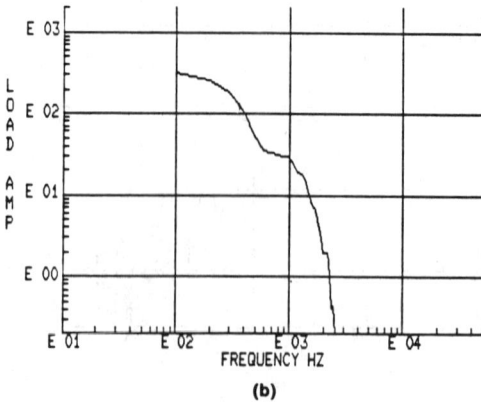

FIG. 15—*Time history* (a) *and frequency spectrum* (b) *for the data in Fig. 13 following application of the low-pass filter.*

vides a very powerful tool. Essentially, one has complete freedom as to how the frequency component, amplitudes, and phase are manipulated. This freedom then leads to the next problem, in that one must chose the optimum way to do this. The optimum approach will depend, to some extent, on the nature of the data to be processed. For this reason, it is not likely there will exist a universal filter for all impact tests. Instead, the filter may have to be selected individually with the nature of the data in mind. For example, a filter method may be appropriate for brittle tests but not for ductile tests.

Implementing the filtering process digitally in the frequency domain has advantages over incorporating analog signal filtering in the test equipment, in

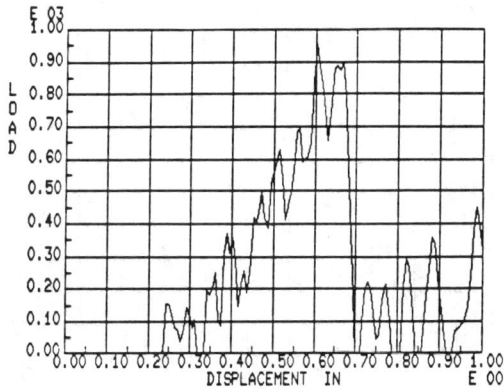

(a)

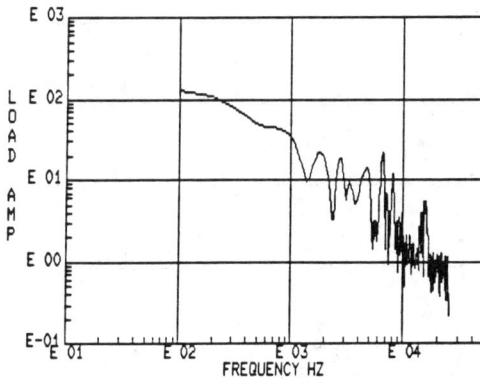

(b)

FIG. 16—*Time history* (a) *and frequency spectrum* (b) *for an impact test of polysterene.*

that the maximum information content of the original data can be retained and the filter process can be modified and tested while still retaining the original data.

References

[*1*] Turner, C. E., "Measurement of Fracture Toughness by Instrumented Impact Test," *Impact Testing of Metals, ASTM STP 466*, American Society for Testing and Materials, Philadelphia, 1970, pp. 93-114.
[*2*] Ireland, D. R., "Procedures and Problems Associated with Reliable Control of the Instrumented Impact Test," in *Instrumented Impact Testing, ASTM STP 563*, American Society for Testing and Materials, Philadelphia, 1974, pp. 3-29.

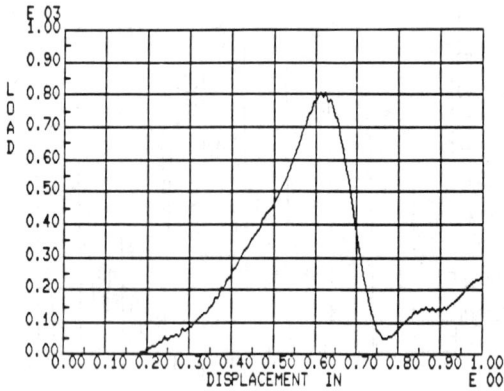

(a)

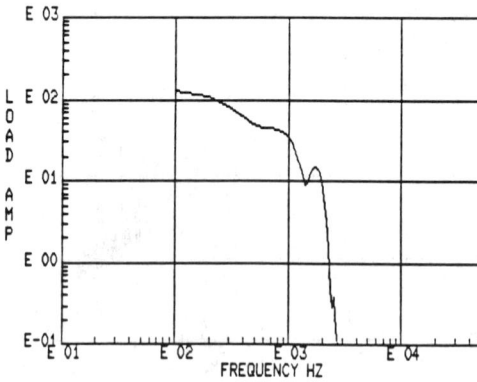

(b)

FIG. 17—*Time history* (a) *and frequency spectrum* (b) *for the data from Fig. 15 after application of the low-pass filter.*

[3] Saxton, H. J., Ireland, D. R., and Servey, W. L., "Analysis and Control of Inertial Effects During Instrumented Impact Testing," in *Instrumented Impact Testing, ASTM STP 563,* American Society for Testing and Materials, Philadelphia, 1974, pp. 50–73.

[4] Brigham, E. O., *The Fast Fourier Transform,* Prentice Hall, Englewood Cliffs, NJ, 1974, p. 75.

[5] Sohaney, R. C. and Nieters, J. N., "Proper Use of Weighting Functions for Impact Testing," *Proceedings,* Third International Modal Analysis Conference, Orlando, Florida, 1985, pp. 1102–1106.

[6] Oppenheim, A. V. and Schafer, R. W., *Digital Signal Processing,* Prentice-Hall, Englewood Cliffs, NJ, 1975, p. 250.

Impact Testing for End-Use Applications

Richard C. Progelhof[1]

Impact Measurements of Low-Pressure Thermoplastic Structural Foam

REFERENCE: Progelhof, R. C., "**Impact Measurements of Low-Pressure Thermoplastic Structural Foam,**" *Instrumented Impact Testing of Plastics and Composite Materials, ASTM STP 936,* S. L. Kessler, G. C. Adams, S. B. Driscoll, and D. R. Ireland, Eds., American Society for Testing and Materials, Philadelphia, 1987, pp. 105–116.

ABSTRACT: The most common technique used to report the impact characteristics of a plastic member found in the literature today is the drop-weight-to-fracture test, the ASTM Test for Impact Resistance of Rigid Plastic Sheeting or Parts by Means of a Tup (Falling Weight) (D 3029-82). An impactor or tup of specified mass is dropped from a known height. Through the use of a staircase testing procedure, the probable energy required to crack 50% of the specimens, F_{50}, is obtained. The use of this experimental technique for thermoplastic structural foam members can result in misleading data.

A series of constant velocity instrumented impact tests were conducted on several sets of test plaques. The impact energy correlated with the local density of the member, but relatively large variations did occur between identical positions on "identical" plaques. The experiments clearly indicate that the basic premise that the drop-weight-to-fracture test must be conducted on "identical" sets of specimens from a single population is not valid for thermoplastic structural foam specimens.

KEY WORDS: impact, impact testing, thermoplastic structural foam, fracture, drop-weight test, ASTM D 3029-82

Thermoplastic structural foam (TSF) refers to a class of foamed thermoplastics that have a nearly uniform cellular core and integral high-density skin. There are two generic processes for forming TSF parts, the single-component and two-component processes. The most common TSF process being used commercially is the single-component low-pressure process, in which molten plastic containing a dissolved gas, such as nitrogen (N_2), carbon dioxide (CO_2), or a fluorocarbon, is plasticized and melted in a single-screw extruder, and the melt is stored temporarily in an accumulator (Fig. 1). The

[1]Professor of mechanical engineering, New Jersey Institute of Technology, Newark, NJ 07102.

Full accumulator

Pressurized accumulator

Injection completed

Expansion completed

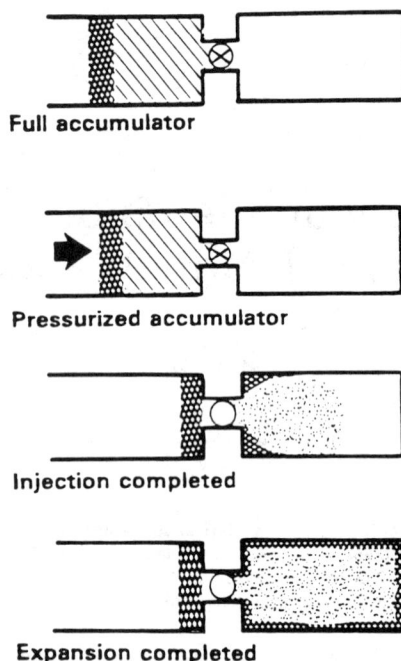

FIG. 1—*Schematic drawing of the low-pressure TSF process.*

foamable resin is then injected at relatively high accumulator pressure into a cold mold with a volume considerably larger than the volume of the unfoamed polymer melt. The low pressure in the mold cavity and the volatile nature of the dissolved gas causes the molten plastic to foam and fill the cavity.

During the expansion process when the plastic is foaming, the gas bubbles in the vicinity of the mold surface are subjected to unbalanced forces and increasing solubility, which tend to collapse the bubbles, thus forming a dense skin. At the same time, the bubble growth rates near the centerline are controlled by the rate of gas diffusion from the polymer melt. Since the gas concentration near a growing bubble is being depleted by the diffusion process, the bubble growth process is self-controlling. This process should result in a relatively uniform bubble structure within the core.

Recent experimental evidence indicates that during the mold-filling process, in which the gas comes out of the solution and forms individual bubbles, there will be some bubble coalesence. As the temperature of the polymer melt drops because of transient conduction to the mold surface, the resultant structure is frozen. The original Union Carbide Co. approach typifies this type of TSF molding process. Thus, the resultant structure is very complex. Across any section of the part there will be a distribution of small bubbles as well as a separate distribution of much larger bubbles due to coalesence. The final bubble structure is dependent on the base resin, blowing agent, and processing conditions.

This process should not be confused with several noteworthy modifications referred to as medium- or high-pressure TSF that were developed to produce a better surface finish and thicker, stronger skin. These processes are the USM expanding mold and the gas counterpressure (GCP) method. The two-component process, as represented by the Imperial Chemical Industry (ICI)/ Battenfeld or Billion process, is a further attempt to improve the surface finish of the product by simultaneously injecting a surface resin and a core resin that contains a dissolved gas.

Previous Impact Studies

It is common practice to relate TSF properties to a reduced TSF average density, which is defined as the ratio of the bulk density of the foamed part, ρ, to the density of the unfoamed resin, ρ_0. Similarly, a reduced property is defined as the ratio of the property for the foamed part to that of the unfoamed resin. Progelhof and Throne [1] showed that there is sound empirical evidence that for many properties such as moduli and strengths, the reduced property, for *uniform density foams,* is proportional to the square of the reduced density

$$\frac{x}{x_0} = \left(\frac{\rho}{\rho_0}\right)^2 = \phi^2 \qquad (1)$$

where

x = foamed property and
x_0 = unfoamed property.

For nonuniform density foams, such as TSF, the relationship depends on other factors, such as the thickness of the skin and the density gradient in the cellular core. Nonimpact TSF-reduced properties seemed to correlate well with reduced density to an arbitrary power, a, which had as a lower bound the value of 2 (for example, the square law of Eq 1).

In the early correlations for impact properties on TSF shapes, the reduced impact value was compared with the reduced density. To include the effect of part thickness, an arbitrary standard thickness, t_0, of 6.35 mm ($1/4$ in.) was used. The initial impact data were developed using either Gardner drop-weight tests or unnotched Izod excess-energy pendulum tests. Based on limited test data with considerable data scatter, but from evaluations on many TSF materials, Throne proposed the following empirical correlation [2]

$$\frac{I}{I_0} = \left(\frac{\rho}{\rho_0}\right)^4 \left(\frac{t}{t_0}\right)^2 \tag{2}$$

where

I = impact energy of the TSF member and
I_0 = impact energy of the solid member.

Probably the greatest error in attempting to correlate any property of a TSF member with a correlation of this type is the absence of the effect of skin thickness, density profile, bubble size, and bubble distribution across the member. Progelhof and Eilers [3] have shown theoretically and verified experimentally to a limited degree that both skin thickness and density profile have profound effects on the flexural and tensile moduli of a TSF beam.

Throne attempted to verify the concept that skin thickness has a predominant effect on impact properties by preparing a set of low-pressure high-density polyethylene (HDPE) TSF specimens of different resins. The skin thickness was measured for each specimen. The Gardner drop-weight impact test data showed no correlation with reduced density but correlated quite well with skin thickness. $F_{50} = 7.85 \exp [1.123 t]$, where F_{50} is the Gardner impact energy in joules, and t is the skin thickness in millimetres. However, these data could not be reconciled with Eq 2 or any other published models for impact.

Hengesback and Egli's [4] data showed a strong influence of the carefully formed high-density skin on impact strength.

In an attempt to clarify the large discrepancies of impact characteristics between single-component and two-component TSF members and to eliminate the use of multiple test specimens as required by the Gardner drop-weight-type test, Progelhof and Throne [5] conducted a series of high-speed puncture impact tests on representative specimens of both single and two-component TSF specimens.

These tests and the subsequent data reported in Refs 6 and 7 were obtained on a Rheometrics impact tester.

Puncture Impact Testing

The impact machine consists of a servohydraulically driven 12.7-mm-diameter dart with a 6.35-mm-radius spherical tip. The test specimen is held in place by a parallel plate-clamping device. The impact target is a 7.62-cm-diameter exposed surface with the dart impacting at the center of the target. The geometric configuration is a circular plate restrained at the edges. The force exerted by the traveling dart against the specimen surface is measured by the piezoelectric force transducer mounted between the base of the dart and the head of the actuator rod of the hydraulic cylinder. The electronic signals from the force and velocity transducer are stored in a microprocessor memory. The data are viewed on a cathode-ray tube (CRT), and a hard copy is obtained from an X and Y plotter. The force deflection can be electronically integrated to obtain the impact energy supplied by the dart as a function of position as it penetrates the TSFs specimen.

The initial screening experiments [5] clearly indicate that the type of failure, brittle or ductile, was directly related to the base resin properties and to the entrapment of gas bubbles in the skin. That is, impact characteristics of two-component TSF members were significantly different from those of single-component TSF members. In all cases, the single-component specimens had significantly lower impact values than a similar two-component specimen.

The tests also clearly indicated that the impact characteristics of lower-pressure single-component TSFs were a function of the local cellular structure in the region of impaction. It was postulated that the impact test results would vary with flow length from the gate or position on the plaque. To test this hypothesis, Progelhof and Kumar [6] conducted a series of high-speed puncture impact tests at different locations on a single plaque of polycarbonate (Lexan FL 500) and a polyphenylene oxide-based resin (Noryl FN 215). The experimental results reported by Progelhof, Kumar, and Throne [8] for both resins showed a relatively large variation in impact performance over the face of each plaque. No single correlation—that is, skin thickness, localized reduced density, and so on—was found to correlate impact energy with cracking or breaking of the test specimens. The data for the Noryl test plaque are shown in Fig. 2.

The preceding tests clearly indicate that impact performance varied with local cellular structure, that is, the position on the molded plaque. Since the impact tests were conducted on only one plaque, the variation due to processing could not be ascertained. Two distinct parameters need investigation: (1) the impact variation for identical processing conditions and (2) the effect of processing variables, injection speed, gas content, melt temperature, mold temperature, and so on, on the impact.

Progelhof, Kumar, and Throne [8] attempted to determine the variations of impact performance for plaques made under identical processing conditions. Three polystyrene (PS) plaques with three different thickness sections (Fig. 3) of approximately the same weight and molded from the same molding conditions were used for the test.

A visual examination of the fractured plaques showed that all of the polystyrene TSF test specimens failed in a brittle manner. The failures were predominantly small circular holes approximately the size of the tup on the impaction side with a larger-diameter irregular hole on the back surface. The

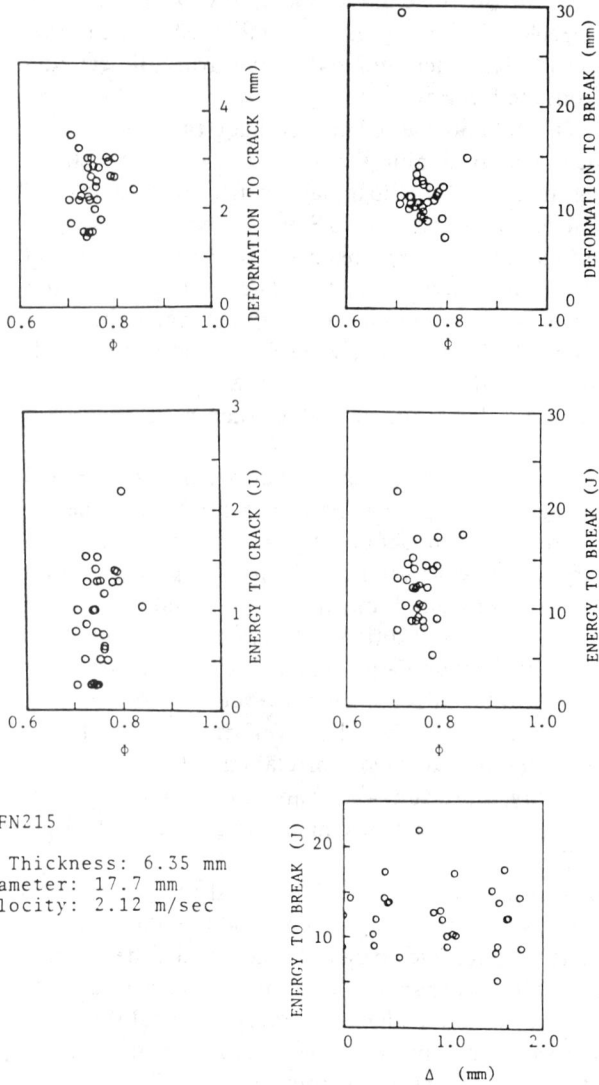

FIG. 2—*Puncture impact data for low-pressure TSF Noryl FN215.*

angle or propagation of the fracture line was at approximately 45° to the axis of penetration.

The data were plotted in the conventional manner, impact energy (initiation or "to break") or deflection (initiation of "to break") versus reduced density. Shown in Figs. 4 and 5 are plots of energy to crack and energy to break versus density, respectively. No correlation was obtained for impact energy to

1.1	1.2	1.3	1.4	1.5	1.6	1.7	1.8	1.9	1.10	1.11	1.12
1.13	1.14	1.15	1.16	1.17	1.18	1.19	1.20	1.21	1.22	1.23	1.24
1.25	1.26	1.27	1.28	1.29	1.30	1.31	1.32	1.33	1.34	1.35	1.36

FIG. 3—*Polystyrene low-pressure single-component TSF test plaque.*

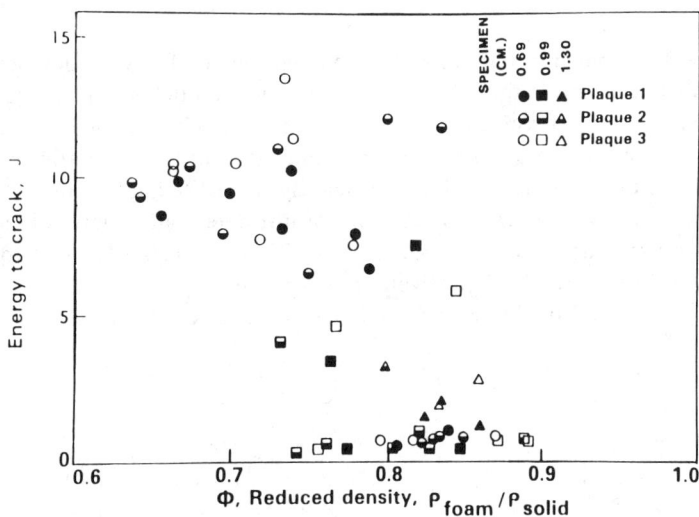

FIG. 4—*Polystyrene low-pressure single-component TSF energy to crack.*

crack versus reduced density. However, the energy to break, in joules, was evaluated by

$$I = 24.22 \left(\frac{\rho}{\rho_0} \right)^4 \left(\frac{t}{t} \right)^2 \tag{3}$$

The average percentage of error in all the data points was 19.2%.

An analysis of the raw data indicates that the impact characteristics to fracture at several locations vary significantly between the three plaques. Con-

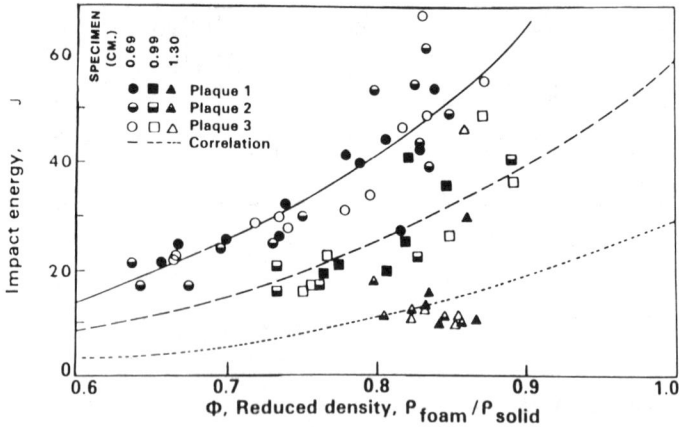

FIG. 5—*Polystyrene low-pressure single-component TSF energy to break.*

sider the data shown in Fig. 6 for Plaque Location 10. The variation between Plaques 1 and 3 and Plaque 2 is significant. The load deflection curves up to the point of first cracking, point *a*, are approximately the same, hence the magnitudes of the energy to crack (area under the curve) are relatively close. Past this point, the curves differ significantly. Both Plaques 1 and 3 show a significant increase in force to the point of ultimate load, whereas Plaque 2 exhibits a lower deflection. The energies to fracture (Table 1) for Plaques 1 and 3, 14.7 and 13.0 J, respectively, are considerably larger than those for Plaque 2—8.5 J. Note, however, that the densities of Plaques 1 and 3 are

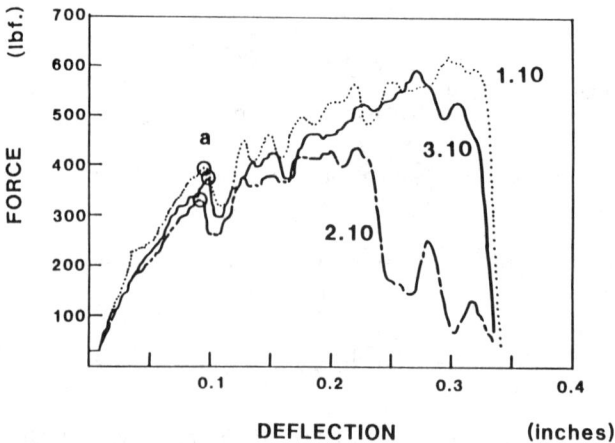

FIG. 6— *Typical PS TSF constant-velocity instrumented puncture impact test results: Plaque Position 10.*

TABLE 1—*Comparison of predicted versus actual impact results.*

Location[a]	Energy, Predicted, J	Energy, Actual, J
1.1	14.0	14.7
2.1	11.28	8.5
3.1	13.6	13.0

[a]1.1 refers to Location 1 in Plaque 1; 2.1 refers to Location 1 in Plaque 2; 3.1 refers to Location 1 in Plaque 3.

almost identical, and that of Plaque 2 is significantly lower. Based on the correlation of all the test data (Eq 3), the predicted energies to fracture for the three plaques are listed in Table 1. It is apparent for the regression analysis that the energy to fracture for the PS TSF test plaques can be correlated, to a reasonable degree of accuracy, with reduced density and part thickness. Hence the variations in impact characteristics over the PS TSF plaques can be directly related to the local density variation in the plaque. The nonuniform impact characteristics to fracture of the three plaques at one specific location can be attributed to the effect of processing, that is, gas evolution and mold filling, on local reduced density rather than to a resin property.

Analysis of Impact Characteristics of Low-Pressure Single-Component TSF

Based on the data presented in the literature [1–8], the author has concluded that the impact characteristics of a TSF structure are dependent on the base resin, localized bubble structure, density profile, number of components, and skin thickness. An X-ray of a typical low-pressure TSF test specimen (Fig. 7) clearly indicates that the specimen *does not have a uniform structure* but varies significantly across the part. This type of bubble structure is prevalent in actual molded parts (Fig. 8). It has been shown that these variations are not necessarily a result of different machine conditions but are an inherent variation in the process. It is, thus, impossible to obtain a uniform set of specimens for a single population.

Conclusions

The Gardner drop-weight test is based on the premise that the test specimens are taken from a single population. When testing solid specimens, if the test is conducted at the identical position on the test specimens and the processing conditions remain unchanged during the fabrication of the test specimens, it can be assumed that the specimens are from a single population. The preceding work clearly indicates that TSF specimens do not satisfy the basic

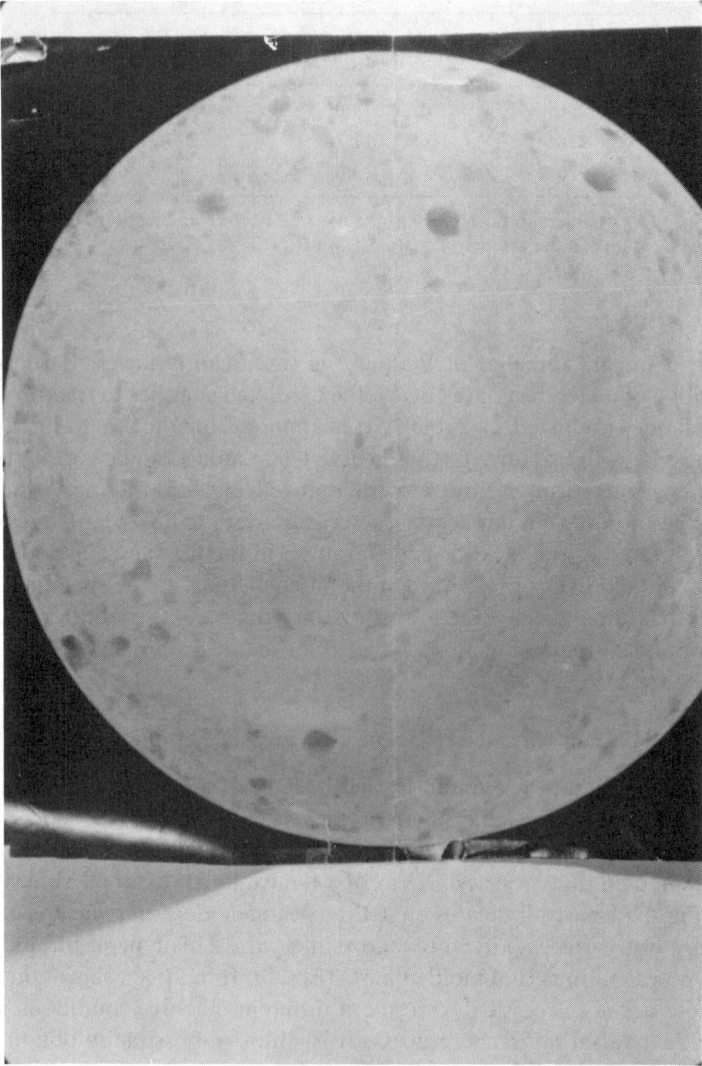

FIG. 7—*X-ray of polystyrene low-pressure TSF plaque.*

criteria necessary for the use of the Gardner test. If the actual specimens represent a bimodal or trimodal distribution, the Gardner test will depict only the probable 50% failure energy, F_{50}, but the standard deviation will not be representative of the actual population.

A computer simulation of this phenomenon by M. Patel for his MS thesis at the New Jersey Institute of Technology (Newark) has been completed. The

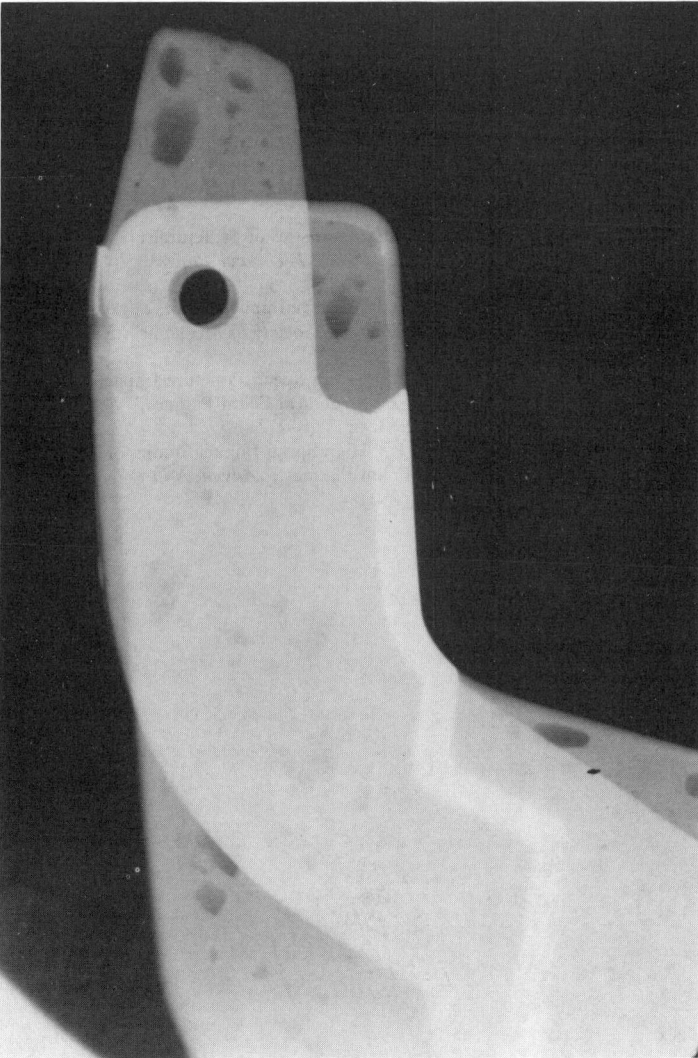

FIG. 8—*X-ray of low-pressure TSF chain saw handle.*

modeling results, which will be reported in the future, appear to confirm this conclusion.

When testing any specimen that is not represented by a single population, a Gardner-type test may give very misleading information to the designer. The author, therefore, recommends that all TSF impact data reported be measured only on a constant-velocity instrumented puncture impact system.

References

[1] Progelhof, R. and Throne, J., "Young's Modulus of Uniform Density Thermoplastic Foams," *Society of Plastics Engineers Technical Papers,* Vol. 24, 1978, pp. 678–684.

[2] Throne, J., "Design Criteria for Thermoplastics Structural Foam," *Plastics Design and Processing,* September 1976, pp. 20–23.

[3] Progelhof, R. and Eilers, K., "Apparent Modulus of a Structural Foam Member," *Society of Plastics Engineers DIVTEC,* Woburn, MA, 27–28 Sept. 1977.

[4] Hengesback, J. and Egli, E., "Structural Foam Molding with Good Surface Finish" *Plastics and Rubber Processing,* Vol. 4, No. 2, June 1979, p. 56.

[5] Progelhof, R. and Throne, J., "Impact Characteristics of Structural Foam, I. High Speed Puncture Tests," *Society of Plastics Engineers Technical Papers,* Vol. 27, 1981, pp. 863–866.

[6] Progelhof, R. and Kumar, S., "High Speed Puncture Impact Studies of Low Pressure Thermoplastics Structural Foam Plaques," *American Society of Mechanical Engineers—PED,* Vol. 5, Sept. 1982, pp. 83–94.

[7] Progelhof, R., Kumar, S., and Throne, J., "High Speed Puncture Impact Studies of Low Pressure Single Component Thermoplastic Structural Foam Plaques," *Society of Plastics Engineers Technical Papers,* Vol. 29, 1983, pp. 270–272.

[8] Progelhof, R., Kumar, S., and Throne, J., "High Speed Impact Studies of Three Low Pressure Styrene Thermoplastic Structural Foam Plaques," *Advances in Polymer Technology,* Vol. 3, No. 1, 1982, pp. 15–22.

Wartan A. Jemian,[1] Bor Z. Jang,[1] and Jyh S. Chou[1]

Testing, Simulation, and Interpretation of Materials Impact Characteristics

REFERENCE: Jemian, W. A., Jang, B. Z., and Chou, J. S., **"Testing, Simulation, and Interpretation of Materials Impact Characteristics,"** *Instrumented Impact Testing of Plastics and Composite Materials, ASTM STP 936,* S. L. Kessler, G. C. Adams, S. B. Driscoll, and D. R. Ireland, Eds., American Society for Testing and Materials, Philadelphia, 1987, pp. 117–143.

ABSTRACT: Research was initiated to measure the characteristics and properties of materials to support a valid simulation for general design purposes. Cushioning systems are used for personal protection in sports and transportation and for the safe transport of fragile materials. Design costs can be reduced by replacing prototype construction with simulation procedures.

Materials impact characteristics are in response to transient dynamic loading involving a sequence of structural processes during the period of contact. In most applications the engineering system responds in an overdamped configuration to eliminate oscillations. The cushioning component, however, is usually a viscoelastic, low-density substance with distinctive characteristics and properties. Cushioning is defined as the redirecting of motion under a controlled level of deceleration. The damping function is separate, being a dissipation of kinetic energy to terminate motion.

A variety of foamed plastics, including polypropylene, polystyrene, and polyurethane formulations were tested in the form of homogeneous blocks with dimensions of 3.8 by 3.8 by 2.5 cm (1.5 by 1.5 by 1.0 in.). Flat-surface impact was employed with drop heights ranging to 122 cm (48 in.). Voltage signals from an accelerometer in the impactor and a force cell below the specimen stage were recorded using a digital storage oscilloscope. These data were processed separately for analysis.

Specimen characterization also included measurement of density and stress relaxation and static testing at rates allowing effective specimen compliance. The objective of the experimental measurements was to determine the density, damping, and structural stiffness to provide the needed information for the deformation process in the governing differential equation

$$M\ddot{u} + C\dot{u} + Ku = R$$

[1]Professor and assistant professor, Mechanical Engineering and Materials Engineering, and graduate assistant, Materials Engineering, respectively, Wilmore Laboratories, Auburn University, Auburn, AL 36849.

where $\ddot{u}$, $\dot{u}$, and u are the acceleration, velocity, and displacement in compression, respectively, and M, C, and K are the mass, damping, and structural stiffness properties of the materials in the test system.

Simulation was performed by finite element (FE) structural analysis developed in prior research. The FE program provided transient, dynamic analysis using nonlinear structural stiffness and damping. The FE model included the corresponding size, shape, and properties of the impactor, specimen, and stage in an appropriate manner. Simulation results were matched with measured responses to establish the validity of the simulation procedure.

Additional experimental drop tests showed pronounced differences between the characteristics of different materials. These observations and specific test procedures allowed conclusions to be drawn relative to the nature of the mechanical and structural responses at different stages of cushioning. These include separate effects of inertial stiffness, dynamic stiffness (due to damping), and structural stiffness occurring when motion is arrested.

KEY WORDS: structural foams, deformation mechanisms, impact, impact testing, cushioning, damping, materials impact response, finite element simulation

Interest in head and neck protection has prompted the development of a series of helmets and other protective devices. A variety of test methods and equipment are being used to simulate the conditions of impact, and models of the head are available to simulate the mechanical nature of impact response [1]. It was found that the severity of head impact is related to the level of acceleration and to the duration of the event. Gadd [2] proposed a severity index (SI), evaluated by integrating the acceleration factor, a (dimensionless quotient of acceleration divided by the standard gravitational acceleration), raised to the 2.5 power over the contact period, where t_s and t_f are the starting and final times.

$$SI = \int_{t_s}^{t_f} a^{2.5}\, dt \qquad (1)$$

Using this method it is possible to distinguish levels of human tolerance.

More recently, Jemian simulated the mechanical response of a torso-head-helmet impactor system using finite element (FE) analysis [3,4]. The method utilizes three-dimensional models with nonlinear materials properties and a stepwise integration scheme that is inherently stable [5]. A special feature is the incorporation of transient analysis. The ability to assign specific damping characteristics to each of the elements was added. The method was applied in connection with the work of an ASTM task force of ASTM Subcommittee FO8.52 on Playing Surfaces and Facilities (a subcommittee of ASTM Committee F-8 on Sports Equipment and Facilities) for a first-order estimation of the severity of impact experienced by a pole vaulter falling into an athletic landing pit. The simulation results were later shown to follow the results of experimental impact tests. Another application was a procedure developed for designing crash helmets, which evaluated the use of the wide range of cushioning materials currently or potentially available [4].

The general characteristics of impact between two materials involve the deformation and possible recovery of both under available mechanical actions. The important events occur while the impacting objects are in contact. The deformation in each material component follows a sequence characteristic of that material and the level of forces. This generally involves compressive loading of the cushioning materials. Deformation mechanisms include processes listed in Table 1. Each process is capable of absorbing and storing energy under static loading conditions. Storage by elastic distortion of interatomic spacing or by the reshaping of extended groups is recoverable. The first is dependent on deformation potential energy, and the second is entropy dependent and sensitive to loading rate. These effects on the free energy, F, of the material are given, in this order, by the terms on the right of Eq 2

$$F = \left(\frac{\partial U}{\partial L}\right)_T - T\left(\frac{\partial S}{\partial L}\right)_T \qquad (2)$$

where U and S are the internal energy and entropy, respectively, L is the specimen dimension in the loading direction, and T is the absolute temperature.

Dynamic loads involve an equilibrium that includes inertial, $M\ddot{u}$, kinetic, $C\dot{u}$, and configurational energy, Ku, terms. The general governing relation is

$$M\ddot{u} + C\dot{u} + Ku = R \qquad (3)$$

where M, C, and K are mass, Rayleigh damping, and stiffness properties of the material, and R is the driving force; u, $\dot{u}$, and $\ddot{u}$ are the displacement,

TABLE 1—*Materials deformation mechanisms.*

Mechanism	Description
Elastic deformation	recoverable deformation, maintaining nearest neighbor connections
Plastic deformation	slip, twinning, and martensite transformation in crystals, crazing, and shear-band formation in polymers
Viscous flow	time-dependent deformation involving interatomic or intermolecular shear
Void formation	coalescence of lattice vacancies formed at deformation bands, inclusions, and internal surfaces in crystals and in the general phase structure in polymers
Cracking	cleavage, void coalescence, and craze enlargement
Brittle fracture	separation without shape change
Rupture	general separation

velocity, and acceleration, respectively. These apply directly to a lumped mass system, such as that described in Fig. 1. In this system no external loads are applied ($R = 0$), and a small transient displacement of the mass results in a continuing oscillation without loss in amplitude if there is no damping ($C = 0$), as shown. The period of natural oscillation, τ, is given by Eq 4

$$\tau = 2\pi\sqrt{\frac{M}{R}} \qquad (4)$$

A portion of the total energy of this system is continually converted from kinetic to potential and back, without any loss. In any real system many processes dissipate a portion of this energy in each cycle. In fact, this dissipation is a continuing process within any portion of a cycle. This damping produces an attenuation of the amplitude of the oscillation. A critical amount of damping in the system, C_c, dissipates motion without oscillation; that is, the system moves unidirectionally to its equilibrium position.

$$C_c = 2\sqrt{MK} \qquad (5)$$

Figure 2 illustrates the form of the motion of undamped, partially damped, and critically damped oscillators with the same masses and elastic stiffnesses. The form of impact deceleration curves of real cushioning systems indicates that they are underdamped; that is, their damping is less than C_c. The period of contact is approximately one-half cycle of oscillation.

The matrix material in a structural foam is usually a viscoelastic polymer

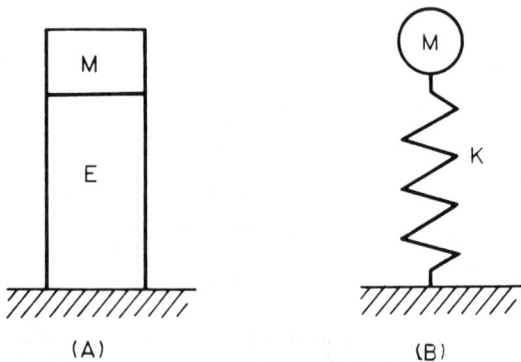

FIG. 1—*Simple, undamped oscillator in the form of* (a) *a materials system with an elastic element with elastic stiffness,* E, *supporting a weight of mass,* M, *on a rigid base, and* (b) *the mechanically equivalent lumped mass analog with the same mass concentrated at the free end of the spring with constant* K = AE/l, *where A and l are the cross-sectional area and length of the elastic element of Part a.*

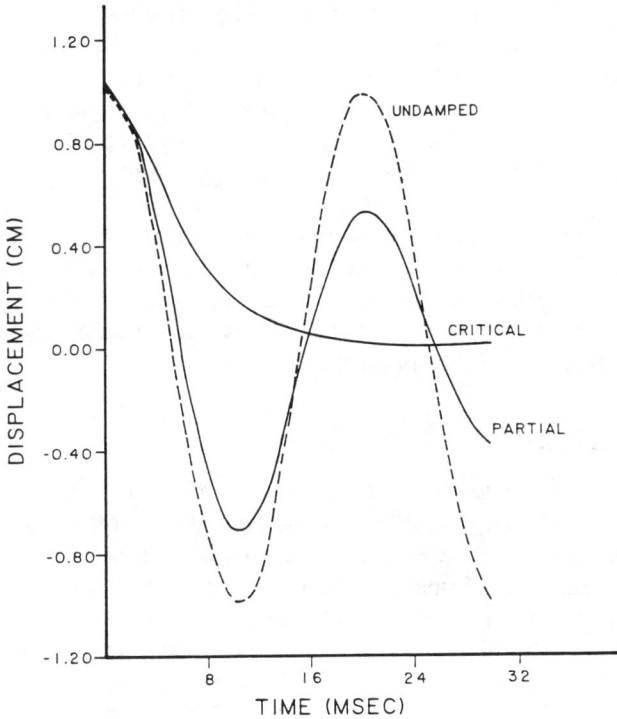

FIG. 2—*The effect of damping on the two-element, simple oscillator after an initial elastic extension and release. Partial damping, not shown in Fig. 1, is one tenth of the critical for this configuration.*

such as polystyrene or polyurethane. The static compression curve is sensitive to loading rate, and a stress relaxation can be observed when the displacement is held constant. The relaxation time, T_r, is the ratio of the viscosity, β, to the elastic shear modulus, μ.

$$T_r = \frac{\beta}{\mu} \qquad (6)$$

Static compression must be measured under conditions comparable to the impact test if the compression characteristics are to be used in simulation, as in Eq 3. Each of the deformation processes in the cushioning material has a characteristic time constant and activation stress. Those processes with small enough time constants contribute to structural compliance (the reciprocal of structural stiffness) at each level of loading, and the remainder dissipate kinetic energy. Actual proportions depend on test conditions.

In an experimental measurement of impact response, the specimen is loaded between two materials, the impactor and stage (anvil), which are relatively undeformable, being limited to minor elastic deformation. The drop test is a frequently used configuration in which the impactor, of known weight, is allowed to fall freely until it is stopped by contact with the specimen. The available energy is the potential energy due to the mass of the impactor and the difference in starting and final elevations. During the impact, the compression displacement, velocity, and deceleration change continuously under the influence of the response of the cushioning material. Changes in deformation mechanism within the specimen are marked by changes in the impact deformation characteristic curve. This paper is concerned with the fine structure of the deformation characteristic.

Experimental Procedure

The test equipment consisted of a guidance system composed of two parallel rods, fluorocarbon-type bushings at the sides of the impactor and an electromagnetic holding and release device for the impactor. The test system is illustrated in Fig. 3. The impactor was an aluminum block with provisions for adding weights and attaching an accelerometer. The specimen was placed on the stage, a metallic block. The stage was positioned immediately above a force transducer, which was supported by a rigid mounting plate. Both the accelerometer and force transducers were quartz piezoelectric devices (Kistler Nos. 8002 and 9212M01). Each was connected to the two-channel Nicolet Model 9031 digital oscilloscope through a separate Kistler Model 5004 dual mode charge amplifier. The falling impactor tripped a microswitch to initiate the scanning oscilloscope approximately 2.5 cm (1 in.) before contact with the specimen. The impact events were recorded on bubble memory. This information was later transferred into a Hewlett-Packard 9836 minicomputer for processing. The oscilloscope provided the required speed and precision for

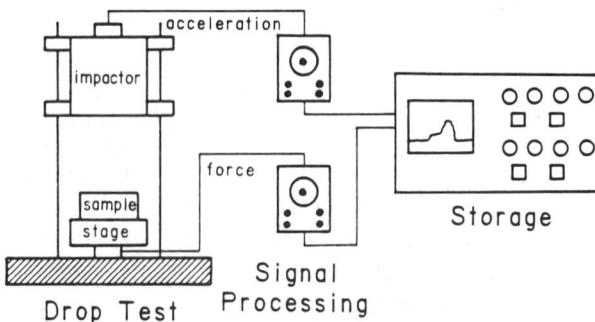

FIG. 3—*Configuration of the drop test system.*

this purpose. Curves presented here were plotted directly from the stored data. All testing was conducted in the laboratory environment.

Earlier testing employed the same drop test equipment but with separate recording of acceleration and force profiles. Also, in the early series, the minicomputer was used only to store data and drive the plotter. Points were read from each plot into a file in a main frame computer, Harris Model 800, with which further processing was conducted. All later work involved the direct transfer of data into the minicomputer for both storage and processing, without intermediate visual and manual procedures.

Static compression characteristics were measured on Instron and MTS test machines. Compression test data from the MTS system was stored directly in the minicomputer. The principal testing was performed in cellular polystyrene specimens purchased locally in the form of 2.5-cm (1-in.) plate. These were cut into 3.8-cm-(1.5-in.) square sections. Several other types of rigid foams were tested. These included polyurethane and polypropylene formulations.

Simulation

A finite element structural analysis program (NONSAP) was adapted to the simulation of the drop test of low-energy compression impact (flat surface) of specimens in the form of rectangular blocks. The simulation involved the period of initial compression and resulted in an impact deceleration time curve that corresponded to that produced by an accelerometer in the dropping weight. NONSAP provides an inherently stable stepwise solution procedure (Wilson Θ-method) with capabilities for nonlinear materials stiffness and separately assignable density. The specification of damping in each element, as required in Eq 4, was added. In the finite element method of structural analysis the specimen is represented by a model divided into regions, called elements, with shapes and sizes defined by nodal points. These are illustrated in Fig. 4, which represents a compression specimen and impactor. The stress and strain fields of the specimen are represented, respectively, by the actions of forces applied to the nodal points, to which stiffnesses have been assigned, and nodal displacements calculated as the unknowns. The FE representation of Eq 3 is

$$[M]\{\ddot{u}\} + [C]\{\dot{u}\} + [K]\{u\} = \{R\} \tag{7}$$

$[M]$, $[C]$, and $[K]$ are matrices of nodal mass, damping, and stiffness assigned to the model. The column vector quantities, $\{\ddot{u}\}\{\dot{u}\}$, $\{u\}$, and $\{R\}$, are the unknown nodal accelerations, velocities, displacements, and known loads, respectively.

The two elements in Fig. 2 have 20 degrees of freedom, as marked by the arrows. This assignment reduces the number of variables, taking advantage

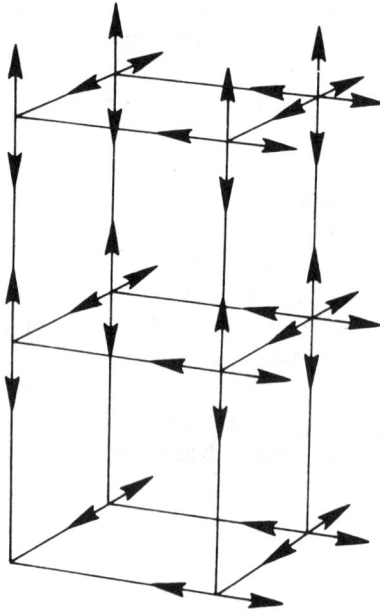

FIG. 4—*The two-element, finite-element drop test model. Each nodal degree of freedom is indicated by a double-headed arrow to indicate the direction of allowed motion.*

of symmetry in the test configuration, and provides for the necessary actions and reactions. The upper element represents the impactor. Its density and size, as defined by nodal point coordinates, account for the weight of the impactor, it is also given appropriate stiffness and damping characteristics. All eight nodal points are assigned the contact velocity, calculated from the drop height, as the starting velocity, v_0, for the simulation.

$$v_0 = \sqrt{2gh} \tag{8}$$

g is the gravitational acceleration, and h is the starting elevation of the impactor relative to the contact surface of the specimen.

The lower element represents the specimen with the appropriate mass, stiffness, and damping. The stiffness is reassigned in each time step according to the slope of the static compression curve at the given compression. A typical stiffness characteristic is shown in Fig. 5. Damping is assigned from a separate measurement. The vertical degrees of freedom are removed or fixed on the bottom surface to correspond to the rigid stage. Additional degrees of freedom are fixed on two sides to take advantage of specimen symmetry. These surfaces bisect the specimen, and they meet along the vertical centerline. Points on these surfaces remain on these surfaces during the impact.

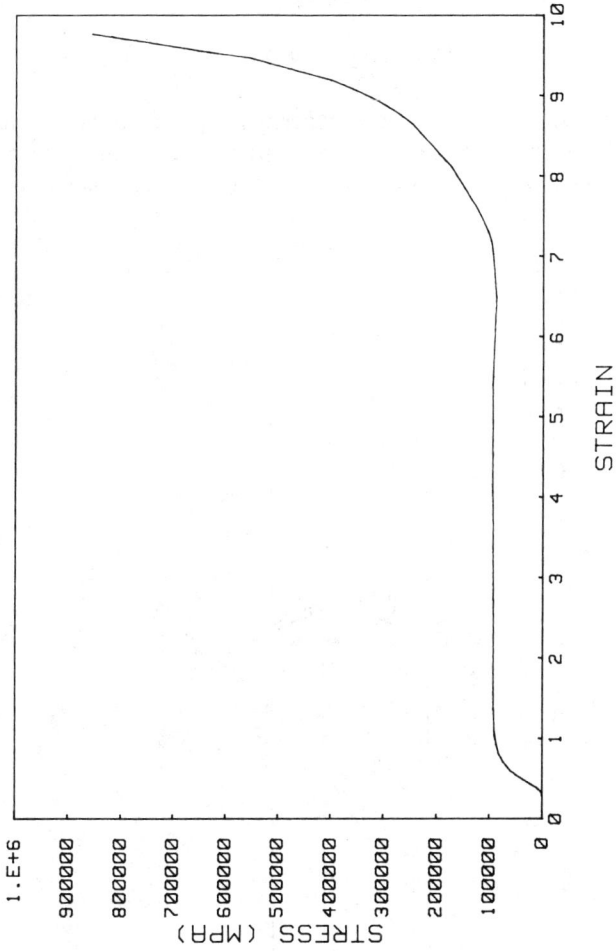

FIG. 5—*Compression stiffness curve used to represent the stiffness of polystyrene foam in FE simulation.*

The steps in the solution procedure of NONSAP are outlined in Fig. 6. At each time step the effective stiffness matrix, $[K]_{eff}$, for the entire structure is reformulated, as is the vector of loads, $\{R\}_{eff}$, on each nodal point. This procedure involves the evaluation of matrixes for the masses, damping, and structural stiffness in the governing differential equation (see Eq 7). Based on the assumption that the acceleration varies linearly over the time step, the effective governing equation is reduced to one of the following form, which is solved for the vector of nodal displacements, $\{u\}$.

$$[K]_{eff}\{u\} = \{R\}_{eff} \tag{9}$$

Starting with known values of the displacement, velocity, and acceleration at time, t, as shown in Fig. 6, the displacement and acceleration at time, $t + T$, where $T = \Theta\Delta t$ and Δt is the desired time interval in the simulation, are

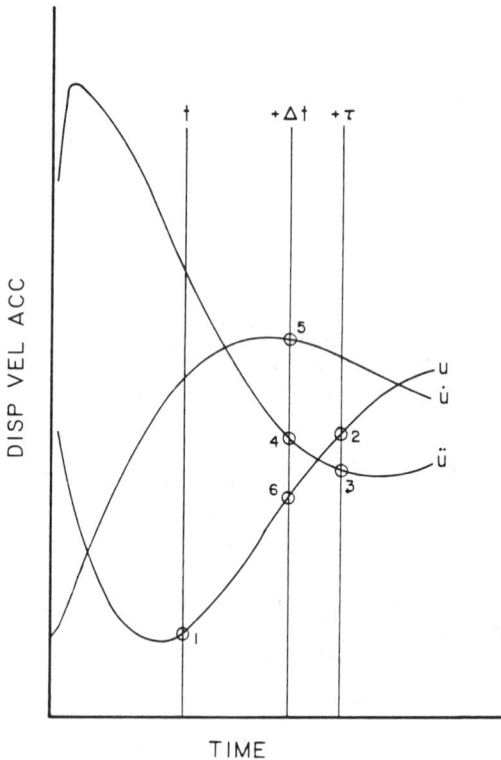

TIME

FIG. 6—*Steps in the FE simulation solution procedure. The effective stiffness matrix and load vector are formulated at Point 1 from the known values of compression, velocity, and acceleration. The corresponding values at the end of the time step, Δt, which are used for the next time step, are computed in the indicated numerical sequence.*

calculated, assuming that the acceleration varies linearly over this extended time interval. Finally, acceleration, velocity, and displacement are back calculated to time $t + \Delta t$ at each nodal point. A value of $\Theta = 1.4$ is found to provide unconditional stability and reduced integration error [5]. Specific desired results are determined from the displacements, velocities, and accelerations of specific nodal points, representing points on the specimen being simulated.

Figure 7 illustrates the time variation of deceleration for the impact of a plastic foam specimen. The displacement and velocity profile curves are also shown. These curves are the result of a finite element simulation and correspond to the measured characteristic of a specimen. The first impact maximum, called G_{m1}, precedes maximum compression. G_{m2} is the peak that corresponds with maximum compression and zero velocity. Under low-energy impact there may not be a peak at this moment of maximum compression. This curve demonstrates the effect of damping in imparting stiffness while the deformation velocity is significant. Several features of the curve, including the oscillations in the final peak, are related to features of the simulation model.

Program development involved a systematic confirmation of program operation by numerical tracing, a series of simulation tests of progressive complexity and ultimately by matching experimental test results. Using this program, it was possible to simulate the effects of material properties and size on various parameters of interest in impact testing. Figure 8 shows the effect of damping on the first impact peak (labeled G_{m1}), which is part of the cushioning characteristic in protective equipment [4]. The simulation model included a moderately compliant pad, to represent the hard-rubber anvil used in early tests. This pad reduces the magnitude of the G_{m1} peak. This impact peak occurs in that period when the compression velocity is at its highest level. It is the contribution of damping, as represented by the $[C]\{\dot{u}\}$ term of Eq 7 to specimen stiffness. This portion of the curve represents a stiffness above that measured in the static compression test.

The inertial effect, involving specimen and impactor masses, produces a stiffness effect in proportion with the acceleration. This generally coincides with G_{m2}. This is in the same region of deformation as the structural stiffness maximum. The simulation results show that these three effects can be associated separately with impact stiffness. Stiffness and density correlate strongly in structural foams and make the strongest contribution to impact stiffness in that portion of the deformation process that should not be utilized for cushioning. This represents "bottoming out" and is associated with the principal structural damage that can occur. Damping is therefore identified as a significant materials property in relation to cushioning.

In other applications, the effect of varying cushion size and properties were studied by simulation. The test configuration, which represented a crash helmet drop test system, included elements for the pad and the helmet liner, which provides a separate element of compliance in the line of action. Only

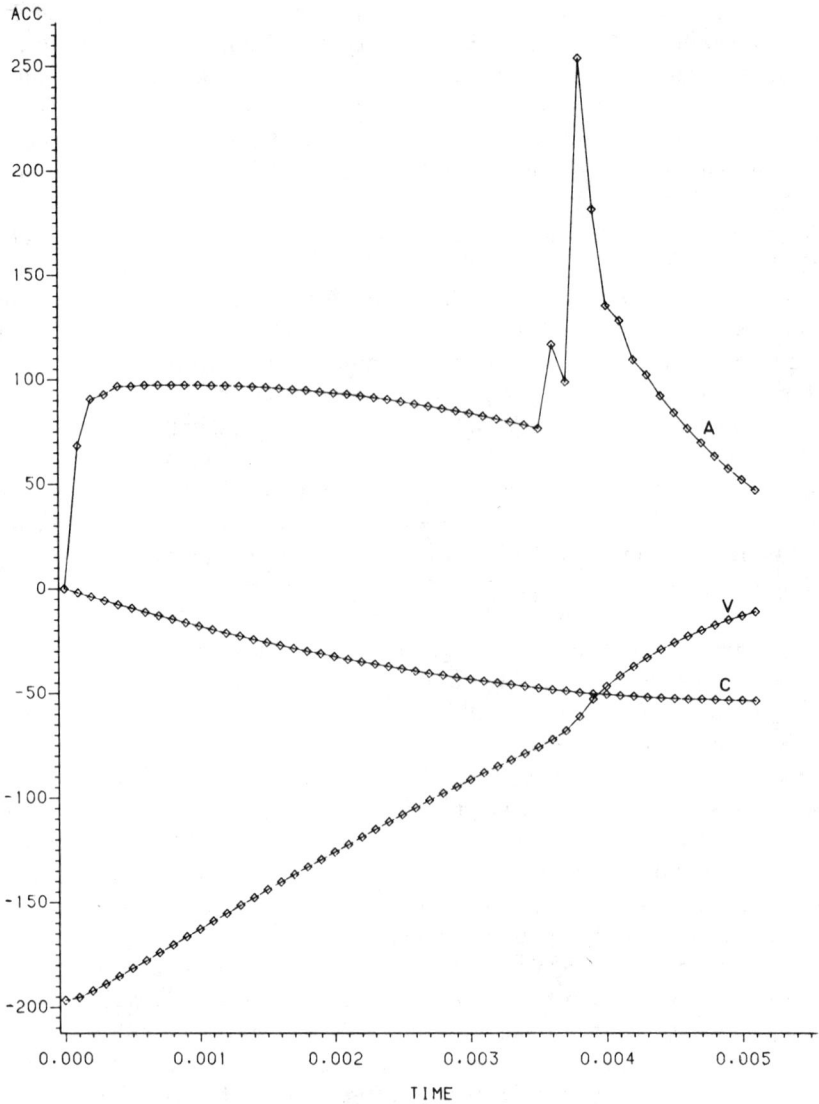

FIG. 7—*The computed impact displacement,* C, *velocity,* V, *and deceleration,* A, *for the two-element model, representing an impactor and polystyrene foam cushioning specimen.*

cushion specimen properties were varied in these simulation runs. All the dimensions and properties of the other structural elements were maintained constant. The results of studies of the effect of damping changes, separately, are shown in Fig. 8. The significant behavior is that although damping produces an initial improvement in cushioning, the extended effect is to increase

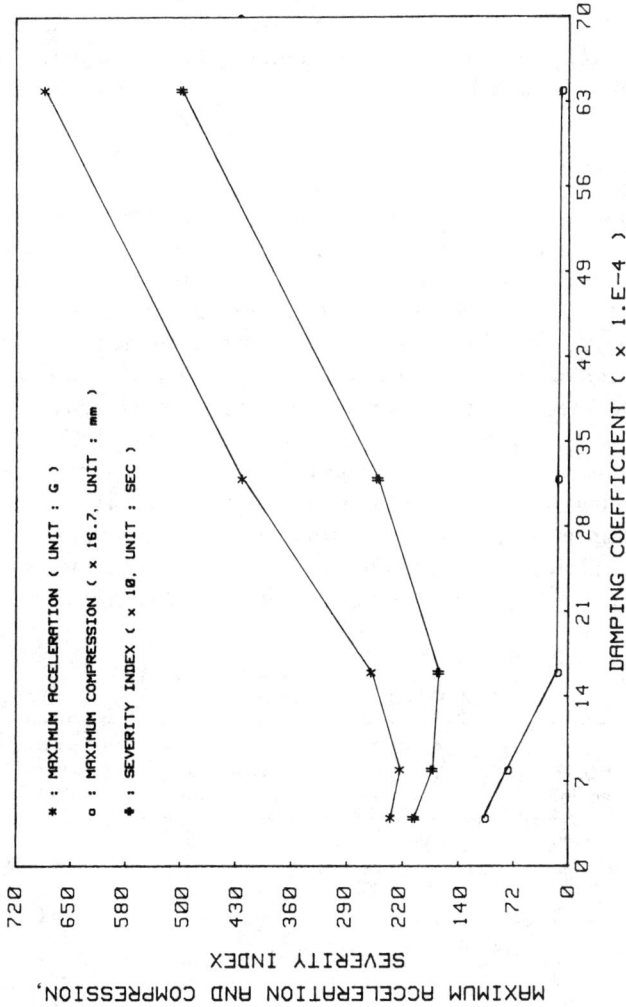

FIG. 8—*The effect of damping on the initial acceleration peak, G_{m1}, severity index, SI, total compression, ϵ, and time of contact, T_{exp}. Simulation results are based on a four-element model patterned after components in a crash helmet drop test. Elements represent the polyurethane cushioning foam, comfort liner, helmet shell, rubber pad, and impactor.*

specimen stiffness. On this basis, cushioning is defined as the reversal of motion under a controlled level of deceleration, while damping is the elimination of motion (the conversion of kinetic energy to heat and radiation, such as elastic waves in the specimen). An increase in damping, as a materials or system property, results in a loss of motion; however, it has only a limited effect on reducing impact severity. The severity index supports this observation, since it follows a similar trend with a minimum at a finite but low level. Total compression is steadily reduced with increases of damping, and the duration of the impact event is extended most rapidly up to the level of damping that corresponds with the minimum stiffness.

Results and Discussion

Experimental studies with individual acceleration and force measurements produced results similar to those in which the results were recorded concurrently. Table 2 lists the test conditions and general results.

Drops 1 and 2 were conducted separately on two specimens cut from the same block of polyurethane foam. Drops 3 and 4 were similarly conducted on two specimens from a common block of polyurethane foam but a different block from that of the preceding specimens. The difference between the available energy and that calculated up to the peak acceleration point, G_{m2}, represents frictional losses of the impactor against air and the guide rods as well as the work of tripping the microswitch. In every case measured, this loss is less than 2%. The impact deceleration curve from Drop 1 is shown in Fig. 9.

Figure 9 is a typical deceleration profile. These data represent the experience of the accelerometer during the impact event. The accelerometer is rig-

TABLE 2—Drop test results.

Drop[a]	Height, cm (in.)	Available Energy, J(ft · lb)	Total Work, J(ft · lb)	Transmitted Work, J(ft · lb)
1, Ua	122 (48)	8.65 (6.38)	8.50 (6.27)	. . .
2, Uf	122 (48)	. . .	. . .	7.58 (5.59)
3, Ua	122 (48)	8.66 (6.39)	8.50 (6.27)	. . .
4, Uf	122 (48)	. . .	. . .	7.58 (5.59)
5, Uc	static	. . .	5.48 (4.04)	. . .
6, Saf	91 (36)	6.60 (4.87)	6.48 (4.78)	4.47 (3.30)
7, Uaf	91 (36)	6.62 (4.88)	6.48 (4.78)	5.97 (4.40)
8, Paf	91 (36)	6.66 (4.91)	6.48 (4.78)	6.44 (4.75)
9, Sc	static			
10, Sf	91 (36)	6.60 (4.87)	6.48 (4.78)	5.53 (4.08)
11, Sf[b]	91 (36)	6.60 (4.87)	6.48 (4.78)	5.55 (4.09)

[a]Key to abbreviations: U, polyurethane foam (Arthur D. Little Laboratories); S, polystyrene panel; P, polypropylene; a, accelerometer signal; f, force cell signal; c, static compression data.
[b]Aluminum stage.

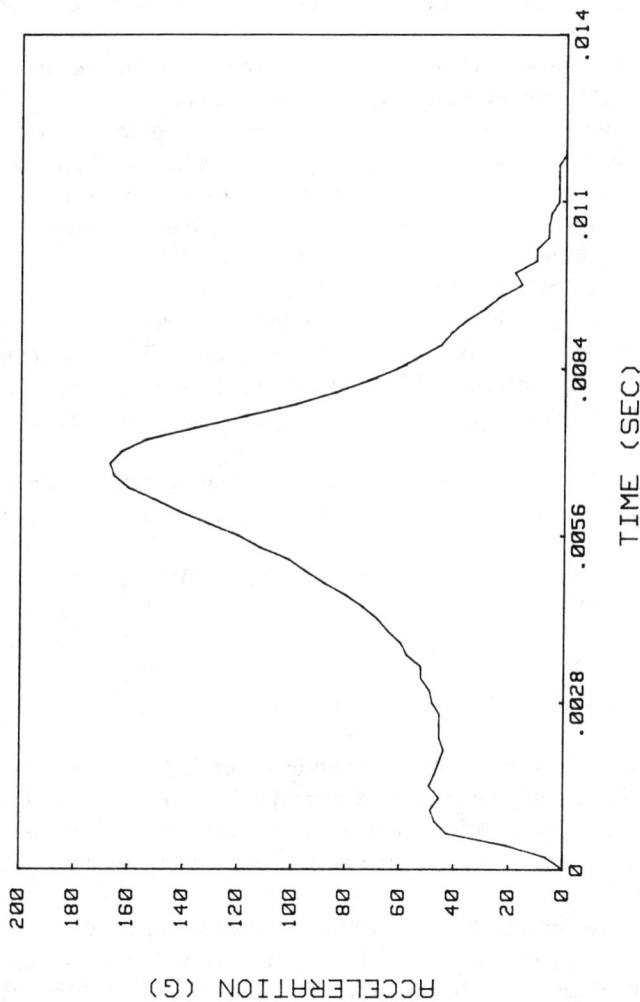

FIG. 9—Impact deceleration profile of polyurethane specimen of Drop 1 (see Table 2).

idly attached to the impactor and follows it reliably. This acceleration, multiplied by the mass of the impactor and integrated over the event, measures the total work of compression. This quantity is computed directly as the change of elevation (initial height minus the elevation at maximum compression of the foam), which is slightly greater than the nominal drop height due to the compression of the specimen. The differences of the available energies listed for any one drop height represent the measured amount of specimen compression.

As the specimen is deformed, it utilizes some of the kinetic energy of the impactor by other processes, such as the production of heat (this was not measured), permanent deformation of the specimen, and radiation of sound. Therefore, not all of the force exerted by the impactor is transmitted to the stage and load cell under the specimen. Figure 10 is a typical force cell profile. It represents the same material represented by the deceleration profile of Fig. 9, but two different specimens were tested out of the same block of polyurethane foam. The integration of the force profile curve with the specimen compression distance shows a correspondingly lower work. This transmitted work may be considered as work performed on an equivalent specimen that does not undergo nonrecoverable deformations. It provides a basis for comparison of the relative magnitudes of the responses at various positions in this system.

Velocity and compression profiles were generated from the deceleration profile by a numerical integration between time steps. The equations are

$$v_{i+1} = v_i + 0.5g \, (a_i + a_{i+1})\Delta t \qquad (10)$$

and

$$u_{i+1} = u_i - v_i\Delta t - [0.25g \, (a_i + a_{i+1})]\Delta t^2 \qquad (11)$$

where g is the standard gravitational acceleration (9.806 64 m s^{-2} [386.088 in. s^{-2}]) and Δt is the time interval, which is uniform for the event. The procedure was tested by integrating a sine function twice and overlapping the starting and final curves. The fit was reassuring, even for relatively large angular increments.

The work transmitted by the specimen to the load cell is displayed as the transmitted force curve shown in Fig. 10. The energy transmitted is less than that accounted for by the motion of the impactor, ΔW_a. The difference is due to the permanent deformation that is produced in the specimen. The energy lost, ΔW_t, in this way represents damping in this quarter cycle. This is identified here as the "quarter cycle damping," β_{qc}.

$$\beta_{qc} = \frac{\Delta W_t}{\Delta W_a} \qquad (12)$$

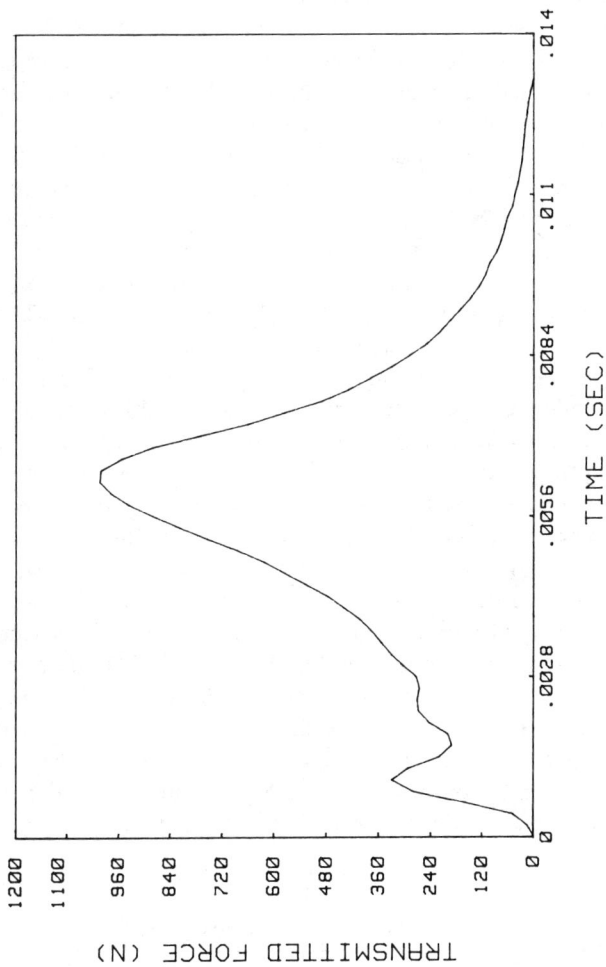

FIG. 10—*Transmitted work profile for a drop test. This is a direct plot of the signal from the quartz piezoelectric load cell under the specimen stage.*

The measured values of quarter cycle damping are listed in Table 3.

Figure 11 is a comparison of the two curves of Figs. 9 and 10 and the corresponding static compression curve, all expressed as force (in newtons) versus time (in seconds). The impact deceleration profile (Fig. 9) was converted using the mass of the impactor. The deceleration and force curves were coordinated at the points of contact and maximum compression. The small mismatch is consistent with the experimental procedure of testing separate specimens and transferring data by visual and mechanical procedures. These results support the concept that at the moment of maximum compression, the velocity is zero and the deceleration is at its peak. The system is momentarily in static equilibrium at this time. The deceleration profile curve shows a pronounced peak following initial contact. The deceleration and transmitted force curves match closely over the remaining event. The static compression curve parallels the others at a lower level. The difference in work between the static compression curve and the accelerometer profile represents the actions of all the potential energy loss processes and may be considered to be the maximum available quarter-cycle damping, β_{qcm}, in the material. The measured value for polyurethane is listed in Table 4.

Additional parameters, shown in Table 5, were calculated from the test data. These include the maximum compression, rebound height, and coefficient of restitution. The maximum compression is taken from the computed compression profile. The rebound height is calculated from the velocity of separation after rebound, by Eq 8, which was used to determine the initial velocity based on the free-fall drop height. The coefficient of restitution is calculated as the negative ratio of the parting to arrival velocities.

Figures 12 through 16 are representative test results, which correspond to a display of the measured parameter in each case. The actual measurement from a piezoelectric transducer is output as a voltage, and the output from a chart record of a test machine is in inches of chart scale. These measured quantities are converted to the recorded dimensions prior to this display. In the initial tests an excessive amount of oscillation and "noise" was found in the test results. A number of drops were made to isolate and eliminate spurious vibrations from the equipment by adding dampers to the guide rails, and

TABLE 3—*Quarter cycle damping.*

Specimen	Drop No.	β_{qc}
Polyurethane	1, 2	0.108 5
Polyurethane	3, 4	0.108 5
Polystyrene	6	0.310
Polyurethane	7	0.079 5
Polypropylene	8	0.006 28
Polystyrene	10	0.146
Polystyrene	11	0.144

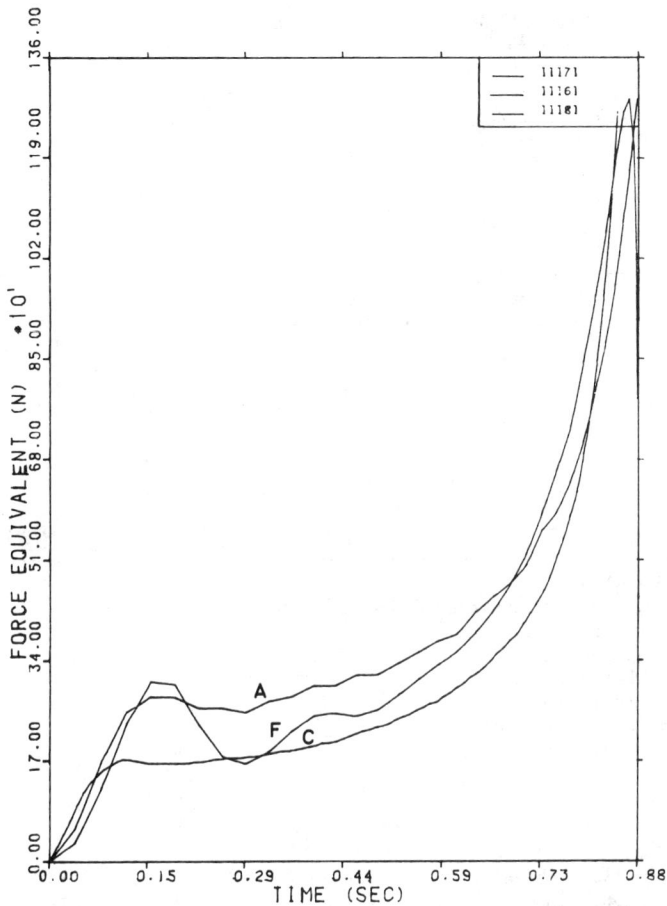

FIG. 11—*Force equivalent comparison of the impact deceleration,* A, *transmitted force,* F, *and static compression,* C, *curves to the time of peak deceleration.*

so on. After sampling a variety of cushioning materials and exploring specific phenomena, it was realized that a certain amount of the noise consisted of materials characteristics and general operating characteristics of the test system. The latter may be altered by system redesign.

The transmitted force profile is found to show an oscillation not generally reflected in the deceleration profile. This is illustrated in Figs. 12, 13, and 14. This is a characteristic of all the recorded transmitted force profile data available. One possible source is an interaction between the specimen, as a compliant element, and some massive part of the system. Another is the irregular plateau region that is characteristic of the deformation of many foams. A

TABLE 4—*Maximum available quarter cycle damping.*

Specimen	Drop No.	β_{qcm}
Polyurethane	3, 4, 5	0.356

TABLE 5—*Computed parameters.*

Drop[a]	Maximum Acceleration, g	Maximum Compression, cm (in.)	Rebound Height, cm (in.)	Coefficient of Restitution
1, Ua	167	2.2 (0.86)	32.08 (12.63)	0.51
3, Ua	185	2.2 (0.88)	20.11 (7.92)	0.41
6, Saf	82	1.8 (0.71)	5.28 (2.08)	0.24
7, Uaf	76	2.0 (0.79)	20.32 (8.00)	0.47
8, Paf	97	2.5 (0.98)	25.27 (9.95)	0.53
10, Sf	74	1.83 (0.72)	1.83 (0.72)	0.14
11, Sf[b]	71	1.85 (0.73)	1.70 (0.67)	0.14

[a]Key to abbreviations:
 U = polyurethane foam (Arthur D. Little laboratories).
 S = polystyrene panel.
 P = polypropylene.
 a = accelerometer signal.
 f = force cell signal.
 c = static compression data.
[b]Aluminum stage.

number of observations were made of the static deformation characteristics and any possible relation to the dynamic, transmitted force profile.

The difference in amplitude between the two curves for each specimen reflects the different sensor sensitivities. The ordinate is voltage from the device. As in the other data reported, the energy represented by the area under each curve is restricted to the range of initial compression.

Polystyrene drop test results are similar to those of polyurethane in that a G_{m1} damping peak is evident, and both materials transmit the oscillating force to the load cell. However, the plateau region of the impact deceleration profile of the polystyrene specimen reflects its irregular static compression plateau region. The polyurethane response is smooth in both tests. This detail is associated with materials deformation processes.

The polypropylene drop test results and the static compression test results, as well, are remarkably symmetrical. The damping effect is observed as a slight increase in initial amplitude in the impact deceleration profile compared with the terminal region of that same curve. The transmitted force profile, however, displays the same type of oscillation as by the other specimens.

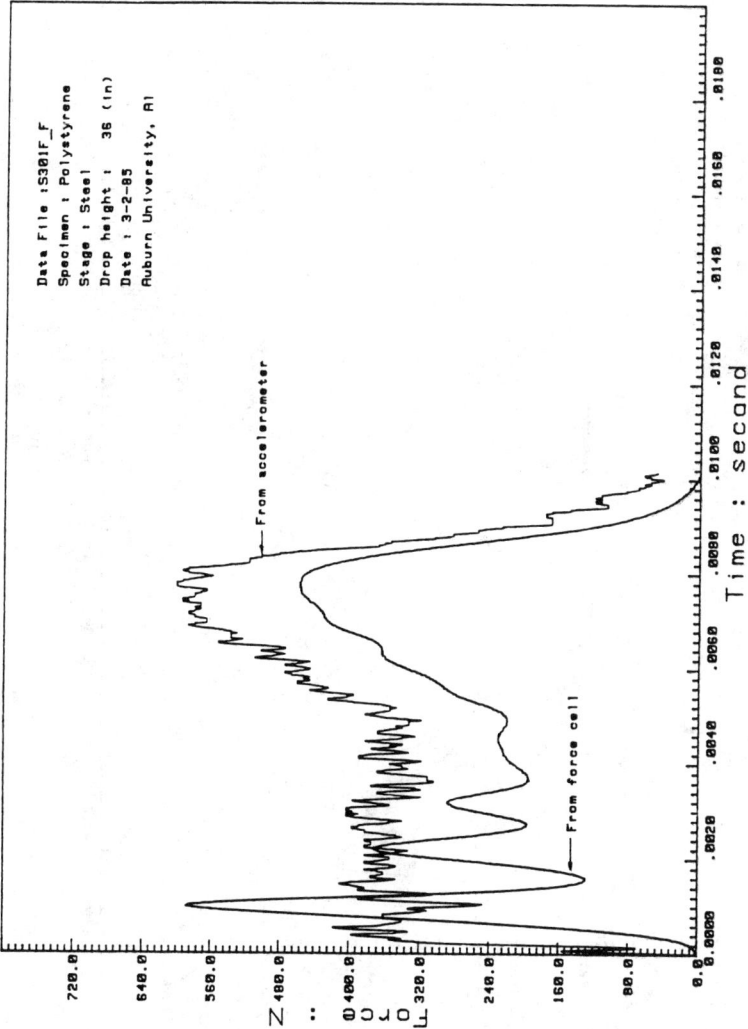

FIG. 12—*Polystyrene foam drop test results (Drop 6, Table 2).*

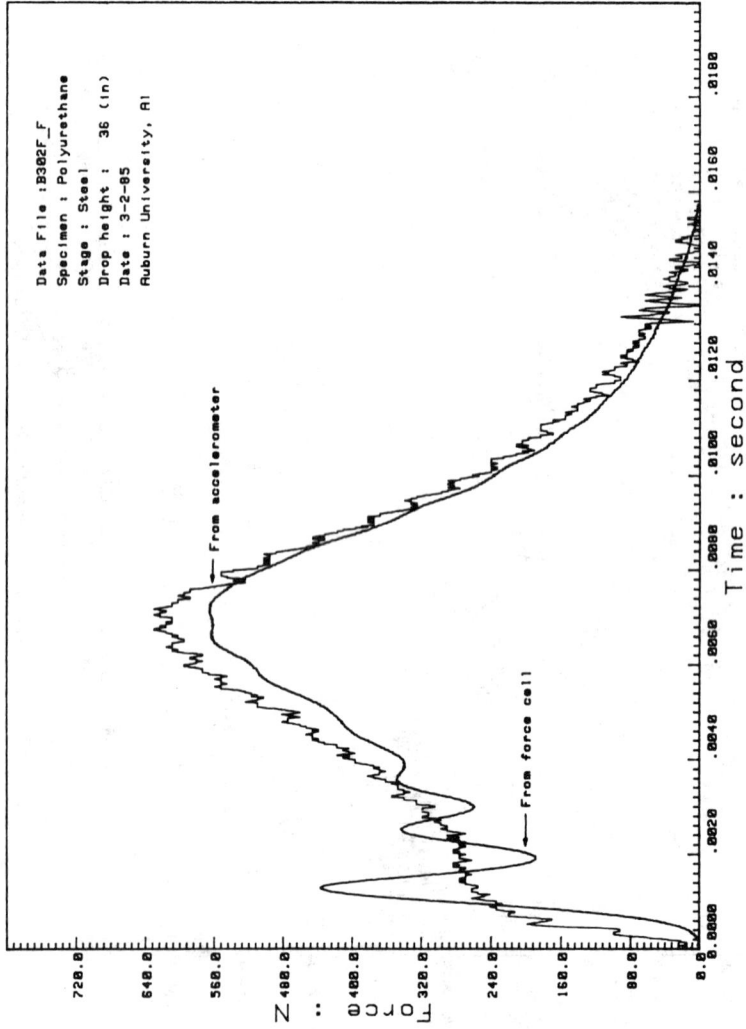

FIG. 13—*Polyurethane foam drop test results (Drop 7, Table 2).*

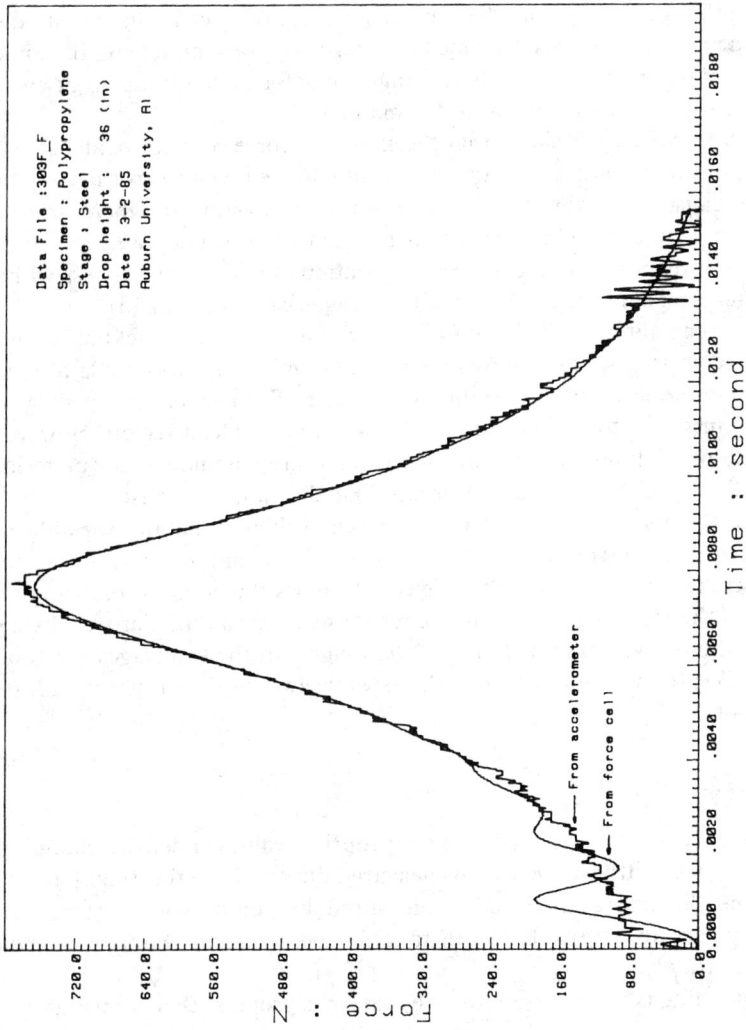

FIG. 14—*Polypropylene foam drop test results (Drop 8, Table 2).*

Specimens from this same block have been repeatedly compressed statically and dynamically with no apparent change in characteristics. During the compression test, the maximum compression attained is of the same order as with the other types of foams. The calculated results indicate that most of the energy in the test is transferred to the load cell, that is, the damping is very low.

In both static and dynamic testing the polystyrene specimens were observed to deform progressively from the contact surfaces. After cell collapse initiated in the surface regions, this deformed region grew by expanding into the undeformed central region. It was found that the deformation initiated at either surface. Not all foams deform in this manner.

The force cell clearly shows that the stage and force cell receive an oscillating or pulsating load. This can be attributed to either of two causes. One possible cause is the time-varying rupture of cells within the foam and the other is a coupled mechanical action between the specimen and the stage. Static compression tests show a cyclic variation of compression stress within the plateau region of the deformation characteristic (see Fig. 15).

Polystyrene and polyurethane foams were found to emit crackling sounds as static compression testing was conducted. An attempt was made to measure the sound level, using a sound level meter. The results tend to show an increase in sound following each load drop in the plateau region. However, since the oscillations in the transmitted force are common to all materials tested, this special relationship was eliminated as a prime cause.

Separate tests of the effect of impactor weight showed that this variable did not affect the period of this oscillation, although the amplitude was increased with increased impactor weight. Figure 16 shows the effect of reducing the weight of the stage by comparing the response using the standard steel stage and an alternative aluminum stage. The weights of the two stages are 186.0 and 466.7 g for aluminum and steel, respectively. This is consistent with the prediction of Eq 4.

Conclusions

1. Impact can be simulated using appropriate values of density, damping, and structural stiffness. Density is measured directly from the weight and dimensions. Structural stiffness is measured by compression testing with proper regard to natural time constants in the material. Damping is measured separately.

2. Damping is a structure-sensitive materials property that contributes directly to material stiffness and to cushioning in a complex manner. It is required to prevent excessive oscillations but can produce an adverse effect.

3. The impacting system is in static equilibrium at the time of maximum deceleration, maximum compression, and zero velocity.

4. A number of special properties and parameters may be calculated from simulation results.

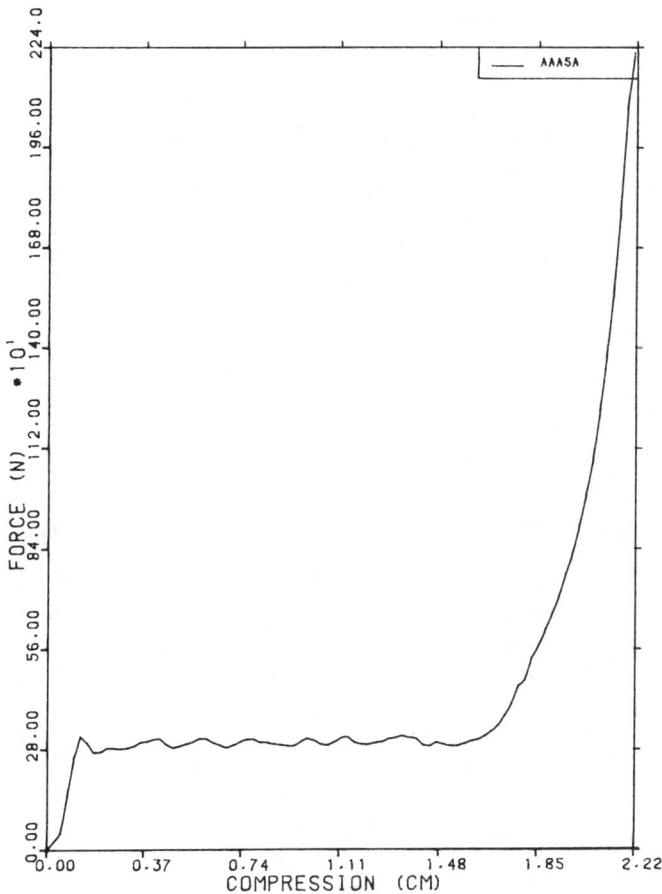

FIG. 15—*Polystyrene foam static compression test results.*

5. Damping capacity may be measured from the difference in work of deformation represented by the acceleration and transmitted force profiles to the point of static equilibrium.

6. The maximum available damping capacity may be calculated from the difference in work of deformation represented by the acceleration and static compression curves to the point of static equilibrium.

7. The force transmitted by the compliant specimen, under impact, may be oscillatory because of system characteristics.

8. Instrumented impact testing is capable of producing design information of use in human safety and transportation of fragile equipment.

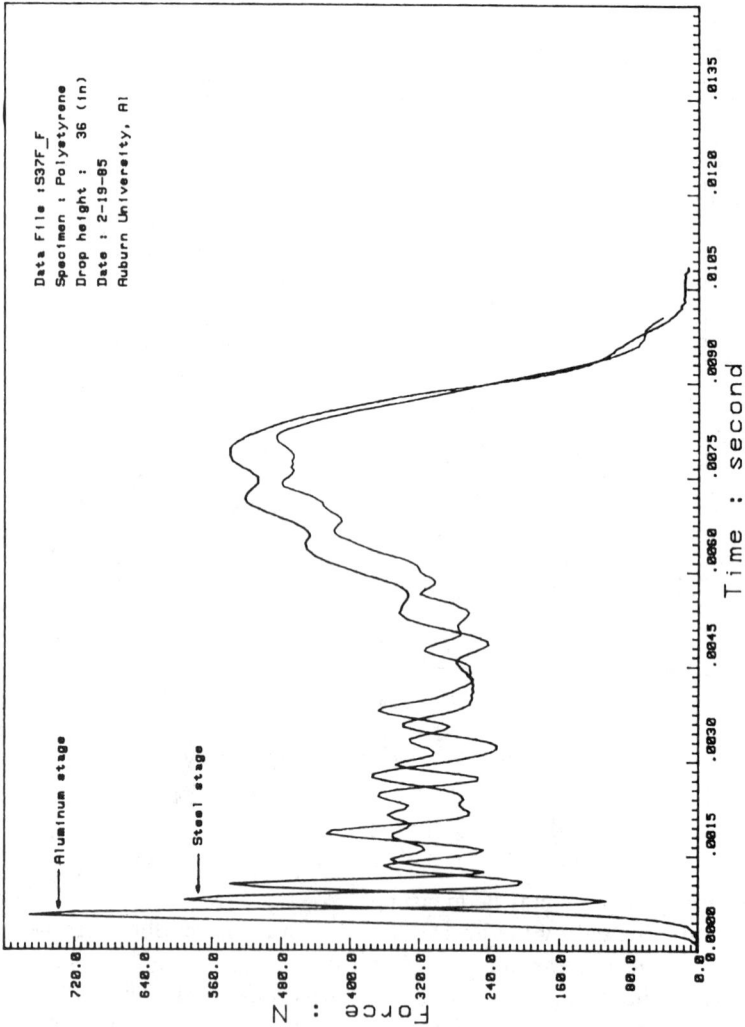

FIG. 16—*Effect of the stage on the transmitted force (Drops 10 and 11. Table 2).*

Acknowledgments

The authors wish to acknowledge an Auburn University grant-in-aid that provided funds for the piezoelectric sensors and charge amplifiers. The polyurethane specimens studied were supplied by the Arthur D. Little Laboratories.

References

[1] Jones, N. and Wierzbicki, T., Eds., *Structural Crashworthiness,* Butterworth, London, 1983.

[2] Gadd, C. W., "Use of a Weighted Impulse Criterion for Estimating Injury Hazard," *Proceedings of the Tenth Stapp Car Crash Conference,* Society of Automotive Engineers, New York, November 1966, pp. 164–174.

[3] Jemian, W. A. and Lin, N.-H., "Computer Modeling of the Body-Head-Helmet System," USAARL Report No. 76-13, Vols. I and II, U.S. Army Aeromedical Research Laboratory, Fort Rucker, AL, 1972.

[4] Jemian, W. A., "Simulation and Testing of the Double Shell Crash Helmet," Final Report to U.S. Army Aeromedical Research Laboratory, Fort Rucker, AL, 1976.

[5] Bathe, J. K. and Wilson, E. L., "Stability and Accuracy Analysis of Direct Integration Methods," *Earthquake Engineering and Structural Dynamics,* Vol. 1, 1973, pp. 283–291.

S. Norm Kakarala[1] *and James L. Roche*[1]

Experimental Comparison of Several Impact Test Methods

REFERENCE: Kakarala, S. N. and Roche, J. L., **"Experimental Comparison of Several Impact Test Methods,"** *Instrumented Impact Testing of Plastics and Composite Materials, ASTM STP 936,* S. L. Kessler, G. C. Adams, S. B. Driscoll, and D. R. Ireland, Eds., American Society for Testing and Materials, Philadelphia, 1987, pp. 144–162.

ABSTRACT: The results of a statistical evaluation of ten impact test methods are presented in this paper. The concept of "good impact test" has two aspects: the quality and the applicability of the test results. The applicability of various test methods for predicting in-service performance is the subject of a companion paper. This paper presents an evaluation of the quality of the test output, as well as an analysis of the correlation between various test outputs.

The ten test methods evaluated—three pendulum, one tensile, four drop dart, one flex, and one driven dart—are described briefly. Three of the test instruments were instrumented. Seven materials, ranging from unfilled thermoplastics to reinforced thermosets, were tested by each of the methods.

Evaluation of a test's output quality is developed in terms of the range of applicability and the degree of material differentiation. Specific factors that serve as criteria include the existence and distinctness of failure, and the spread and scatter of measured results. The test methods are ranked from "highly desirable" to "severely restricted."

Evaluation of the correlation of material ranking by the various test methods is presented. In general, results from different impact tests do not correlate. However, there are instances of strong or fair correlation. The degree of correlation is explained in relation to the similarity of test stress states and measured characteristics. It is meaningful to classify impact tests according to stress states and measured characteristics. Material selection for impact performance should be based on *test methods* that simulate the stress states, controlling variables, and failure limits of the intended application.

KEY WORDS: composite materials, reinforced plastics, impact behavior, impact testing, impact stress, failure mode, failure existence, failure distinctness, material differentiation

Even a brief review of the literature on impact behavior of composite materials reveals the enormous complexity of the subject. Many composite material systems are anisotropic and are highly nonuniform in composition on a

---

[1]Supervisor and senior project engineer, respectively, Advanced Engineering Staff, General Motors Technical Center, Warren, MI 48090-9040.

macroscopic scale, typically consisting of matrix, fillers, and reinforcement flakes or fibers. Impact response in composites reflects a failure process involving crack initiation and growth in the viscoelastic matrix, fiber breakage and pullout, delamination, and disbonding. Even the description of failure can be complex.

Until recently, the impact behavior of composite materials has been characterized by the same test methods originally developed for metals—namely, the Izod, Charpy, and drop-weight tests. Several major concerns have arisen in connection with these test methods. The scatter in results is often extremely high, even for carefully executed tests. Material performance may be dramatically different from that given by another test method. The test result ranking of materials is often contrary to comparative performance in actual application.

This paper examines the nature and the causes of these inconsistencies in material rankings between test methods and proposes a methodology for comparison of impact test methods. There are three primary objectives in this study: (1) description of the stress states and failure modes for the various impact tests, (2) determination of the range of applicability and the degree of material differentiation with the various impact tests, and (3) determination of the consistency of material impact ratings between test methods.

Experimental Details

Experimental Plan

The focus of this study is on impact test methods. The experimental plan included ten test methods commonly used for material screening, product design evaluation, and production quality control. Seven materials were selected to represent a broad spectrum of commercial materials and the full range of automotive applications. An objective of the experimental design was that the conclusions from the study would not be material specific.

Impact Test Methods

The ten test methods selected for this study can be grouped into five classes: pendulum, tensile, drop dart, flex, and driven dart. The test method details are summarized in Table 1. An understanding of the mechanics of specimen loading and failure in the various impact tests is necessary for comparison of test methods. An overview of each of the test methods is provided, with a sketch to illustrate the test configuration.

The impact geometry, specimen shape, and design of the test fixture determine the induced stress state and failure mode. The stress states and failure modes in each of the test methods are indicated. This information is essential for a discussion of the correlations between test methods.

TABLE 1—*Test methods evaluated.*

No.	Test Name (ASTM Standard)	Test Velocity, m/s	Stress State	Failure Criterion	Measured Characteristic
1	Izod notched (D 256-81 Method A)	3.45[a]	in-plane bending	break	total energy
2	Izod reverse notched (D 256-81 Method E)	3.45[a]	in-plane bending	break	total energy
3	Charpy (D 256-81 Method B)	3.45[a]	in-plane bending	break	total energy
4	Tensile, MTS (D 1822M-83 modified)	3.45	uniaxial tension	break	tensile strength tensile elongation
5	Falling weight (D 3029-84 Method B)[b]	3.60 → 0	biaxial bending/tension	cracking	yield energy[c]
6	Gardner standard	variable → 0	normal shear	cracking	yield energy
7	Gardner 1.25-in. ring	variable → 0	biaxial bending	cracking	yield energy
8	Gardner anvil	variable → 0	normal compression	cracking	yield energy
9	Flex, Dynatup	3.60[d]	uniaxial bending	yield-break	yield deflection yield energy
10	Driven dart, Rheometrics	3.60[d]	biaxial bending	yield-break	ultimate force total energy

[a]Velocity decreases during impact by 5 to 60%.
[b]ASTM Test for Impact Resistance of Rigid Plastic Sheeting or Parts by Means of a Tup (Falling Weight) (D 3029-84).
[c]Yield refers to proportional limit.
[d]Velocity decreases during impact by 0 to 10%.

Pendulum Tests—The impactor and specimen configurations of the three pendulum tests (Izod notched, Izod reverse notched, and Charpy) are shown in Fig. 1. Procedures recommended in the ASTM Test for Impact Resistance of Plastics and Electrical Insulating Materials (D 256-81) were followed in conducting these tests. For the two Izod tests, the specimens are clamped in the vertical position like a cantilever beam with the notch toward the striker (or on the reverse side for the reverse-notched test). In the Charpy test, the specimen is notched and supported horizontally against the stops at either end, and the pendulum impacts the specimen directly opposite the notch.

In the three tests, a free-swinging pendulum with a rounded-tip mount is used as an impactor. A single value of total absorbed energy is obtained by measuring the energy left in the pendulum after completion of the impact fracture. The primary stress state in these pendulum tests is in-plane bending, and the failure mode is total breakage of the specimen.

Tensile Impact Test—Type L specimens, as recommended by the ASTM Test for Tensile-Impact Energy to Break Plastics and Electrical Insulating Materials (D 1822-84), were tested on a hydraulically driven high-speed tensile tester (Model 819, MTS Systems Corp.). The test setup is shown in Fig. 2.

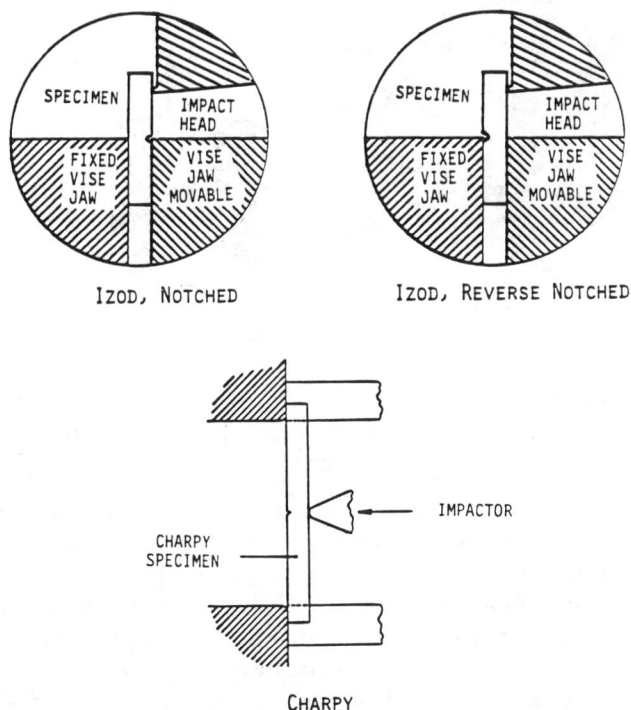

IZOD, NOTCHED   IZOD, REVERSE NOTCHED

CHARPY

FIG. 1—*Test configurations in pendulum tests.*

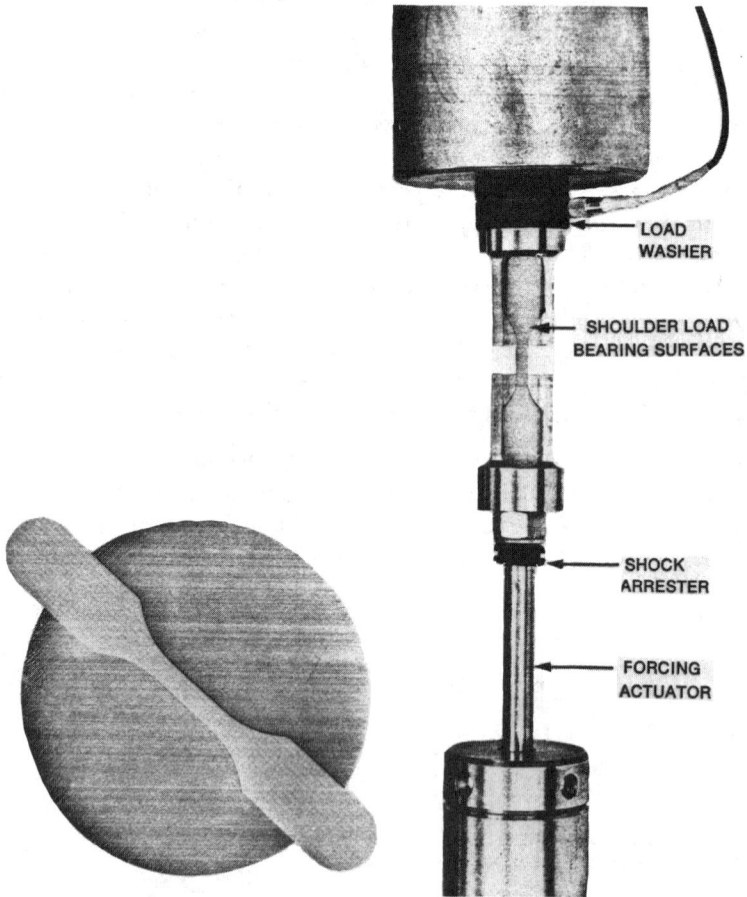

FIG. 2—*Specimen and apparatus for the tensile test.*

Tensile strength and elongation at break were calculated from the force-deflection traces. Hence, the "elongation at break" does not truly correspond to the extensometer measurements of true tensile elongations.

Drop Dart Test—The specimen supports and the impactor geometries for the four drop dart tests—falling weight, Gardner standard, Gardner 1.25-in. ring, and Gardner anvil—are illustrated in Fig. 3. The tests differ mainly in the way the specimens are held and the ratio between the unsupported area of the specimen and the cross-sectional area of the dart. The stress states in these tests are listed in Table 1. The staircase method described in the ASTM Test for Impact Resistance of Polyethylene Film by the Free Falling Dart Method [D 1709-15 (1980)] was used in calculating the energy to initiate cracks on these tests.

2.00 in dia

5.00 in dia

FALLING WEIGHT

0.625 in.

Impact hammer
0.625 in. dia ball

Crack

0.64 in.

1.2 in.

GARDNER, STANDARD

0.625 in.

Impact hammer
0.625 in. dia ball

1.2 in.

2.0 in

GARDNER, 1.25 in RING

0.625 in.

Impact hammer
1.250 in. dia ball

1.2 in.

2.0 in

GARDNER, ANVIL

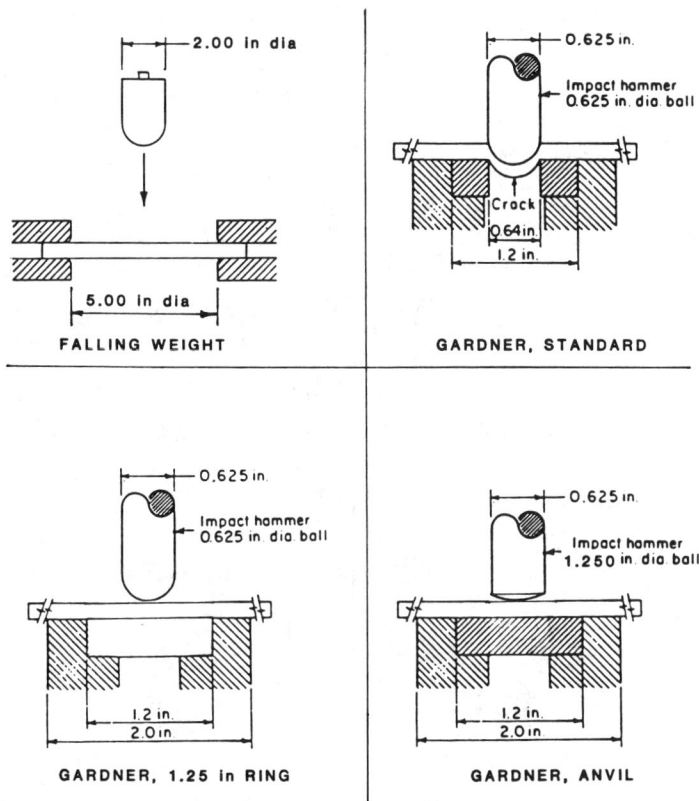

FIG. 3—*Test configurations in the four drop dart tests.*

Flex Impact Test—The Dynatup Model 8000 instrumented drop-weight impact system was used for the flex test. Flex specimens meeting the requirements of the ASTM Test for Flexural Properties of Unreinforced and Reinforced Plastics and Electrical Insulating Materials (D 790-81) were tested with a span between supports of 50.8 mm (2 in.), as shown in Fig. 4. The striking tup [12.7 mm (0.5 in.) in diameter] is instrumented. Force-deflection and absorbed energy-deflection traces were recorded during the impact event. The measured characteristics on this test are the deflection and energy at yield, force at break, and total energy. The term "yield" is used in this paper to represent the material characteristic that is commonly known as "proportional limit."

Driven Dart Test—The Rheometrics high-speed RIT-8000 system was used for the driven dart test. A flat specimen is clamped over a 76-mm (3-in.)-diameter ring. A hemispherical probe (25.4 mm in diameter) is driven at

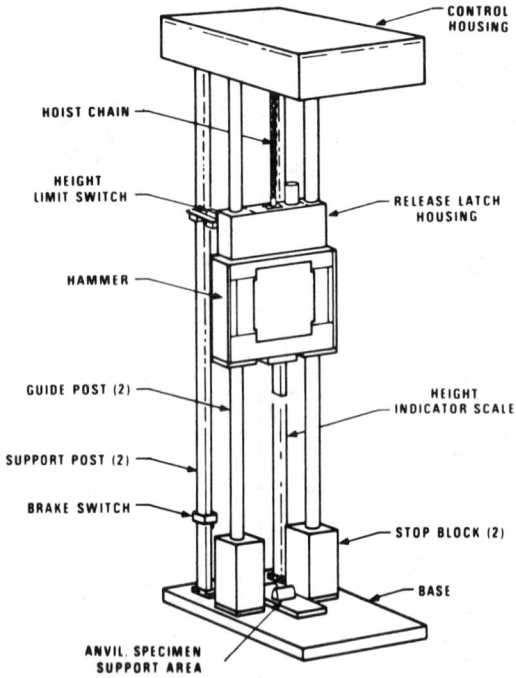

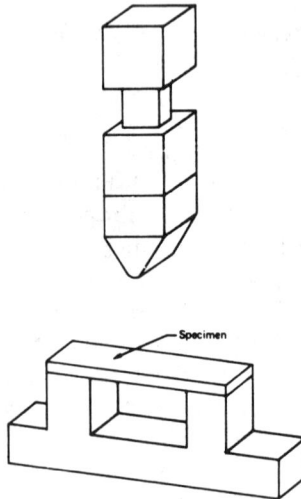

FIG. 4—*Apparatus and test configuration for the flex test.*

3.60-m/s velocity and penetrates through the specimen, as shown in Fig. 5. The velocity and load transducers on the probe record complete load-deflection and energy-deflection traces during the impact event. The four measured characteristics—yield deflection, yield energy, ultimate force, and total energy—were obtained from the plotted traces.

Materials Tested

Seven materials were selected to represent a variation in impact behavior ranging from brittle to ductile. Each of the materials selected represents an important material class. These materials are known to display distinctively different impact responses. Details of the materials used in this study are given in Table 2.

Large sheets of the seven materials were procured. The specimens for different impact test methods were prepared according to the specific ASTM standards.

The material variables were considered only to the extent necessary to provide a reasonable range of test responses. Material evaluation was not an objective of this study. The specimens tested were produced in accordance with existing production specifications. However, it cannot be assumed that they are typical of their material class.

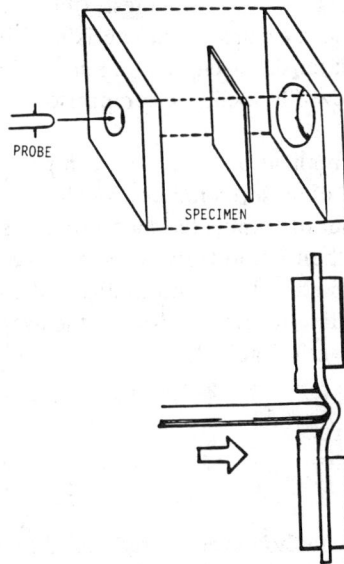

FIG. 5—*Impact stress state in the driven dart test.*

TABLE 2—*Materials tested.*

Test Method No.	Material			Automotive Applications
	Type	Resin	Reinforcement	
1	SMC-R28	polyester	28% glass fiber	front end panels, body panels
2	SMC-R65	vinyl ester	65% glass fiber	bumper supports
3	RIM	polyurethane	none	fascias
4	RRIM	polyurethane	20% glass flake	fenders, door skins
5	nylon RIM	nylon w/ 20% polyol	20% milled glass	fenders, door skins
6	ABS	acrylonitrile-butadiene-styrene	none	instrument panel
7	PP	polypropylene	none	interior trim

Data Reduction

The impact tests produced different measured characteristics—force, deflection, and energy—at specified failure limits. That means the measured characteristic has two dimensions—the monitored property, measured at a set failure limit. The instrumented impact tests (tensile, flex, and driven dart) produced more than one measured characteristic (MC). For example, the tensile test has two MCs: tensile strength and tensile elongation. The MCs on all the tests were normalized to facilitate comparison of the impact test methods.

For each test, the normalization procedure compared the MC of each material to the average MC of all seven materials. The average MC for each material was calculated from the replicated measurements. The average MC for all seven materials was then found from the seven averages of the replicated measurements on each material. The normalized MC for each material was then determined by dividing its average MC by the average MC of all materials. The results are shown in Table 3.

Test Method Evaluation

Evaluation Criteria

Performance of a test method is evaluated based on the extent of applicability of the test and how well the test differentiates the materials. The "range of applicability" of the test depends on the existence and distinctness of fail-

TABLE 3—Matrix of normalized measured characteristics.

No.	Test Method Name	Material Tested, normalized MC						
		SMC-R28	SMC-R65	RIM	RRIM	Nylon RRIM	ABS	PP
1	Izod notched	1.95	1.95	1.29	0.30	0.70	0.60	0.22
2	Izod reverse notched	0.93	1.63	NB[a]	0.38	1.17	0.59	1.28
3	Charpy	1.80	2.55	NB	0.28	0.69	0.38	0.29
4	Tensile,							
	Ultimate tensile strength	0.83	2.68	0.31	0.86	0.56	1.08	0.69
	Total elongation	0.08	0.11	5.39	0.22	0.57	0.29	0.35
5	Falling weight	0.11	0.07	2.10	0.07	0.84	0.17	1.78
6	Gardner standard	0.16	0.04	2.50	0.19	1.19	0.95	1.96
7	Gardner, 1.25-in. ring	0.21	0.08	NB	0.27	3.49	0.95	NB
8	Gardner anvil	0.74	0.14	1.24	1.76	1.24	1.30	0.59
9	Flex,							
	Yield deflection	0.54	0.85	NB	1.68	0.73	1.20	NB
	Yield energy	1.03	1.39	NB	1.66	0.42	0.51	NB
	Ultimate force	1.43	1.72	NB	0.79	0.47	0.59	NB
	Total energy	0.90	1.43	NB	0.55	0.87	1.26	NB
10	Driven dart							
	Yield deflection	0.38	0.44	2.92	0.28	1.06	0.62	1.30
	Yield energy	0.21	0.28	3.85	0.04	0.69	0.40	1.53
	Ultimate force	0.85	1.00	1.53	0.25	0.95	0.98	1.44
	Total energy	0.75	0.86	2.79	0.36	0.54	0.31	1.39

[a]NB = no break.

ures for the materials tested. The "degree of material differentiation" is determined by the ratio between the spread of reported results for all materials and the scatter in the data for each material tested. The test method performance is considered highly desirable if the test has general applicability and provides a high level of differentiation between materials.

Range of Applicability

The range of applicability for the ten test methods studied is reported in Table 4.

Existence of Failure—An essential feature of an impact test is that the test specimens must have failures. If for some materials the specimens do not show any deterioration (cracking or breaking) when tested, then that test will have limited applicability. Reaction injection molding (RIM) and polypropylene (PP) materials, which have relatively low moduli, did not break in a bending stress state. There are four tests that have some limitations for lower modulus materials, as shown in Table 4.

Distinctness of Failure—One other limitation to the general applicability of an impact test involves indistinct measured characteristics. This was observed only in an instrumented impact test, while yield behavior was being

TABLE 4—*Existence and distinctness of measured characteristics.*

No.	Name	No Break For	Indistinct For	Rating[a]
	Test Method			
1	Izod notched			7
2	Izod reverse notched	RIM		6
3	Charpy	RIM		6
4	Tensile,			
	Ultimate tensile strength			7
	Total elongation			7
5	Falling weight			7
6	Gardner standard			7
7	Gardner, 1.25-in. ring	RIM, PP		5
8	Gardner anvil			7
9	Flex,			
	Yield displacement	RIM, PP	all except RRIM	1
	Yield energy	RIM, PP	all except RRIM	1
	Ultimate force	RIM, PP		5
	Total energy	RIM, PP		5
10	Driven dart			
	Yield deflection		RIM, PP	5
	Yield energy		RIM, PP	5
	Ultimate force			7
	Total energy			7

[a]See text for an explanation of the rating scale.

recorded. Table 4 indicates that both driven dart and flex tests have this limitation for low-modulus materials.

Applicability—Applicability ratings were based on the number of materials exhibiting distinct specimen failures on each of the tests, as shown in Table 4. A rating of 7 implies that the test has no limitations for all seven materials tested. Lower ratings on other tests indicate limitations due to either lack of specimen failures or indistinct measurements.

Material Differentiation

The normalized measured characteristics shown in Table 3 were used to determine the degree of material differentiation. The two characteristics that determine material differentiation are the spread of MCs and the amount of data scatter. Scatter in the data for each material relative to the range of MC on a given test should be small, for clear distinction of impact rankings between materials. Consequently, large spread and small scatter are the necessary and sufficient conditions for a good quality impact test.

Spread of Measured Characteristics—The difference between the maximum and the minimum normalized MCs for the materials tested is called the normalized range, or spread, of the results for that test. The normalized range results for different impact tests are shown in Table 5.

Data Scatter—The data scatter is determined by the percentage of error in replicated measurements, which is based on the t-statistic at the 95% confidence level. The average percentages of scatter for all seven materials in a given test are listed in Table 5. The data scatter is greater than 20% for most of the impact tests. Drop dart impact tests use the "staircase method" for data analysis, which gives a single statistical average number as the reported result. Thus, the concept of data scatter is not used for drop dart impact tests.

Degree of Differentiation—One way of quantifying the degree of material differentiation is by comparing the ratio between the spread and the scatter of the data. Before calculating the spread-to-scatter ratio, the normalized range for each test is converted to an average percentage of spread to provide consistency of units. The conversion is accomplished by dividing by the number of specimens and multiplying by 100. The ratio numbers for all the impact tests evaluated are given in Table 5. The tests with higher normalized range values yield a greater extent of differentiation between materials. The flex test, with a range of less than two, is the least effective in the impact ranking of materials.

The scatter bands of the normalized MCs for three impact tests are shown in Fig. 6. Comparisons were made on a log scale to provide proportional representation of scatter at large and small MC values. The driven dart test and

TABLE 5—*Spread and scatter of measured characteristics.*

Test Method		Normalized Range	Average Spread, %	Average Scatter, %	Spread-to-Scatter Ratio
No.	Name				
1	Izod notched	1.73	24	11	2.2
2	Izod reverse notched	1.25	21	22	1.0
3	Charpy	2.27	38	15	2.5
4	Tensile,				
	Ultimate strength	2.37	34	10	3.4
	Total elongation	5.31	76	18	4.2
5	Falling weight	2.03	29	N/A[a]	
				(staircase method)	
6	Gardner standard	2.46	35	N/A	
7	Gardner, 1.25-in. ring	3.41	68	N/A	
8	Gardner anvil	1.62	12	N/A	
9	Flex,				
	Yield displacement	1.14	14	147	0.1
	Yield energy	1.24	18	156	0.1
	Ultimate force	1.25	25	16	1.6
	Total energy	0.88	18	37	0.5
10	Driven dart				
	Yield deflection	2.64	38	18	2.1
	Yield energy	3.81	54	41	1.3
	Ultimate force	1.28	18	19	1.0
	Total energy	2.48	35	33	1.1

[a]N/A = Not applicable.

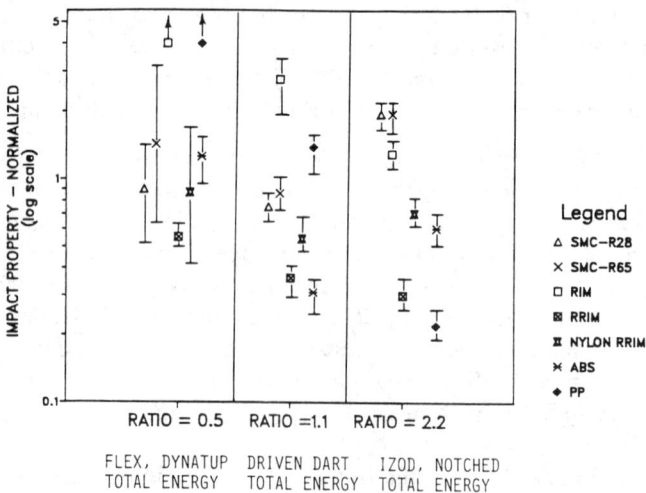

FIG. 6—*Degree of material differentiation.*

the Izod notched test provide more differentiation between materials than the flex impact test.

The ratios for the three tests are shown in Fig. 6. This comparison illustrates that tests with higher speed-to-scatter ratios are more effective in differentiating impact behavior of materials.

Test Method Performance

Test method performance can be represented in a matrix, as shown in Table 6. Here the test methods are qualitatively classified into three levels for each of the two evaluation criteria. The classification limits shown are somewhat arbitrary, but useful.

The tests are classified as "limited" or "narrow" in range of applicability when two or more materials exhibit no failures. Similarly, the tests with spread-to-scatter ratios of less than 1 are considered less than acceptable in material differentiation.

According to this rationale, the flex test and some drop dart tests showed severely restricted performance for impact ranking of materials. Interestingly, the notched Izod test, the standard Gardner test, and the instrumented tensile test were found to be highly desirable in evaluating impact behavior of materials.

Correlation Between Test Methods

Correlation Criteria

The degree of correlation between test methods was determined by the linear regression coefficients. The correlation coefficient defines the level of consistency in material impact ratings between test methods. The correlation between test methods is considered statistically insignificant when the coefficient is less than 0.8. Results from different test methods are judged to have fair to close correlation as the coefficient exceeds the value of 0.9.

Test Method Correlations

The correlation coefficient between two tests is determined by relating the normalized measured characteristics of the seven materials from one test to the corresponding measurements of the other test. There are more than 100 correlation coefficients between the 17 MCs listed in Table 3. However, most of the coefficients are less than 0.7, which indicates the lack of general correlation between impact tests.

Specifically, some impact tests do have fair to close correlations in ranking materials. The instances of correlation were explored and are discussed in the following sections. An hypothesis is proposed that correlation between test

TABLE 6—Matrix of test method desirability.[a]

Range of Applicability	Poor $(R < 1)$	Acceptable $(1 < R < 2)$	Good $(R > 2)$
General $(N = 7)$	Gardner anvil	falling weight, driven dart, ultimate force, driven dart, total energy	Izod notched, tensile, ultimate strength, tensile, total elongation, Gardner standard
Limited $(N = 5; N = 6)$	flex, total energy	Izod reverse notched, flex, ultimate force	Charpy, Gardner, ring, driven dart, yield deflection, driven dart, yield energy
Narrow $(N < 5)$	flex, yield deflection, flex, yield energy		

[a]Ranking of test method desirability levels:

	Poor	Acceptable	Good
General	D	B	A
Limited	D	C	B
Narrow	D	D	D

A = highly desirable.
B = desirable.
C = acceptable, but limited.
D = severely restricted.

[b]Material differentiation: R = ratio of spread to scatter; N = number of materials with distinct failures.

methods occurs only if the stress states are the same for the two tests and is further improved if the measured characteristics also match.

Pendulum Tests—The stress state for all three pendulum tests is in-plane bending with one significant variation. The Izod notched and Charpy test specimens contain a stress concentration at the point of maximum stress, while the Izod reverse notched specimen does not. The measured characteristic for all three tests is total energy.

Results from Izod notched and Izod reverse-notched tests do not correlate, as is shown in Table 7. Stress is concentrated at the notch tip of the specimen in the Izod notched test, and stress is more evenly distributed in the Izod reverse-notched test. The difference in stress distribution between these two tests explains the variation in material rankings. The Charpy test also involves stress concentration. It has close correlation with the Izod notched test and no correlation with Izod reverse-notched test, in accordance with the proposed hypothesis.

Drop Dart Tests—All drop dart test methods show close correlation, except for the Gardner anvil test, as shown in Table 8. The measured characteristic for all four tests is an energy approximately at the yield point. The Gardner anvil test represents a normal compression stress state compared with the biaxial bending/tension in the other three tests.

TABLE 7—*Correlation between pendulum tests.*

	Correlation Coefficient Between Tests		
Test Method	Izod Notched	Izod Reverse Notched	Charpy
Izod notched	1.00	0.47	0.96
Izod reverse notched		1.00	0.62
Charpy			1.00

TABLE 8—*Correlation between drop dart tests.*

	Correlation Coefficient Between Tests			
Test Method	Falling Weight	Gardner Standard	Gardner, 1.25-in. Ring	Gardner Anvil
Falling weight	1.00	0.96	0.99	0.00
Gardner standard		1.00	0.86	0.13
Gardner, 1.25-in. ring			1.00	0.29
Gardner anvil				1.00

Driven Dart Test Response Features—Among the four response features of an instrumented driven dart test given in Table 9, yield deflection, yield energy, and total energy show close correlation. However, ultimate force does not correlate with any other response feature. These instances of correlation are surprising and suggest an overriding dependence of correlation on similarity of stress states.

Drop Dart and Driven Dart—The yield energy and total energy responses from the instrumented driven dart test are compared with the drop dart test results in Table 10. The stress state for the driven dart test is biaxial bending/tension, which is similar to the stress states for the drop dart tests, except for the Gardner anvil test. The drop dart tests, excluding the Gardner anvil test, show close correlation with the yield energy but reduced correlation with the total energy response from the driven dart test. The failure criterion used in the drop dart tests is crack initiation, which corresponds to the yield energy response from the driven dart test, resulting in close correlation between them. The total energy corresponds to a complete break, which is a different failure criterion.

This comparison supports the hypothesis that both stress states and mea-

TABLE 9—*Correlation between driven dart test response features.*

| Driven Dart Test Response | Correlation Coefficient Between Driven Dart Test Responses | | | |
	Yield Deflection	Yield Energy	Ultimate Force	Total Energy
Yield deflection	1.00	0.99	0.76	0.93
Yield energy		1.00	0.77	0.96
Ultimate force			1.00	0.76
Total energy				1.00

TABLE 10—*Correlation between drop dart and driven dart test response features.*

| Drop Dart Test Methods | Correlation Coefficient for Driven Dart Test Response | |
	Yield Energy	Total Energy
Falling weight	0.90	0.85
Gardner standard	0.89	0.78
Gardner, 1.25-in. ring	0.90	−0.23
Gardner anvil	0.07	−0.12

TABLE 11—*Correlation between tests with various stress states.*

Test Method	Izod Notched	Izod Reverse Notched	Tensile, Ultimate Strength	Gardner Anvil	Flex, Total Energy	Driven Dart, Total Energy
			Correlation Coefficient Between Tests			
Izod notched	1.00			−0.59	0.51	0.18
Izod reverse notched		1.00		−0.89	0.63	0.69
Tensile, ultimate strength			1.00			
Gardner anvil				1.00	0.75	−0.32
Flex, total energy					−0.75	−0.12
Driven dart, total energy					1.00	0.41
						1.00

sured characteristics should be matched between tests to obtain the best correlation of material rankings.

Tests with Various Stress States—The correlation coefficients of six tests with various stress states are compared in Table 11. As expected, no close correlations exist between these common impact tests. Although the tensile test ultimate strength and driven dart test ultimate force are equivalent measured characteristics, they do not correlate.

In summary, correlations between test methods depend primarily on a similarity of stress states, and the correlations do improve when measured characteristics also match between the tests.

Conclusions

The impact performance for a wide range of automotive plastics and composites can be measured using several currently available test methods. The Izod notched, tensile, and standard Gardner test method results had general applicability and showed a higher degree of differentiation between materials. On the contrary, the Gardner anvil and Dynatup flex test methods demonstrated severely restricted performance.

In general, results from different impact tests do not correlate. However, test methods with similar stress states and measured characteristics have good correlation. The correlation between test methods depends primarily on a similarity of stress states, and the correlations improve with similarity of measured characteristics.

Classification of impact tests according to stress states and measured characteristics is useful for comparison of results from different test methods. Consequently, material screening and selection for impact performance should be made with a test method with stress state and measured characteristics that match the stress state, controlling variable, and failure limit of the intended application.

Stephen Burke Driscoll[1]

Variable-Rate Impact Testing of Polymeric Materials—A Review

REFERENCE: Driscoll, S. B., "**Variable-Rate Impact Testing of Polymeric Materials— A Review,**" *Instrumented Impact Testing of Plastics and Composite Materials, ASTM STP 936,* S. L. Kessler, G. C. Adams, S. B. Driscoll, and D. R. Ireland, Eds., American Society for Testing and Materials, Philadelphia, 1987, pp. 163–186.

ABSTRACT: Instrumented impact testing has been discussed since the early 1970s but has still not been readily accepted and employed in characterizing polymeric materials systems. This paper explores this topic by reviewing some of the more important technical papers presented at various professional society meetings. The second part of this paper identifies areas of interest in research methodology, including testing parameters such as material composition, geometry, fabrication, loading history, and temperature.

KEY WORDS: impact testing, variable-rate impact testing, plastics, composites, testing parameters, impact test problems, impact test analysis, impact test applications

It would be appropriate to begin this paper by quickly reviewing the classic approach to impact testing of plastics and composites. In all cases we are concerned about the quality and reproducibility of data generated using actual parts, which are impacted over the full range of environmental in-use conditions to simulate a multistress field situation, with failure usually occurring in the weakest direction (Fig. 1). Driscoll has written in detail on the relative advantages and disadvantages of different modes of impact testing [1]. Table 1 compares the advantages and disadvantages of the three major types of impacting techniques:

(*a*) ASTM Test for Impact Resistance of Plastics and Electrical Insulating Materials (Pendulum-type Izod/Charpy) (D 256-81),

(*b*) ASTM Test for Tensile-Impact Energy to Break Plastics and Electrical Insulating Materials (D 1822-83), and

---

[1]Professor, Department of Plastics Engineering, University of Lowell in Massachusetts, Lowell, MA 01854.

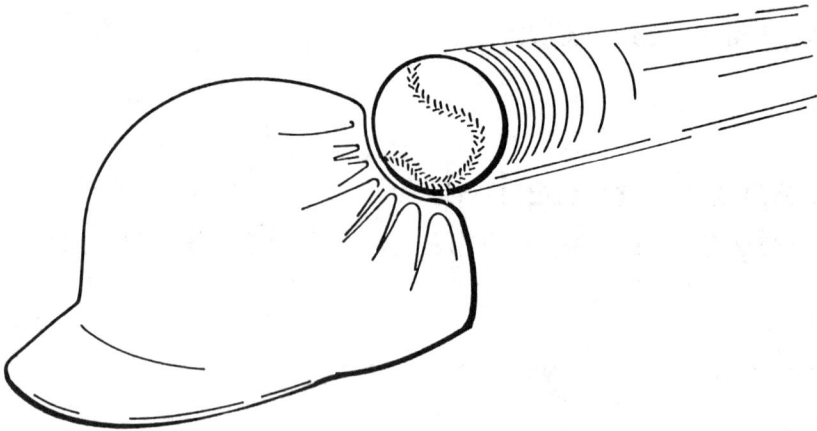

FIG. 1—*Baseball helmet being impacted—a multistress field.*

TABLE 1—*Data on traditional impacting test methods.*[a]

Impact Method	Temperature	Rate	Realistic Geometry
Pendulum impact	X	O	O
High rate tensile	X	X	O
Drop weight	X	O	X

[a]Key: X = good; O = poor.

(c) ASTM Test for Impact Resistance of Rigid Plastic Sheeting or Parts by Means of a Tup (Falling Weight) (D 3029-82).

Briefly, the Izod and Charpy tests are quite inadequate since both are only single-point measurements and provide limited information, which is not always representative of actual in-service usage. More critical is the notch sensitivity aspect, which influences energy absorbing behavior. Should the notch be molded in or machined? What is the influence of the rate of notching and of the thickness sensitivity of the material?

Another important consideration is the quality of the notch. Impact strength will increase as the "sharpness" of the notch decreases. Studies have indicated that the notch need not be too deep to provide adequate stress concentrations. Force/deflection instrumentation has been designed to enhance the value of the data generated by the freely falling pendulum-type test apparatus.

What is the effect of impact at other than the prescribed rate of 3.3 m/s (11 ft/s, 7920 in./min, or 7.5 miles/h). In the mid-1960s, this concern for rate sensitivity was addressed through the development of instrumented high-rate tension testers. Integrating the area under the force/deflection curve generated the energy absorbed by the breaking specimen. Although this information certainly represented a new generation of "quality data," it was limited by the uniaxial stresses imposed on the specimen.

The protocol of ASTM Method D 3029-82 does accommodate a greater range of test geometries but generates limited data since this is a multiaxial stress field impact event. However, when the event is properly instrumented for force/deflection analysis, it is possible to divorce the subjectivity of failure analysis from the reality of the impact event. Defining the criteria for failure is difficult. Instrumentation does provide more complete information—for example, on whether the failure mode is brittle, ductile, tear, or punched hole fracture (Figs. 2 and 3). One final reservation on drop-weight testing is the limitation on impacting rate unless the laboratory ceiling is unusually high. Most conventional rooms cannot accommodate a drop tower of more than 2.4 to 3.0 m (8 to 10 ft). Consequently, the impacting rate is below 5 m/s, which could miss the critical rate-dependent "window" of some polymeric material systems. Faster impacting rates are possible using various methods for accelerating the impacting tup.

Historical Review of Instrumented Impact Testing

Progressing beyond the design of the impact test tower, it is important to account for the instrumental problems associated with the data acquisition/reduction analysis. This brief section will detail some of the historical aspects of instrumented impact testing of plastics and composites.

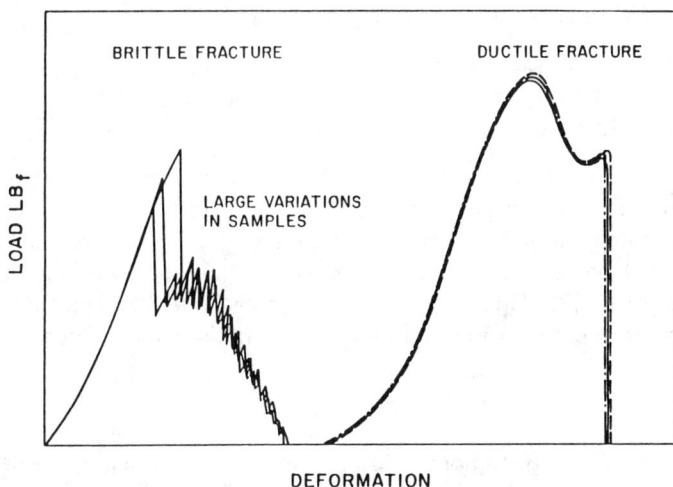

FIG. 2—*Brittle versus ductile impact behavior.*

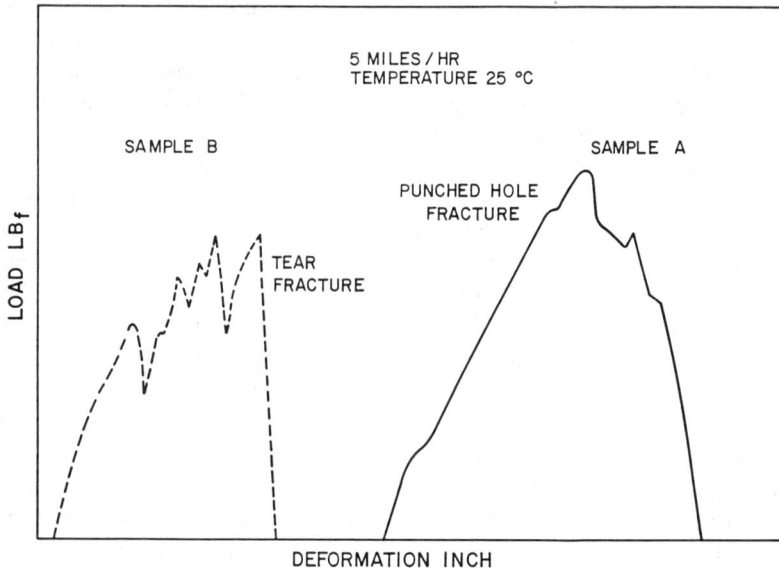

FIG. 3—*Tear versus punched-hole impact behavior for ABS.*

In June 1973 Ireland reported that "in 1970 there were fewer than five laboratories in the United States using instrumented impact tests, about 25 in 1972, and more than 50 by 1973" [2]. Today it would be difficult to guess the number of commercially available and proprietary instrumented systems being used, but certainly there are many more than there were ten years ago. In his paper, Ireland addressed three critical areas:

(*a*) calibration of the dynamic load cell,
(*b*) control of the instrumented tup signal, and
(*c*) data reduction.

A decade later we are still concerned (and often confused) about these same three topics. It is the author's hope that this volume will provide greater insight and more helpful information in formulating solutions. Ireland also coauthored with Saxton and Server a paper on the "initial discontinuities of the load-time trace" [3]. This "inertial load" was evaluated using a Charpy impact configuration for various aluminum and titanium alloys (Figs. 4 and 5).

Ductility Index

Still later in 1973 Beaumont et al evaluated a series of epoxy resin-based composites containing single-component and hybrid graphite, S-glass, boron, and Kevlar [4]. An instrumented Charpy apparatus having a span-to-

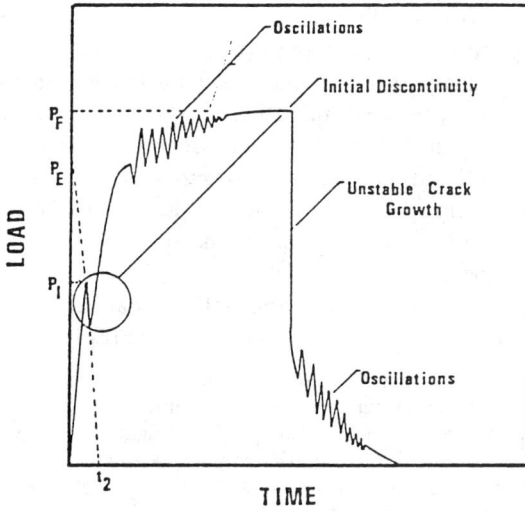

FIG. 4—*Schematic of load-time behavior for a tough material.*

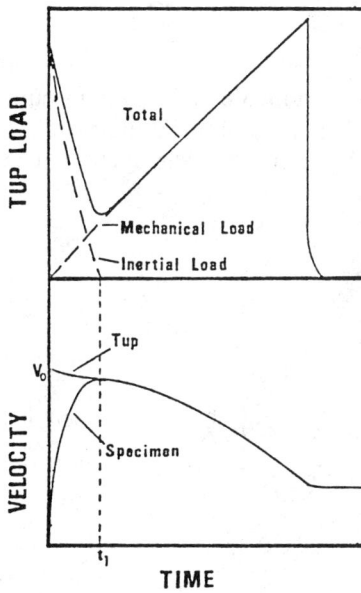

FIG. 5—*Schematic of load-time and velocity-time behavior for a tough material.*

depth ratio of 4 to 1, characteristic of a short-beam shear test, was used. Although these authors stated that "to date, work reported in the open literature has shown significant deficiencies of the standard Charpy test in characterizing the impact response of materials . . . it has been shown that the instrumented Charpy test can provide valuable insight into failure mechanisms and differentiate between initiation and propagation energies." Further, the authors suggested that the total impact energy measured during the test, E, is the sum of the initiation and propagation energies, E_i and E_p. It was stated that Charpy impact energy is not an inherent material property but rather is dependent on the dimensions of the specimen. (More will be said about this parameter later.)

Figure 6 is a schematic representation of the load history in an impact event. Referring to this illustration, Beaumont et al [5] have stated:

> the total Charpy energy, E, does not provide much information about the fracture behavior of a material . . . a brittle, high strength material will have a large initiation energy, E_i, and a small propagation energy, E_p. Conversely, a low strength ductile material will have a small E_i and a large E_p. Therefore, even though the Charpy energy for the two materials may be the same, their behavior is quite different.

This observation prompted these authors to suggest a "ductility index" (DI), which is the ratio of the propagation to initiation energies

$$ \text{DI} = \frac{E_p}{E_i}. $$

Consequently, using an instrumented Charpy test would provide a convenient way of measuring both E_i and E_p. Table 2 illustrates typical impact data for various composites, with emphasis on the ductility index.

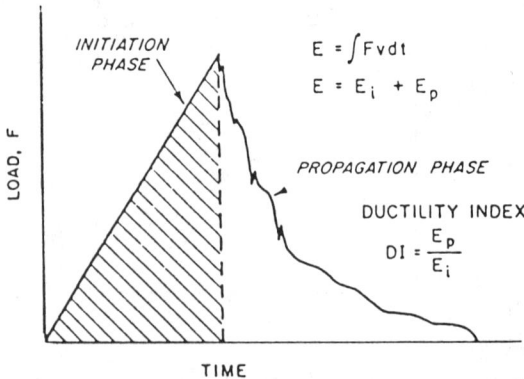

FIG. 6—*Schematic of the load history in an impact event illustrating the ductility index.*

TABLE 2—*Instrumented impact test data—ductility index* [4].

Reinforcing Fiber	Apparent Flexure Strength, ksi (MN/m²)	Total Energy per Unit Area, ft · lb/in.² (J/m²)		Ductility Index
		Dial	Oscilloscope	
E-glass	73 (500)	114 (2.4 × 10⁵)	114 (2.4 × 10⁵)	0.4
Kevlar 49	142 (980)	124 (2.6 × 10⁵)	114 (2.4 × 10⁵)	23 (1.6)ᵃ
HMS-graphite	125 (860)	3.8 (8 × 10³)	3.8 (8 × 10³)	0.0
20% Kevlar 49, 80% HMS-graphite	170 (1170)	34.3 (7.2 × 10⁴)	30.5 (6.4 × 10⁴)	6
41% Kevlar 49, 59% HMS-graphite	141 (970)	46.7 (9.8 × 10⁴)	42.9 (9 × 10⁴)	4

ᵃThe first value is based on the onset of nonlinearity. The number in parentheses is based on maximum stress.

In 1975 Broutman and Rotem presented a paper at the ASTM symposium on Foreign Object Impact Damage to Composites [5]. The paper noted:

> the energy absorption or toughness of homogeneous isotropic materials has been measured by various techniques . . . the results for brittle materials could be correlated with fracture mechanics theories such as the Griffith theory . . . however, [when] these theories which have been developed for single-phase materials are applied to fiber composites, the behavior cannot be predicted . . . and the energy absorption can certainly not be calculated by application of the "rule of mixtures" to the two or more phases.

Regarding this work [5], an instrumented falling-weight/three-point bending test station was constructed using photo diodes to trigger velocity-measuring devices. Impacted specimens were also evaluated using a conventional, noninstrumented Charpy apparatus. It was noted that the fracture stress was not dependent on the weight or on the specimen width. However, when the specimen width, span, and drop weight were doubled, the force/deflection curves sometimes had the same slope but never double the energy absorbed. Further, it was noted that the test series data indicated that the adoption of test methods commonly used for metals was not valid for nonmetallic composites. The Charpy test was found to be not suitable for glass fiber/epoxy composites since these materials may not be as sensitive to notches in either direction to the laminae. More specifically, these laminates are rate sensitive and the energy absorbed is test-apparatus sensitive.

Testing Parameters

Broutman and Rotem also referred to another of the author's concerns for impact testing parameters—the velocity profile during the impact event. They noted that, for similar specimens impacted using different weights, the dependence on the rate of loading is greater for the smaller weight. The lower the ratio between energy applied and absorbed, the higher the dependence on the rate of loading. When the energy applied is only slightly greater than the energy needed to fracture the specimen, the velocity slowdown is considerable, and a significant change in the rate of loading will occur during the test. However, when the available energy is much greater than the energy needed to fracture the specimen, there is only a slight change in the velocity of the impacting tup. Consequently, test results should indicate the dependence of the energy absorbed for a fixed rate of loading.

Rate Sensitivity

Since glass-fiber-reinforced epoxies are rate sensitive, a common impact test such as the Charpy cannot give complete information because it supplies a constant amount of energy regardless of the energy-absorbing capabilities of the specimen. Thus, the difference between the energy supplied and the energy absorbed is not a constant. Since the mechanism of absorbing energy is delamination between layers and between fibers, the type of glass does not change the composite's toughness. However, the surface treatment of the glass fibers may change the ability of the material to absorb energy.

In 1977 Adams presented a paper, "Impact Response of Polymer-Matrix Composite Materials," at the ASTM-sponsored symposium on Composite Materials: Testing and Design [6]. He stated:

> the understanding of the mechanisms governing failure under impact loading remains largely unknown . . . low-level impacts of standard Charpy and Izod specimens now are being considered since these lead to somewhat simpler stress states, although still not as simple as those achieved if a uniaxial tensile impact loading were utilized . . . relatively little has been done in the tension impact testing of fiber-reinforced composites.

Thickness Dependency

Adams specifically cited problems associated with testing various specimen thicknesses [6], mentioning:

> the magnitude of the ratio of shear stress to normal stress increases in direct proportion to the ratio of specimen thickness to specimen length . . . thin specimens tend to fail in flexure (either tensile fracture or compressive buckling), while thick specimens fail in shear (one or more delaminations running along the length of the beam) . . . and, in general, thicker specimens exhibited higher impacting energies, although not always in the same ratio from one material to another.

Table 3 illustrates some of Adams's test observations. Table 4 details more recent data on the instrumented impact behavior of aluminum and steel sheeting of different gage thicknesses [7]. Table 5 illustrates the thickness

TABLE 3—*Influence of specimen thickness on the total impact energy of composites* [6].

Material	Total Impact Energy, kJ/m^2	
	Thin (1.3 to 2.3 mm)	Thick (7 to 9 mm)
Graphite/epoxy	66	98
Graphite/epoxy-S-glass	27	232
Graphite/epoxy-120 glass	74	94
Graphite/epoxy-aluminum mesh	36	75
Graphite/epoxy-titanium foil	208	125

TABLE 4—*Case study: impact data for steel and aluminum for automotive use [2.2 m/s (5200 in./min) 12.7-mm-diameter probe (0.5 in.); 76.2-mm-diameter support ring (3.0 in.)].*

Thickness, mil (mm)	Slope, lb/in.	Force, lbf	Deflection, mil (mm)	Energy, in. · lbf (J)^a
Steel specimens				
34.5 (0.88)	3410	1670	680 (17.3)	470 (4154.8)
47.0 (1.19)	3990	2070	692 (17.6)	620 (5480.8)
59.0 (1.50)	4970	3140	777 (19.7)	1100 (9224.0)
Aluminum specimen				
62.5	4410	2070	649 (16.5)	570 (5038.9)

^{a}1 J = 8.84 in. · lbf.

TABLE 5—*Case study: impact data for printed circuit board laminates.*^{a}

	Single Ply		Double Ply, 50.8 mm (2-in.) Support Ring
	(76.2-mm (3-in.) Support Ring	50.8 mm (2-in.) Support Ring	
Slope	580	920	1520
Yield			
Force, lbf	150	170	230
Deflection, mil (mm)	938 (23.8)	489 (12.42)	468 (11.89)
Energy, in. · lbf (J)	10 (88.4)	10 (88.4)	20 (176.8)

and geometrical dependency of printed circuit board fabrications [8]. Table 6 contrasts the thickness sensitivity of injection-molded structural foam plaque studies by Driscoll, Grolman, and Venkateshwaran [9]. Figure 7 illustrates the instrumented impact behavior of a butadiene-styrene block copolymer at different thicknesses.

TABLE 6—Instrumented impact data for injection molded (8000 in./min) structural foam plaques.[a]

Test Area and Properties[b]	Polypropylene			Polystyrene			Polycarbonate			Modified PPO		
	CP	HSF	CSF	CP	HSF	CSF	CP	HSF	CSF	CP	HSF	CSF
Area C												
Slope, lbf/in.	4637	4343	4142	2739	2875	2581	4348	4224	3970	4135	4098	3914
Yield energy, in. · lbf	663	504	309	88	55	119	697	472	457	62	62	59
Ultimate energy, in. · lbf	1024	783	644	380	323	377	1056	853	790	583	456	469
Area C												
Slope, lbf/in.				3067	2622	2225	4380	4435	2969	4549	4258	3358
Yield energy, in. · lbf				120	51	45	625	449	142	77	48	60
Ultimate energy, in. · lbf				378	311	227	981	962	442	570	353	416
Area D												
Slope, lbf/in.	3749	3758	3479	3257	2726	2467	4408	4296	3281	4554	4300	3624
Yield energy, in. · lbf	683	591	279	51	43	74	662	451	287	64	48	51
Ultimate energy, in. · lbf	1135	911	719	400	279	326	1010	832	593	476	396	445
Area A												
Slope, lbf/in.	3829	4206	4203	7331	6714	7077	3160	3500	3111	10550	9518	10275
Yield energy, in. · lbf	600	421	462	201	283	307	995	600	560	553	106	129
Ultimate energy, in. · lbf	914	698	826	788	591	693	1538	1069	1049	1008	726	799

[a]Key to abbreviations:
  CP = counter-pressure molding technology.
  HSF = heavy conventional structural foam molding technology.
  CSF = conventional structural foam molding technology.
[b]Metric conversion factors:
  1 J = 8.84 in. · lbf.
  1 lbf/in. = 0.1752 kN/m.

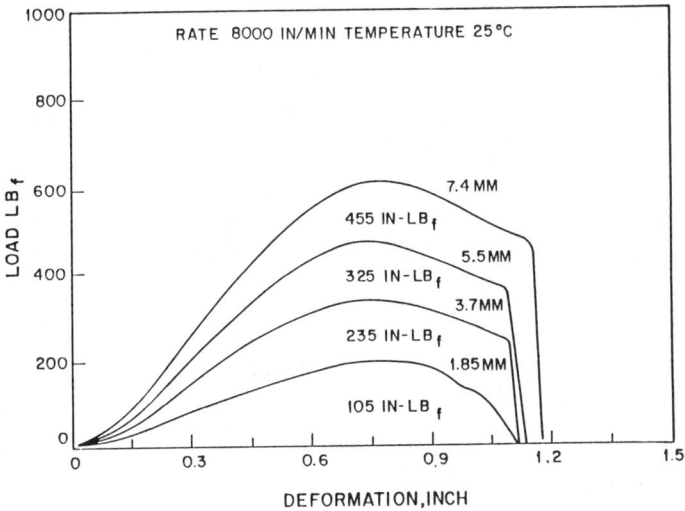

FIG. 7—*Effect of thickness on the impact properties of Kraton.*

Rate Dependency

In discussing the rate dependency [6], Adams continued by stating:

> perhaps the most obvious variable to be considered in an impact test is the impact velocity . . . because, if a polymer-matrix composite were not sensitive to the rate of loading, then an impact test would not be necessary; a simple static test would be sufficient! Surprisingly, however, relatively little attention has been given to experimental studies of effects of rate of impact!

Figure 8 clearly shows the rate dependency of a series of four composite materials impacted at various rates. At the lower impact speeds, Formulations 1, 2, and 3 exhibited similar trends while Formulation 4 absorbed less energy. At the Izod impact rate of 3.3 m/s (8000 in./min), Compounds 3 and 4 were quite similar. Above 5.0 m/s (12 000 in./min), Composite 4 exhibited superior energy-absorbing behavior compared with Compound 3. This is certainly quite a reversal in performance and underscores why impact (and all other) testing should be conducted over a reasonable range that represents typical end-use requirements.

Adams concluded his paper by reiterating the importance of the ductility index [4]. He illustrated this point by mentioning that a large fraction (50% or more) of the total impact energy absorbed by a graphite/epoxy specimen was absorbed before the peak force was reached. However, a Kevlar 49/epoxy composite absorbed only a small fraction of the total energy prior to the impact peak. The Kevlar 49/epoxy formulation exhibited a ductility index more

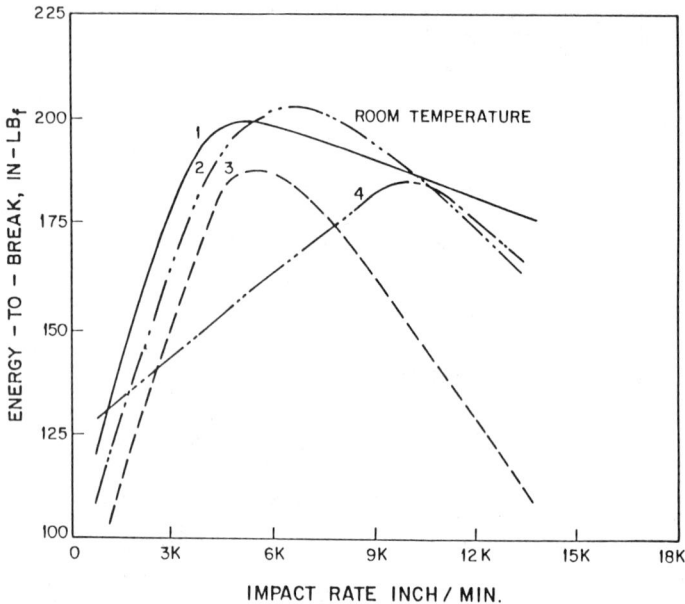

FIG. 8—*Effect of impact rate on the impact strength of four different commercial sheet molding formulations.*

than ten times higher than that of the graphite/epoxy system (15.8 versus 1.5 DI) and a total impact energy more than six times higher (672 versus 109 kJ/m^2). However, the peak energy absorbed by the Kevlar 49/epoxy was slightly less than that absorbed by the graphite/epoxy—and more significantly, the maximum impact force sustained by the Kevlar 49/epoxy composite was only 60% of that of the graphite/epoxy (7.9 versus 12.9 kN). Thus, what constitutes "good" impact resistance must be carefully reviewed and defined for any given application.

In 1979 ASTM Committee D-20 on Plastics sponsored a symposium on Physical Testing of Plastics—Correlation with End-Use Performance. At that June meeting, Ireland presented an excellent paper, "Instrumented Impact Testing for Evaluating End-Use Performance," in which he detailed a series of case studies involving different polymers and fiber-reinforced composites. Examples were cited for using instrumented impact techniques for analyzing the deformation and fracture process of a polymer [10].

Correlation of Different Testing Methods

In that same 1979 symposium, Rieke, of Dow Chemical U.S.A., presented interesting data in a paper, "Intrumented Impact Measurements on Some Polymers" [11]. He stated:

in spite of its practical and technical importance, impact resistance is an ill-defined property generally associated with energy to break in a number of standardized tests normally measured at room temperature . . . and often the results of these various tests do not correlate well with each other . . . some materials will perform well in one test but might not necessarily perform well in another. . . . This is understandable if one recognizes that the impact resistance of a plastic is not only a function of several material properties but also a function of geometry and fabrication conditions . . . that [it] is a property of the specimen.

To illustrate the importance of these variables, a testing program was conducted by Rieke [11] on a series of thermoplastic materials using an instrumented drop tower [65.9 kg (145 lb) dropped from 30.15 cm (12 in.), corre-

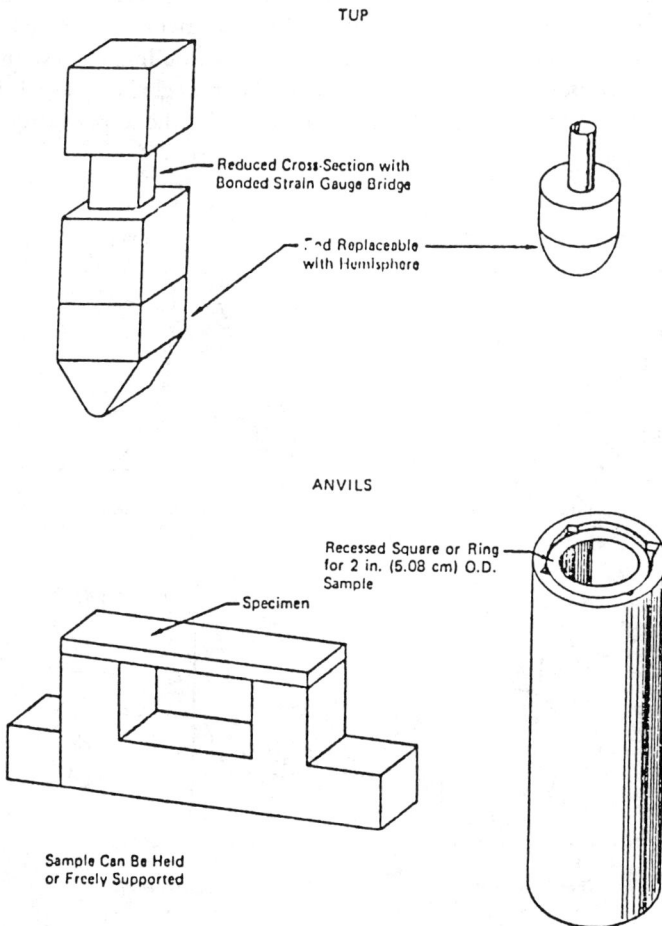

FIG. 9—*Tups and anvils for use with a drop tower.*

sponding to an impacting rate of 240 cm/s (95 in./s)]. Impacting was accomplished using a cylindrical anvil with a 3.475-cm (1.25-in.)-diameter hole and 0.1588-cm (0.0625-in.)-diameter hemispherical tup impacting a 5.08-cm (2-in.)-square specimen. This is similar to the protocol for the ASTM Test for Resistance of Organic Coatings on the Effects of Rapid Deformation (Impact) (D 2794-82). The second geometry was a three-point bending (Charpy-type) configuration (Fig. 9).

Excellent data were generated for extruded polyvinyl chloride (PVC) sheeting of 0.102-cm (0.040-in.) thickness. Typical ductile/brittle trends were observed, and it was noted that "the slope of the loading curve, the time to load and unload, and the character of the fracture pattern . . . should be monitored for unusual features" (Fig. 10).

A series of compression-molded high-impact acrylonitrile-butadiene styrene (ABS) terpolymers was molded into square plaques of various thicknesses, 0.094 cm (0.037 in.) to 0.490 cm (0.193 in.). Below −17°C (0°F), the tup-impacted specimens exhibited quite similar brittle impact responses (at all thicknesses). Below ambient temperature, the impact response exhibited a thickness effect, albeit not a linear one. Not only did the impact strength in-

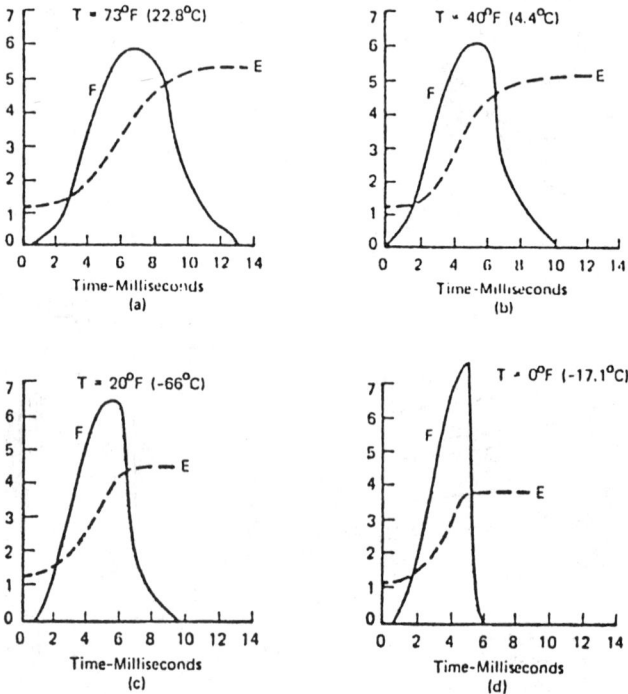

FIG. 10—*Representative force and energy traces for PVC.*

crease with thickness, but thicker specimens tended to absorb more energy per unit thickness than the thinner materials. At elevated temperatures the thickness dependency became less apparent, indicating the need for more research work in this area.

Influence of Test Geometry

At the 1979 Society of Plastics Engineers Annual Technical Conference (SPE ANTEC) in New Orleans, Takemori and Yee (General Electric Co.) investigated the effects of specimen thickness and impact probe on the support ring diameter ratio for various thicknesses of polycarbonate (PC) and polymethyl methacrylate (PMMA) sheeting [12]. Figures 11 and 12 illustrate the relationship between force and thickness, as well as that between energy and thickness.

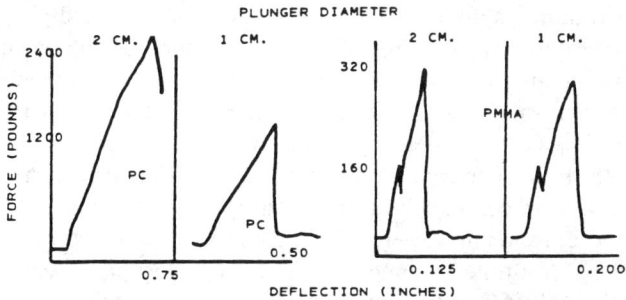

FIG. 11—*Force-deflection behavior for PC and PMMA.*

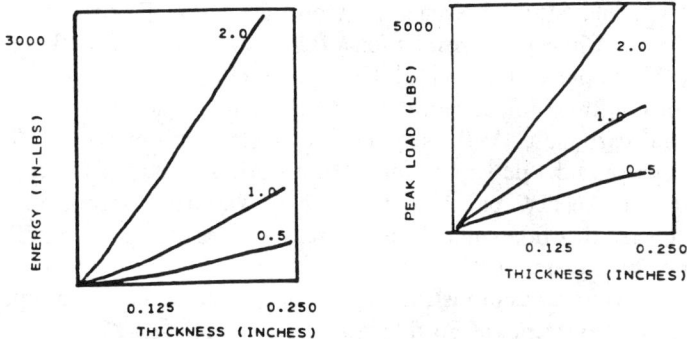

FIG. 12—*Energy versus thickness for PC and PMMA.*

Design Development

Fernando (Northern Petrochemical) presented a series of case studies at the Society of Plastics Engineers National Technical Conference (SPE NATEC) in September 1983 [13]. His study detailed the use of instrumented impact designing to aid in the design development of a series of molded products: hard hats, battery cases, and battery lids. He noted that:

> impact data generated with an instrumented impact machine have been helpful in identifying the critical regions in molded articles. Microscopic examination of these regions has provided useful information in understanding the cause of poor impact behavior. In some instances mold modifications are essential to eliminate the inferior performance and, together with proper material selection, a substantial improvement in impact behavior is seen . . . for example, the high melt flow rate (MFR) impact grade materials exhibit comparable or slightly better impact behavior compared to standard resins . . . these high MFR grades allow sufficient lowering of the melt temperature to reduce the cycle time . . . without affecting the impact performance.

Progelhof (New Jersey Institute of Technology) and Throne (Amoco Polymers) have authored a number of fine papers on instrumented puncture testing of structural foams [14]. Progelhof presents more recent data in this volume. Continuing in the area of practical applications of instrumented impact testing of structural foam, Driscoll and Gacek, at the 1984 SPE ANTEC meeting in New Orleans, contrasted the impact properties of gas-counter pressure with those of conventional structural foam-molded electronic housing [15]. Tests were conducted at various drop heights, using a Rheometrics instrumented drop weight tower (RDT-5000) to simulate actual business office conditions. Multiple specimens were cut from key part areas, tested according to the ASTM Test for High-Speed Puncture Properties of Rigid Plastics (D 3763-79), and compared for mold-fill functional performance behavior (Fig. 13 and Table 7). The data indicated that the mold-fill pattern was critical and that data repeatability was excellent within a site but not for the entire geometry.

At that same SPE conference, Tryson of General Electric reviewed his puncture and Charpy test results for a family of alloyed polycarbonate engineering thermoplastic blends [16]. Goolsby and Miller (University of Texas at Arlington), at this same meeting, reviewed the instrumented fracture toughness of polycarbonate. At the symposium on which this publication is based, Goolsby presented a paper on more recent work on acetals and acrylics [17]. A third SPE ANTEC paper coauthored by Nimmer, Moran, and Tryson (General Electric Co.) explored the impact response of impact-modified polycarbonate box beams [18].

A more recent series of instrumented Charpy studies has been reported by Golovoy and van Oene, of Ford Motor Co. Research [19–22].

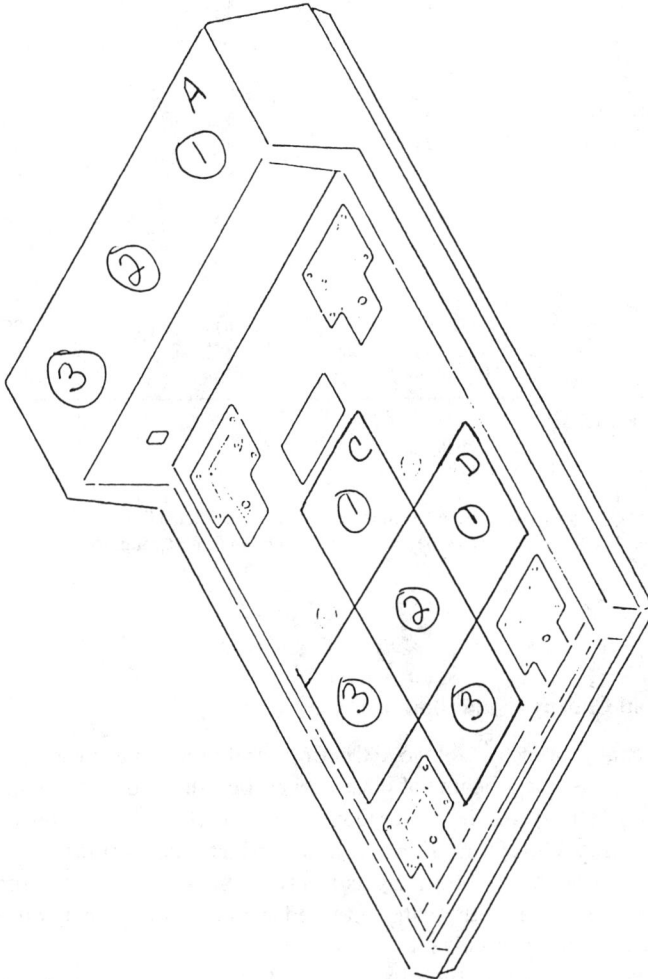

FIG. 13—*Structural foam-molded electronic housing showing sites for instrumented impact testing.*

TABLE 7—*Instrumented impact energy as a function of impacting speed (for counter-pressure foam products).*[a]

Material and Impacting Rate, (in./min)	Slope, lbf/in.	Yield Energy, in. · lbf	Ultimate Energy, in. · lbf[b]
Polypropylene			
6 000	4 333	542	619
8 000	4 067	590	898
10 000	3 637	595	932
Polystyrene			
6 000	7 804	225	616
8 000	6 424	104	843
10 000	5 952	315	825
Polycarbonate			
6 000	2 540	820	1 590
8 000	2 790	1 070	1 630
10 000	4 530	900	1 280
Modified PPO			
6 000	9 454	417	605
8 000	7 808	597	982
10 000	11 469	544	1 074

[a]Metric conversion factors:
  1 in./min = 0.4233 × 10^{-3} m/s.
  1 lbf/in. = 0.1752 kN/m.
  1 J = 8.84 in. · lbf.
[b]The ultimate energy was determined at the maximum deflection exhibiting a detectable force. Attempts to identify the ultimate as a percentage of reduction in maximum force were not satisfactory.

Instrumented Charpy Evaluation of Composites

A paper coauthored by Golovoy, Cheung, and van Oene investigated the impact behavior of short beams of glass and carbon fiber composites using an instrumented falling weight tower configured in the ASTM Methods for Notched Bar Impact Testing of Metallic Materials (E 23-82) for the Charpy impact test. All of the carbon fiber composites broke completely upon impact, and the glass fiber composites retained their overall integrity but exhibited extensive delamination [19].

Golovoy et al supported the observation that when the ratio of the span between the anvils to the specimen thickness is small, for example, less than 10, shear effects are very important. Testing, therefore, at a larger span-to-thickness ratio, usually 16 and above, will induce a tensile failure dominated by fiber orientation and volume. The influence of impact velocity on the fracture energy of different composites was examined over the range of 1.28 to 4.4 m/s. Results are noted in Table 8. Fracture initiation energy was found to be independent of impact velocity. Fracture propagation energy increased from

TABLE 8—*Impact data for unsaturated polyester XMC high-density sheet molding compound (span = 3 cm; width = 1.25 cm; thickness = 0.32 cm)* [19].

Impact Velocity, m/s	Fracture Energy, J/cm²	
	At Initiation	At Propagation
1.28	3.75	30
2.18	3.75	43
3.08	3.50	45
3.50	3.50	49
4.40	3.50	51

30 to 51 J/cm². The deflection was independent of the impact velocity, but the fracture initiation time was inversely proportional to it.

Golovoy et al further commented that the initiation energy was proportional to the square of the interlaminar shear strength, which was expected since the load does rise linearly up to fracture initiation. Fracture propagation includes both tensile and shear failure, which is simply by successive delamination along planes parallel to the midplane.

In a second study, the impact velocity (2.2 and 4.5 m/s) was correlated with a variable, length-to-diameter l/d, ratio of 3.1 to 23.4 [20]. These glass fiber/epoxy composite results are reported in Table 9 and confirm the rate insensitivity for these materials.

The third Ford Motor Co. research project investigated the impact behavior of glass-reinforced polypropylene using a three-point drop-weight system at temperatures between −15 and 85°C at a constant impacting speed of 2.2 m/s (5 mph). The data reported in Table 10 show that, while both the tensile

TABLE 9—*Summary of impact data of unidirectional Scotchply, 0.64 cm thick at 21°C* [20].

Span, m × 10⁻²	Force, N	W_i, MJ/m³ ª	W_p, MJ/m³ ᵇ
At 2.2 m/s			
2	6980	2.4	. . .
10	2700	3.3	1.5
15	1906	3.4	. . .
At 4.5 m/s			
2	7070	2.1	. . .
10	3570	3.3	0.3
15	2519	3.5	. . .

ªW_i = initiation energy.
ᵇW_p = propagation energy.

TABLE 10—*Instrumented impact data for Azdel polypropylene at 2.2 m/s [21].*

Temperature, °C	W_i, MJ/m³	W_p, MJ/m³	$W_i + W_p$, MJ/m³	Span, cm
−15	1.04	5.70	6.74	2
	0.97	2.34	3.31	3
	0.72	0.28	1.00	6
+21	1.30	5.70	7.00	2
	1.09	1.96	3.05	3
	0.76	1.07	1.83	6
+85	. . .	. . .	7.10	2
	. . .	. . .	3.80	3
	. . .	. . .	1.67	6

and shear modulus decreased with increasing temperature, the shear modulus exhibited greater sensitivity. The fracture initiation and propagation energies were relatively independent of temperature.

The influence of temperature on the impact behavior of the previously investigated glass fiber/epoxy composite was investigated in a fourth study [22]. Again, the ASTM Method E 23-82 (Charpy) protocol was followed at 2.2 m/s at temperatures from −20 to +150°C. The impact energy per unit of de-

TABLE 11—*Summary of data for impacted 0.64-cm-thick glass-reinforced polypropylene [22].*

Temperature, °C	Force, N	$W_i + W_p$, MJ/m³	Span, cm
−20	6474	13.6	3
−20	6628	11.8	4
−20	5742	7.3	6
−20	4401	5.4	8
−20	3437	4.9	10
21	6357	14.0	2.5
21	6070	9.3	4
21	4339	4.1	7
55	5578	15.0	2.5
55	5468	11.0	3
55	4703	6.7	5
55	2236	2.8	12
100	4851	14.0	2.5
100	4641	10.8	3
100	3927	7.1	4
100	3297	3.9	7
100	2422	3.2	10.0
150	3428	18.7	2.5
150	3000	14.1	3
150	2773	10.1	4
150	1708	5.6	7

formed volume was found to be very sensitive to the length/thickness ratio of the specimens. At an l/d ratio of 4, the impacted specimens exhibited extensive delamination (shear failure), and the energy absorbed by the composite was about 14 MJ/m³ at temperatures between -20 and $+100°C$, and 18 MJ/m³ at $+150°C$. At l/d ratios above 16, the failure was mostly in tension, and the impact energy approached a constant value of about 3.5 MJ/m³, independent of temperature (Table 11).

Velocity Profile

Recently Driscoll and Grolman conducted instrumented impact analysis of some proprietary thermoplastic alloys. Polymer blends offer the materials engineer the opportunity to develop very quickly and inexpensively a new generation of high-performance materials. The scheduled orderly review of ASTM Method D 3763-79 has prompted several comments on the concern for velocity slowdown during the impact event. The test data were generated to compare different alloys for impact behavior and velocity profile. Figures 14, 15,

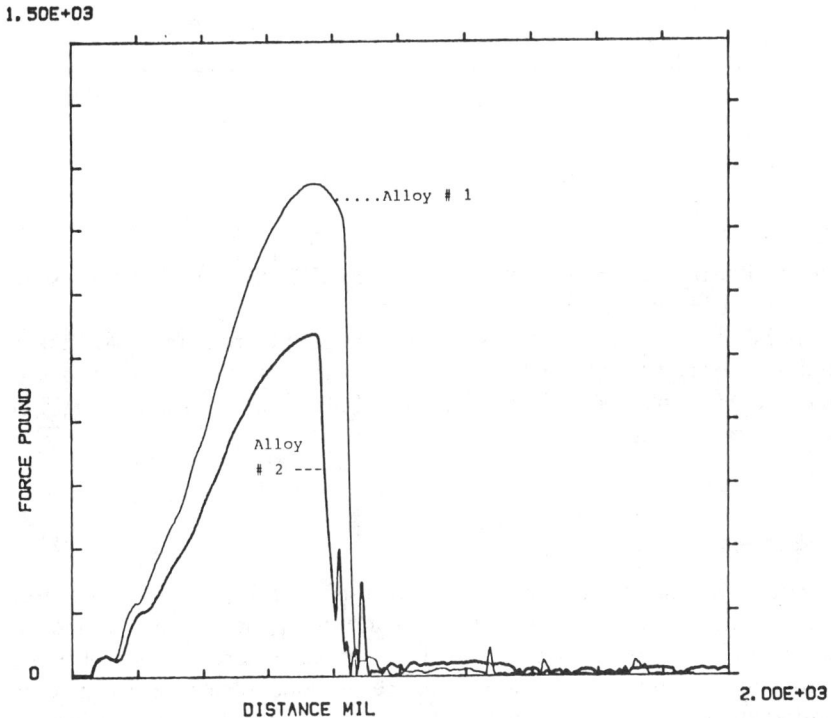

FIG. 14—*Force versus deflection for two alloys.*

1.00E+03

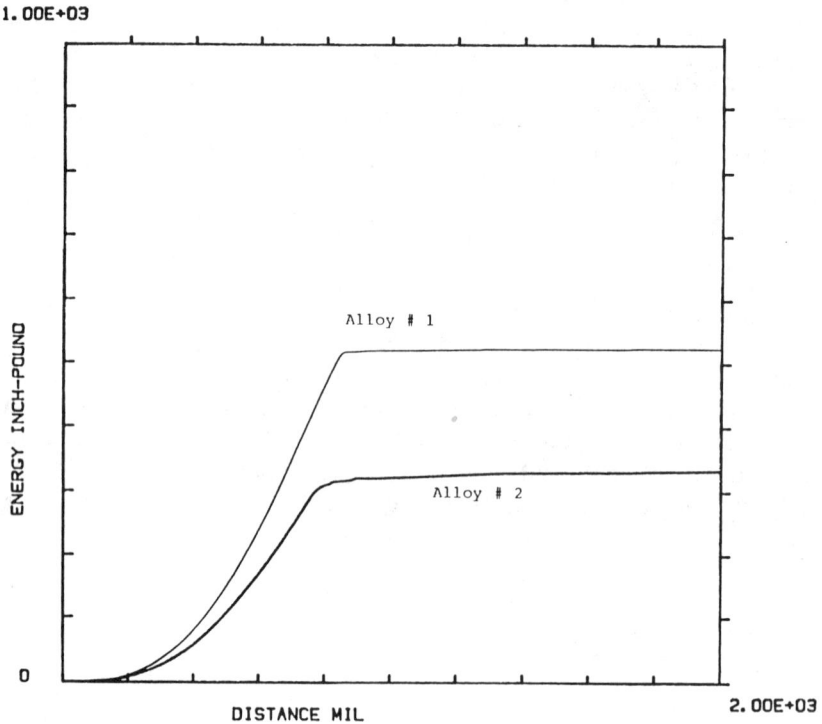

FIG. 15—*Energy versus deflection for two alloys.*

and 16 illustrate the force/deflection, energy/deflection, and velocity slow-down behavior of two candidate alloys.

As will be stated several times in this volume, the mechanism for failure and the performance of a material must be directly related to a specific design and end-use application. What constitutes failure for one product might be superior performance in another application.

Conclusions

These are brief examples of the concern of today's engineers for better-quality, more reliable impact data. Several other papers on the topic of instrumented impact data appear in this volume. A common thread woven throughout is the concern for the impact performance of base resins and composite materials, specifically with respect to (1) composition, (2) design and fabrication, and (3) the consequence of processing, since the manufacturing

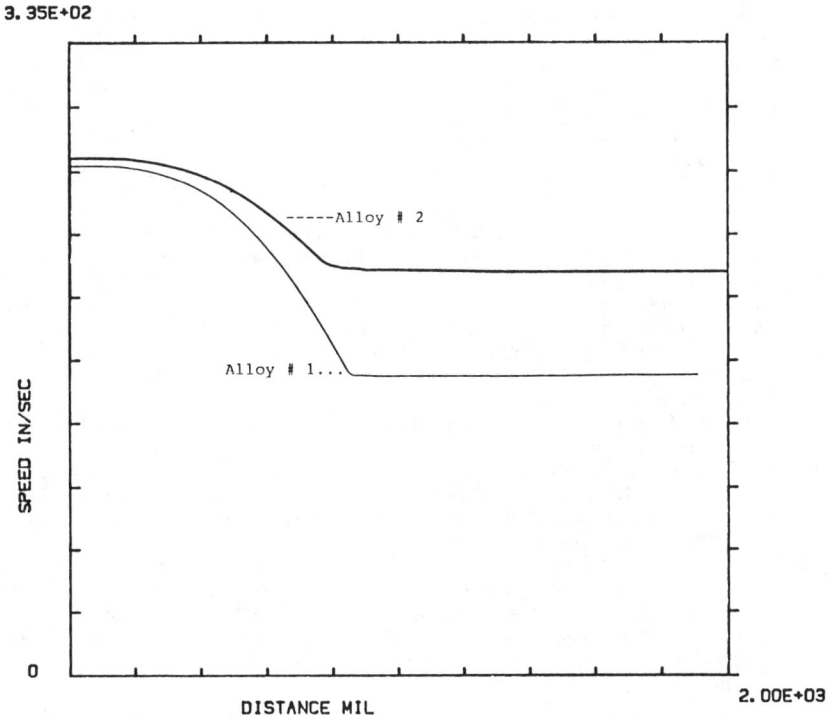

FIG. 16—*Velocity profile for two impacted alloys.*

variables will definitely affect the impact behavior. The fourth and fifth parameters are the influence of test coupon geometry and the specific testing conditions.

References

[1] Driscoll, S. B., "Variable Rate Impact Testing of RIM Products," *Processing and Testing of Reaction Injection Molding Urethanes, ASTM STP 788,* American Society for Testing and Materials, Philadelphia, 1982, pp. 86-93.

[2] Ireland, D. R., "Procedures and Problems Associated with Reliable Control of Instrumented Impact Test," *Instrumented Impact Testing, ASTM STP 563,* American Society for Testing and Materials, Philadelphia, 1973, pp. 3-29.

[3] Saxton, H. J., Ireland, D. R., and Server, W. L., "Analysis and Control of Inertial Effects during Instrumented Impact Testing," *Instrumented Impact Testing, ASTM STP 563,* American Society for Testing and Materials, Philadelphia, 1973, pp. 50-73.

[4] Beaumont, P. W. R., Riewald, P. G., and Zweben, C., "Methods for Improving the Impact Resistance of Composite Materials," *Foreign Object Impact Damage to Composites, ASTM STP 568,* American Society for Testing and Materials, Philadelphia, 1973, pp. 134-158.

[5] Broutman, L. J. and Rotem, A., "Impact Strength and Toughness of Fiber Composite Materials," *Foreign Object Impact Damage to Composites, ASTM STP 568,* American Society for Testing and Materials, Philadelphia, 1973, pp. 114-133.

[6] Adams, D. F., "Impact Response of Polymer-Matrix Composite Materials," *Composite Materials: Testing and Design, ASTM STP 617,* American Society for Testing and Materials, Philadelphia, 1977, pp. 409-421.

[7] Driscoll, S. B., "Instrumented Impact Testing," Rheometrics Technical Report, Piscataway, NJ, June 1981.

[8] Rapuano, J. C., "Rheology of Epoxy Resin Solution Coating Systems for Printer Circuit Laminates," M.S. thesis, University of Lowell, Lowell, MA, May 1986.

[9] Venkateshwaran, L. N., "Evaluation of Foam Concentrates for Thermoplastic Structural Foams," M.S. thesis, University of Lowell, Lowell, MA, December, 1985.

[10] Ireland, D. R., "Instrumented Impact Testing for Evaluating End-Use Performance," *Physical Testing of Plastics—Correlation with End-Use Performance, ASTM STP 736,* American Society for Testing and Materials, Philadelphia, 1981, pp. 45-58.

[11] Rieke, J. K., "Instrumented Impact Measurements on Some Polymers," *Physical Testing of Plastics—Correlation with End-Use Performance, ASTM STP 736,* American Society for Testing and Materials, Philadelphia, 1981, pp. 59-76.

[12] Takemori, M. and Yee, A., "Puncture Testing of Plastics: Effect of Test Geometry," *Society of Plastics Engineers Annual Technical Conference,* Vol. 25, 1979, pp. 638-641.

[13] Fernando, P. L., "Evaluation of Molded Articles Using an Instrumented Impact Machine," *Society of Plastics National Technical Conference,* September 1984, pp. 161-168.

[14] Progelhof, R. and Throne, J., "Puncture Impact Testing of Structural Foam," *Advances in Polymer Technology,* Vol. 3, No. 1, pp. 15-22.

[15] Driscoll, S. B. and Gacek, W., "Gas Counter Pressure Structural Foam Molding Vs. Conventional Low Pressure SF Molding: A Comparison of Mechanical Properties," *Society of Plastics Engineers Annual Technical Conference,* Vol. 30, 1984, pp. 217-222.

[16] Tryson, G. R., "Designing for High Rate and Low Temperature Performance of Engineering Thermoplastics," *Society of Plastics Engineers Annual Technical Conference,* Vol. 30, 1984, pp. 545-548.

[17] Goolsby, R. D. and Miller, J. M., "Instrumented Impact Fracture Toughness Testing of Polycarbonate," *Society of Plastics Engineers Annual Technical Conference,* Vol. 30, 1984, pp. 561-564.

[18] Nimmer, R. P., Moran, H., and Tryson, G. R., "Impact Response of a Polymeric Structure—Comparison of Analysis and Experiment," *Society of Plastics Engineers Annual Technical Conference,* Vol. 30, 1984, pp. 565-568.

[19] Golovoy, A., Cheung, M. F., and van Oene, H., "The Impact Behavior of Glass and Carbon Fiber Composites," Ford Research Technical Report SR-83-138, Ford Motor Co., Dearborn, MI, 1983.

[20] Golovoy, A. and van Oene, H., "Impact Energy Absorption of Unidirectional Glass Fiber-Epoxy Composites," Ford Research Technical Report SR-84-114, Ford Motor Co., Dearborn, MI, 1984.

[21] Golovoy, A., "Instrumented Impact Testing of Glass Reinforced Polypropylene," paper presented at the Annual Meeting of the American Institute of Chemical Engineers, Los Angeles, CA, 1984.

[22] Golovoy, A. and van Oene, H., "The Influence of Temperature on the Impact Behavior of Unidirectional Glass Fiber-Epoxy Composite," Paper 7-D, presented at the 40th Society of the Plastics Industry, Inc., Reinforced Plastics/Composites Conference, Atlanta, GA, 1985.

James S. Peraro[1]

Prediction of End-Use Impact Resistance of Composites

REFERENCE: Peraro, J. S., **"Prediction of End-Use Impact Resistance of Composites,"** *Instrumented Impact Testing of Plastics and Composite Materials, ASTM STP 936*, S. L. Kessler, G. C. Adams, S. B. Driscoll, and D. R. Ireland, Eds., American Society for Testing and Materials, Philadelphia, 1987, pp. 187–216.

ABSTRACT: The ability to predict material behavior under impact loading is very important in the design and manufacture of products. To predict product performance adequately, it is important to simulate the conditions under which the material is used.

Traditional impact tests often fail to provide the data required to evaluate and predict behavior under impact stress. An instrumented multivariable high-rate impact tester is designed to provide this information. The instrument is capable of testing materials of almost any configuration at velocities from 0.0127 to 12.7 m/s (30 to 30 000 in./min) and in environments simulating service conditions. Electronic instrumentation and a sophisticated computerized microprocessor simplifies the gathering, processing, and calculation of the impact property data. The graphic and calculated values permit a detailed analysis of impact performance.

The high-rate impact tester used at Owens-Corning Fiberglas Corp. is described, with examples of test programs in which the instrument capabilities were utilized to simulate the impact forces on boat, pipe, and tire constructions. Modifications made to the ram, or mode of operation, better simulate end-use conditions.

The results adequately predict impact behavior experienced under service conditions within a laboratory environment. This provides the scientist with a means of observing the impact phenomenon under laboratory conditions. As new materials are developed, characterization can be accomplished rapidly and at minimum cost. Screening of candidate materials can be accomplished without the time-consuming and expensive process of producing a product and conducting in-service tests.

The information developed can be invaluable for improving existing products and developing new products. By observing the test and evaluating the traces obtained, one can identify potential weaknesses in a product, which can be corrected by redesign or substitution of alternative materials.

KEY WORDS: composite materials, glass fiber, impact strength, puncture resistance, damage, delamination, pipe, tires, glass size, sandwich laminates, impact testing

[1]Senior scientist, Owens-Corning Fiberglas Corp., Technical Center, Granville, OH 43023.

187

The ability to determine material behavior under impact loading is essential in the design and manufacture of many products. To predict product performance, it is important that the evaluation be conducted under simulated end-use conditions.

Of particular interest to materials engineers is whether a polymer will behave in a brittle or ductile manner at the strain rate and temperature typical of its intended service. Part design and shape can also influence this brittle or ductile performance.

To survive impact, a polymer structure must be able to absorb the impact energy at the strain rate and temperature it experiences. Materials that are impact-resistant at one set of conditions can behave in a brittle manner if impacted at other conditions of temperature or strain rate, or both.

Impact resistance is affected by more factors than simply the applied force. These factors include (1) chemical composition, which includes the degree of matrix crystallinity, (2) orientation of the matrix polymer chain, (3) temperature, (4) geometry of the part, (5) rate of impact loading, (6) orientation of the part in relation to the impact direction, and (7) the support mode. Processing conditions can also have a major effect on the impact resistance of a molded part in that directional properties can develop from processing methods. Impact stress is usually multiaxial, and failure will generally occur along the path of least resistance (that is, along knit or flow lines). For these conditions, the ultimate forces developed during impact may be considerably lower than the ultimate strength of the polymer, as determined by traditional static test methods.

Many times, traditional impact tests fail to provide sufficient data to evaluate and ultimately predict the impact performance of a material or molded part. A meaningful evaluation begins with the ability to test the material under simulated end-use conditions. Also required is the ability to evaluate the design and, ultimately, the molded part or structure under multiaxial stress at the strain rates and temperatures typical of actual service.

Impact Tester

The Rheometrics high-rate impact tester (HRIT) (Fig. 1) is designed to provide detailed engineering impact data on materials, prototypes, and finished and assembled products. Materials can be reinforced, unreinforced, rigid, or flexible.

The unit utilizes a horizontal hydraulic ram (Fig. 2), which can be programmed to impact the specimen at any desired velocity between 0.0127 m/s and 12.7 m/s (0.5 and 500 in./s). The instrumented impacting ram has a 12.7-mm (0.5-in.)-diameter steel rod with a 12.7-mm (0.5-in.)-diameter hemispherical tip. However, larger ram diameters, tipped with other geometric shapes, can also be used.

A microprocessor data-handling system (Fig. 3) permits a detailed analysis

FIG. 1—*Rheometrics high rate impact tester.*

FIG. 2—A 0.0127-m-(½-in.) diameter ram and specimen clamp assembly.

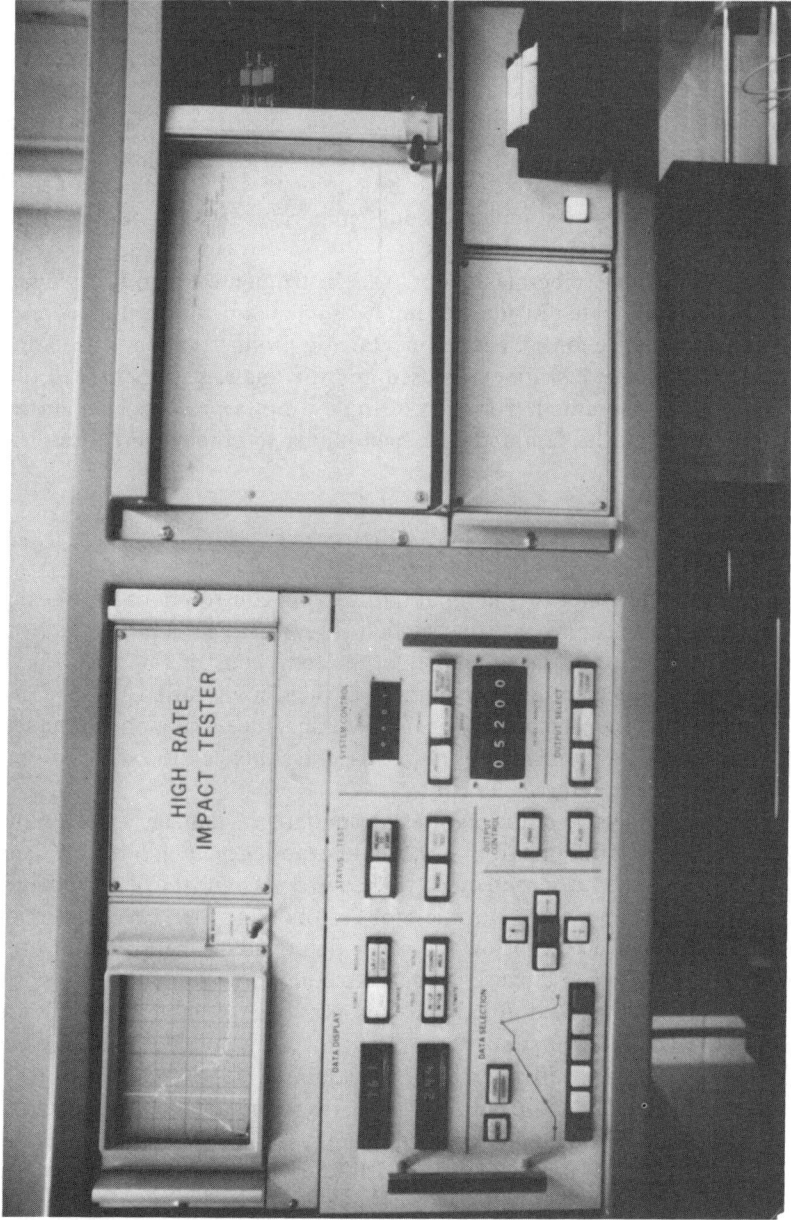

FIG. 3—*Electronics package to record and analyze data obtained during impact testing.*

of the impact performance. After the specimen is impacted, the force-displacement profile is displayed on a CRT. The microprocessor is then used to enlarge the trace for greater clarity, select points on the curve for analysis, and, finally, calculate the associated engineering parameters, which include the apparent modulus, yield load, ultimate load, and energy at yield point and at breaking. The test parameters and calculated properties are recorded on the printer, and the load deflection curve is plotted on an X-Y recorder.

The impact tester has the ability to test a variety of specimen geometries including flat plaques, films, molded parts, and even large assemblies mounted to the machine. It can also be equipped with an environmental chamber to permit testing under various temperature and humidity conditions that simulate service conditions.

At Owens-Corning Fiberglas Corp., this instrumented high-rate impact machine is being used to provide meaningful impact data for the prediction of material performance for both experimental and production products. A few examples of how this instrument is used and the analysis of the test results obtained will be presented. For each of the examples presented, modifications were made to the ram and specimen holder to obtain the desired test conditions.

Skin/Core Laminates

Six experimental skin/core laminates of varying constructions were evaluated for potential use in boat construction (Fig. 4). Each laminate had the same core material, but the skins were of different constructions, with both the type of reinforcement and the number of plies being varied.

Two panels were constructed with four-ply skins; the remaining four panels had six-ply skins. Within each set of panels, the skins had the same volume fraction of fibrous reinforcement.

The primary objective of the test was to simulate the damage a boat, moving at a velocity of 2.235 m/s (5 mph), would experience upon hitting a fixed object such as a dock. A second objective was to determine the depth of penetration and the force required for the probe to puncture the skin of the laminate. If the impact occurs below the water line, water penetration into the core of the laminate could cause irreparable damage to the hull.

The HRIT was modified utilizing a 25.4-mm (1-in.)-diameter ram and a 76.2-mm (3-in.)-diameter support ring (Fig. 5). A bump type of impact was employed to penetrate only the skin of the laminate (Fig. 6).

A ram displacement of 7.62 mm (0.3 in.) was found adequate to penetrate some, but not all, of the laminates when the ram velocity was held constant at 2.235 m/s (5 mph). At this displacement, differences in the impact resistance were observed.

The impact results were compared for specimens within the same set, that is, with the same number of reinforcing plies in the skins. For the four-ply

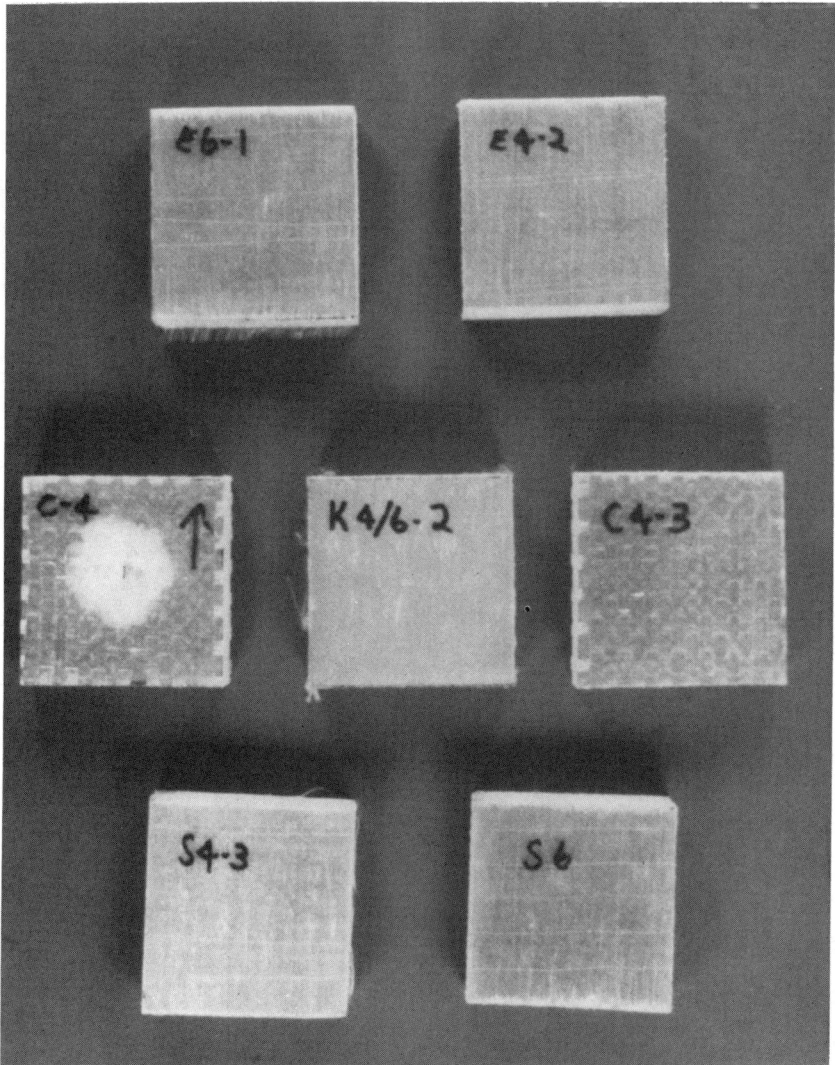

FIG. 4—*Sections of the skin-core panels evaluated. The specimen on the left side of the middle row shows damage after impact.*

specimens, the specimen with skins reinforced with S glass gave higher impact loads than the specimen reinforced with E glass (Fig. 7). It can be seen from the shape of the curves that the E glass laminate fractured before the maximum ram penetration was achieved, while the S glass laminate deflected but did not fracture.

Comparing the impact profiles for the laminates with six plies of reinforce-

FIG. 5—A 25-mm-(1.0-in.) diameter probe with a section of skin-core laminate in place prior to testing.

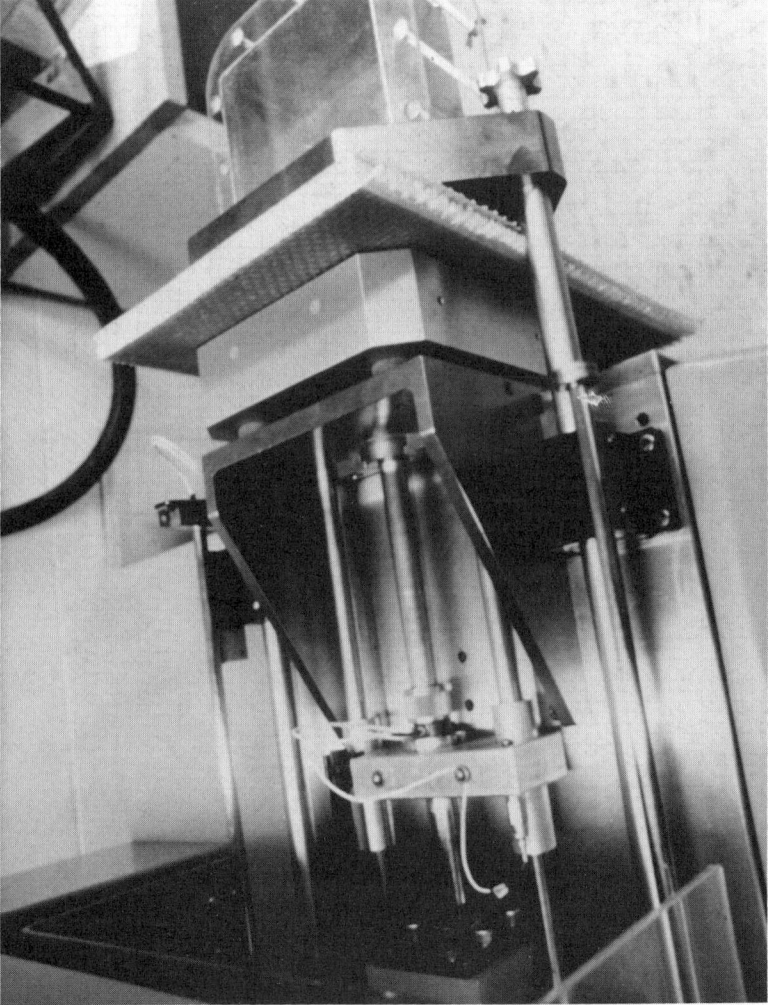

FIG. 6— A 25-mm-(1.0-in.) diameter probe advanced to impact with the specimen being held by clamp.

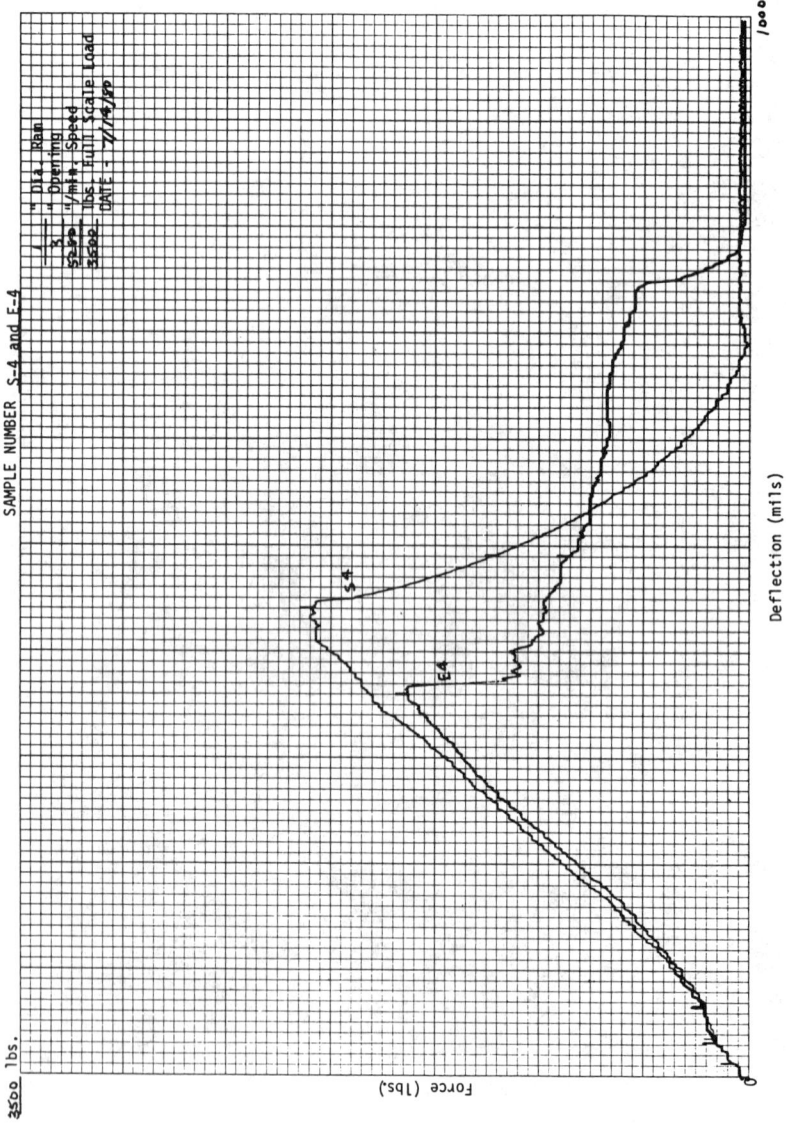

FIG. 7—*Bump impact traces for boat laminates with four-ply skins [25-mm-(1.0-in.) diameter probe: speed—2.235 m/s (5280 in./min)].*

ment, the laminates with S glass skins developed the greatest impact resistance, followed by unidirectional E glass, conventional E glass, mat/woven roving, and unidirectional Kevlar, respectively (Fig. 8). As with the four-ply skins, the S glass laminates deflected but did not fracture. The remaining three skin constructions fractured prior to reaching the maximum ram penetration.

The tabulated data confirm the graphic results (Table 1). The S glass laminates develop the greatest ultimate force and ultimate energy for both four-ply and six-ply skin constructions. When the skin fractures, additional energy is absorbed and is reflected in the higher total energy values for unidirectional E glass (Specimen E-6) and conventional mat/woven roving (Specimen C-4) specimens. Kevlar gave the lowest impact values for both ultimate force and energy.

Filament Wound Pipe

Impact tests were conducted on 57.15-mm (2.25-in.)-diameter filament wound pipes (Fig. 9) to determine the influence of size formulation on impact resistance. Two glass rovings, coated with different sizes, one more flexible than the other, were used in this study. Each roving was used as the reinforcement for one set of filament wound pipe specimens. As with the experimental boat laminates, a bump test was performed, this time with the ram velocity held constant at 0.51 m/s (1200 in./min). However, for each test, the depth of penetration for the ram was progressively increased. The impact properties were calculated and the visible damage was recorded after each impact.

The tubular specimens were cut in half lengthwise, and a wooden support assembly was constructed to hold the semicylindrical specimen (Fig. 10). The wooden supports, set 76.2 mm (3 in.) apart, were attached to the specimen.

The assembled specimen was mounted vertically in the HRIT with the inside surface of the pipe exposed for impact by the 12.7-mm (0.5-in.)-diameter hemispherical tipped ram (Fig. 11). Because of the curvature of the specimen, the ram had to travel a distance of 36.195 mm (1.425 in.) before making contact with the inside surface of the pipe.

A series of ram displacements were selected to produce progressively increasing amounts of damage to the pipe (Fig. 12). The damage ranged from no visible crack to penetration of the ram through the pipe wall. For each ram displacement setting, impact profiles were developed from which the ultimate force, ultimate energy, and total energy were calculated (Table 2).

The results show that for ram displacements up to 40.0-mm (1.575-in.) or 3.81-mm (0.150-in.) deformation, the control (Specimen A), made from glass fibers with the rigid size, produced higher-impact properties than Specimen B, which utilized the more flexible size (Figs. 13 and 14). During impact, the more flexible size permits greater specimen deflection without fracturing the pipe. The interfacial region between the fibers and the resin matrix is more

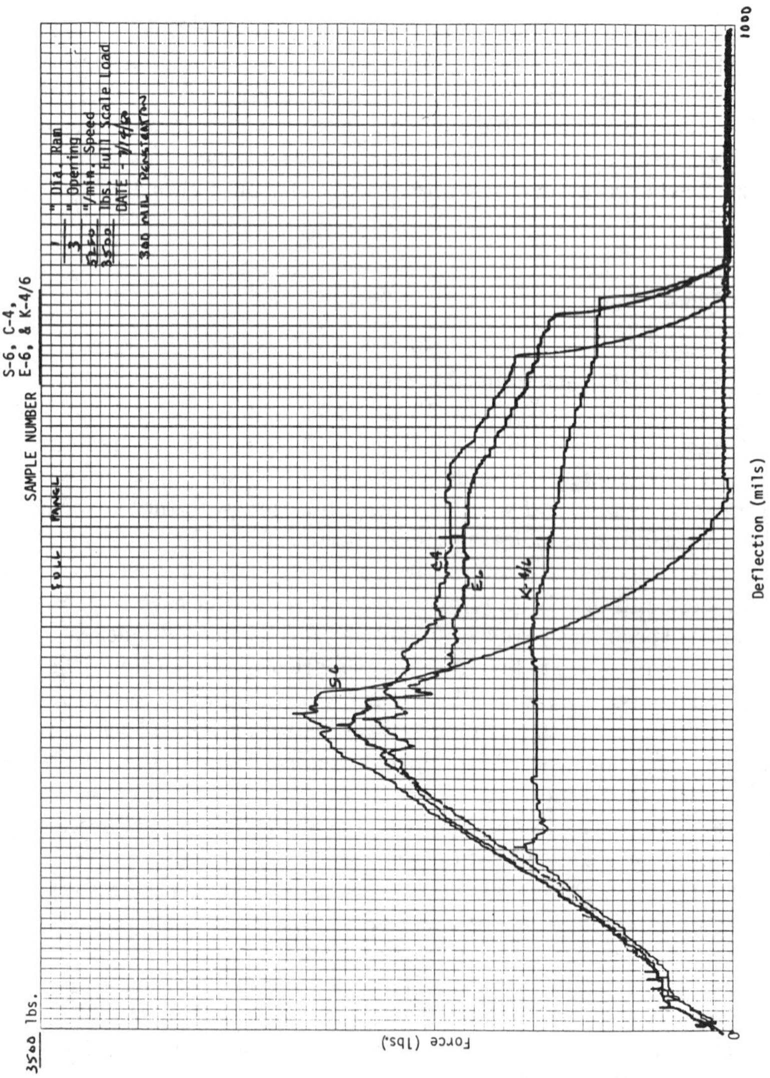

FIG. 8—*Bump impact traces for boat laminates with six-ply skins [25-mm-(1.0-in.) diameter probe: speed—2.235 m/s (5280 in./min)].*

TABLE 1—*Bump impact of skin/core laminates used in boat construction [velocity = 2.235 in./s (5280 in./min); probe diameter = 25 mm (1.0 in.)].*[a]

Specimen	Ultimate Force, N (lb)	Ultimate Energy, J (lb · in.)	Total Energy, J (lb · in.)
E-4	7099 (1596)	30.6 (271)	45.0 (398)
S-4	9132 (2053)	48.2 (427)	57.5 (509)
E-6	8865 (1993)	35.4 (313)	62.6 (554)
S-6	9564 (2150)	37.5 (332)	54.9 (486)
C-4	7918 (1780)	32.2 (285)	64.1 (567)
K-4/6	4728 (1063)	11.8 (104)	40.0 (354)

[a]Ram velocity for all the specimens was 2.235 m/s (5280 in./min).

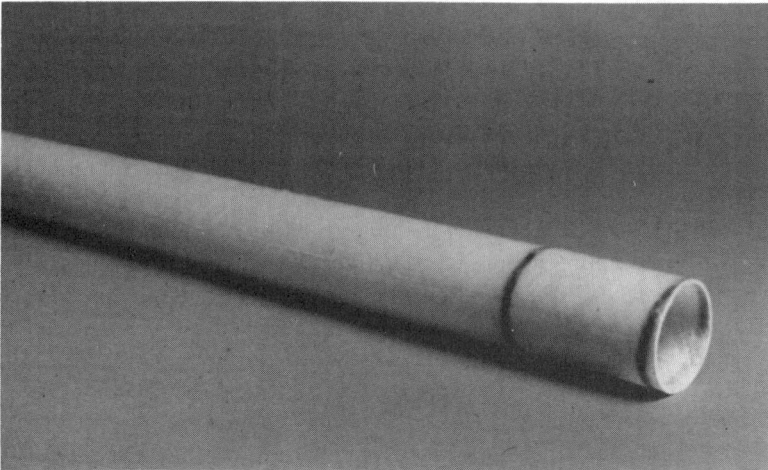

FIG. 9—*Filament wound pipe with a 57.15 mm (2.25 in.) diameter.*

"rubbery," allowing for the greater deformation. This results in the generation of lower-impact forces for Specimen B than for Specimen A. However, visual examination of the impacted specimens shows that for the same ram displacement, Specimen A, being stiffer, appears to have sustained more damage than Specimen B for small deformations.

FIG. 10—*Semicircular section of a filament wound pipe mounted in wooden supports. The distance between the wooden supports is 76.2 mm (3.0 in.).*

When the ram displacement was increased to 40.64-mm (1.600-in) or 4.45-mm (0.175-in.) deformation, the ram ruptured the pipe wall (Fig. 15). At this displacement, the more flexible pipe produced higher impact properties than the more rigid pipe (see Table 2). Therefore, considerably more damage was inflicted on Specimen B than on Specimen A.

Specimen A shows localized damage, while Specimen B shows delamination occurring between the glass fibers and the resin matrix. The more rigid Specimen A deflects less, resulting in shear failure in the glass fibers. The more flexible Specimen B, having the more rubbery interface and ultimately the weaker bond between the glass and resin, delaminated the glass fibers from the resin matrix. The glass fibers maintained their integrity for a greater penetration of the ram before catastrophic failure occurred. In this case, the failure mode appears to be tensile rather than shear.

Finally, puncture tests were conducted at velocities of 0.013 m/s (30 in./min), 3.47 m/s (8200 in./min) (Figs. 16 and 17), and 8.47 m/s (20 000 in./min). The results confirm that for large ram deflections the more rigid material produces lower impact properties than the flexible pipe at all speeds (Table 3).

The author concludes that for applications in which the pipe is subjected to small deformations, the more rigid Specimen A will be capable of withstanding greater impact forces, but with greater localized damage. If the potential exists for high-impact forces and large deformations to the point of pipe rupture, the more flexible material, Specimen B, will provide greater impact resistance, However, the damage sustained by the flexible material will be significantly greater.

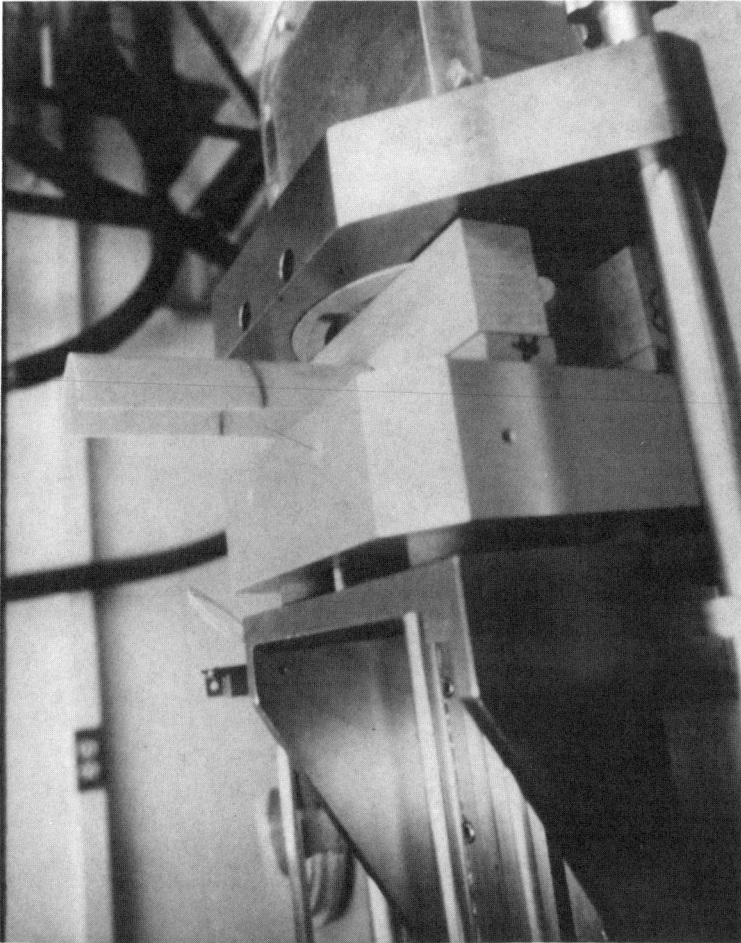

FIG. 11 — *Filament wound pipe specimen and wooden supports mounted in a clamp assembly of high rate impact tester.*

FIG. 12—Damaged section of a filament wound pipe after bump impact.

TABLE 2—*Bump impact of filament wound pipe, 57.15 mm (2.25 in.) [velocity = 0.15 m/s (1200 in./min); probe density = 25 mm (1.0 in.)].*[a]

Specimen	Return Point Selected, mm (in.)	Ultimate Force, N (lb)	Ultimate Energy, J (lb · in.)	Total Energy, J (lb · in.)
A	36.20	351	0.00	0.00
	(1425)	(79)	(0)	(0)
	37.08	1299	0.56	1.01
	(1460)	(292)	(5)	(9)
	38.10	2246	1.69	2.82
	(1500)	(505)	(15)	(25)
	39.37	3403	4.07	6.55
	(1550)	(765)	(36)	(58)
	40.00	4315	6.21	10.28
	(1575)	(970)	(55)	(91)
	40.64	4893	8.59	17.74
	(1600)	(1100)	(76)	(157)
B	36.20	142	0.00	0.00
	(1425)	(32)	(0)	(0)
	37.08	400	0.00	0.11
	(1460)	(90)	(0)	(1)
	38.10	1557	0.90	1.58
	(1500)	(350)	(8)	(14)
	39.37	1957	1.81	2.94
	(1550)	(440)	(16)	(26)
	40.00	3283	3.84	6.10
	(1575)	(738)	(34)	(54)
	40.64	6352	19.89	30.73
	(1600)	(1428)	(176)	(272)

[a]Ram velocity for both specimens was 0.51 m/s (1200 in./min).

Tire Puncture Resistance

A study was established to develop a test for evaluating the impact properties of tires. The primary objective was to characterize and rank tire constructions as measured by puncture resistance. A second objective was to measure the impact resistance of tires when subjected to different types of puncture road hazards under various environmental conditions.

Puncture resistance of tires can be simulated very effectively using the instrumented HRIT. Present techniques of evaluating tire puncture resistance requires constructing the tires, mounting the tires on a vehicle, and then road testing the tire over a puncture-producing hazard. The road test is typically a pass/fail test.

With the instrumented impact test, tires, sections of tires, or laboratory-prepared specimens simulating tire body constructions can be evaluated (Fig. 18). By observing the penetration of the probe through the specimen under laboratory conditions, additional information regarding the behavior under impact loads, the failure mode, as well as a measure of the impact resistance can be obtained.

The impact forces and energy generated by the probe penetrating the tire

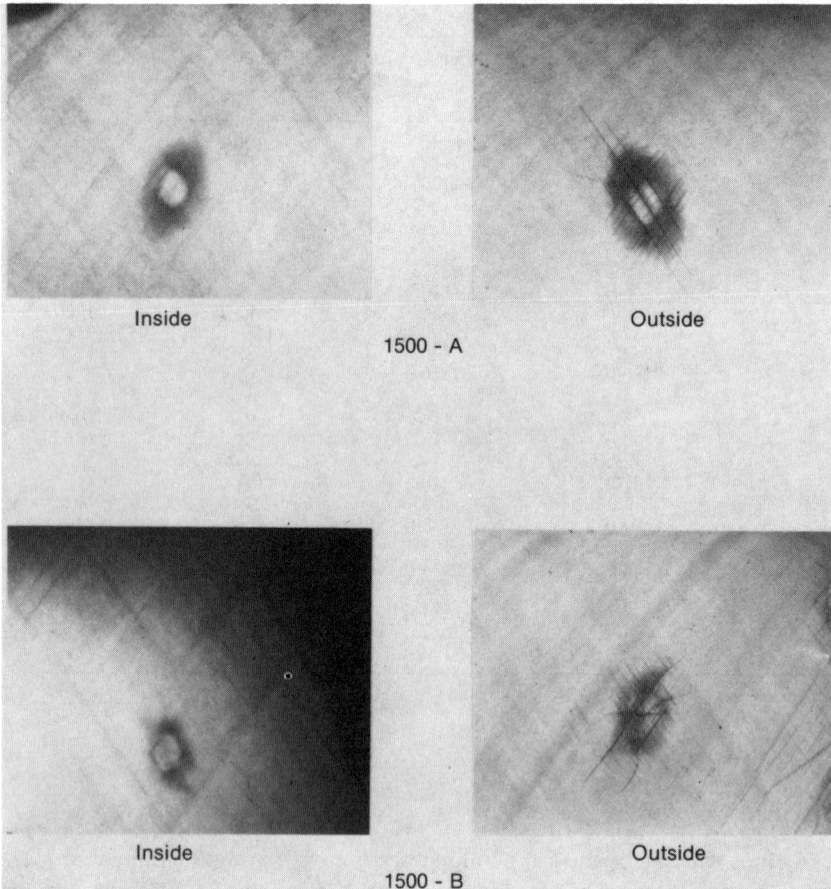

FIG. 13—*Photographs of damage developed by bump impact at a velocity of 0.51 m/s (1200 in./min) and a deformation of 1.91 mm (0.08 in.).*

composite can be measured and recorded. The velocity of the probe can be varied to evaluate the effect of speed on the impact resistance of the tire. The effect of tip geometry (round, pointed, and so on) on impact resistance can be evaluated easily. Environmental conditions (cold, hot, wet, dry, and so on) can also be simulated. It should be noted that, in this study, only puncture-type road hazards were simulated. Other road hazards, such as broken glass or potholes, and their effect on the impact resistance were not simulated.

The initial tests, conducted on sections of both production and experimental tires, were inconclusive because of the nonuniform thickness (or cross section) tire tread pattern. The probe did not penetrate through the same thick-

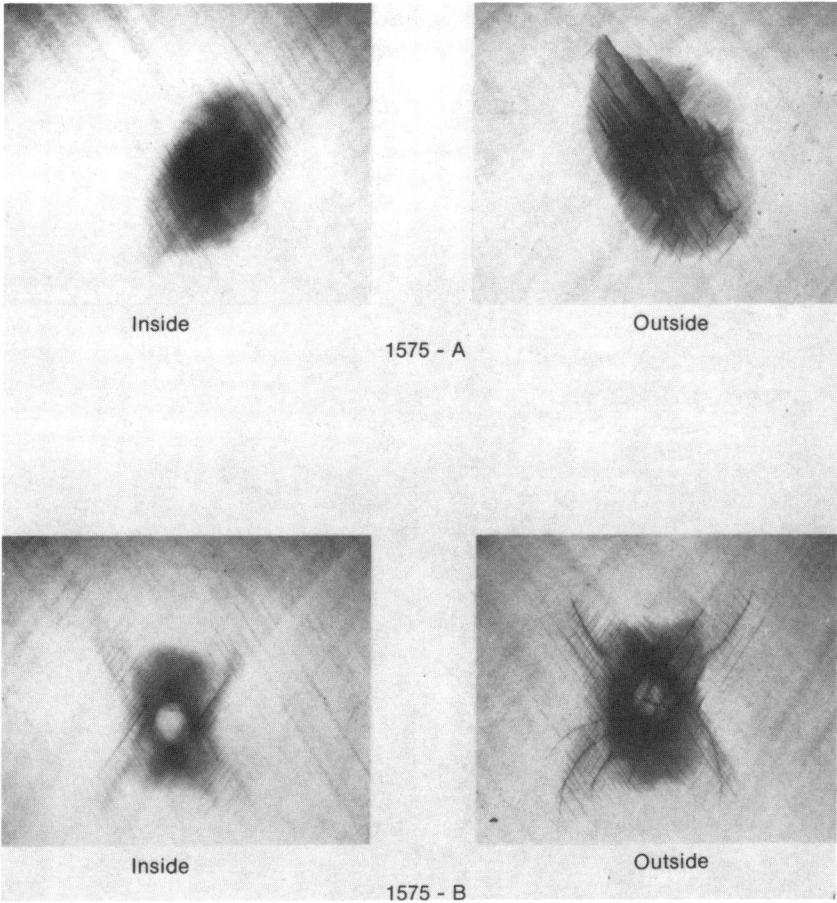

FIG. 14—*Photographs of damage developed by bump impact at a velocity of 0.51 m/s (1200 in./min) and a deformation of 3.81 mm (0.150 in.).*

ness of the tire on each impact. Laboratory specimens were made of tire belt constructions without the unreinforced tread, to eliminate the influence of tread pattern. Each specimen was mounted against a support plate with a hole through which the probe passed (Fig. 19).

Initially, a 76.2-mm (3-in.)-diameter opening in the support plate was used. The scatter in the data was so large that differences in tire construction could not be detected. Additional tests were conducted using other size openings in the support plate, ranging from 12.7 mm (0.5 in.) to 63.5 mm (2.5 in.) in diameter, in an attempt to reduce the variability in the data. After numerous trials, the 12.7-mm (0.5-in.)-diameter opening was found to minimize the

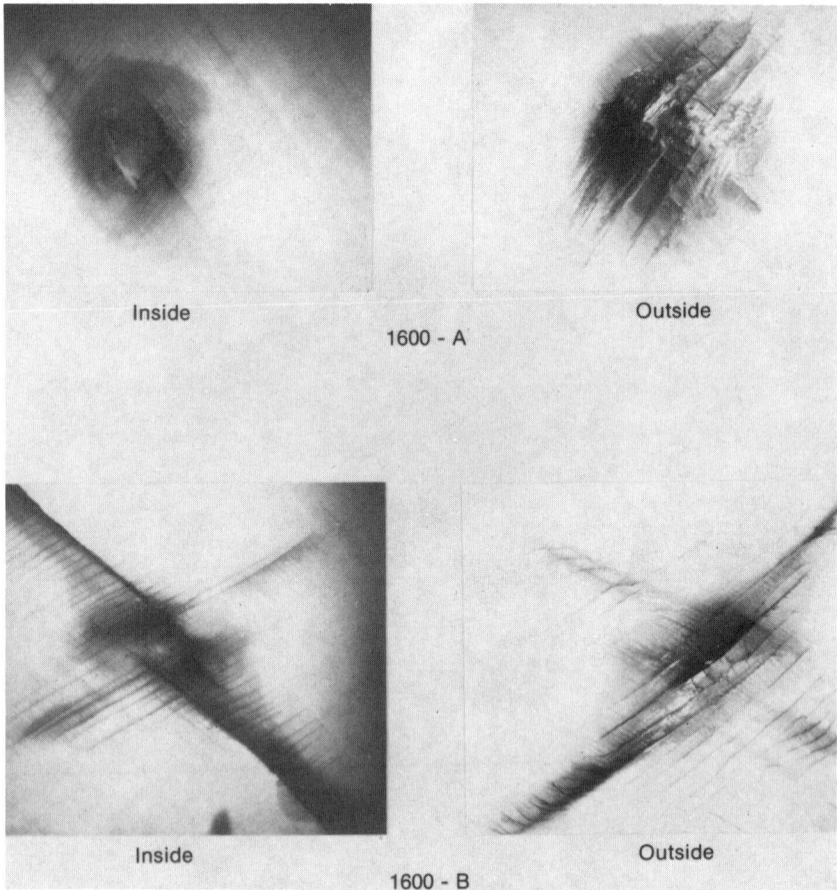

FIG. 15—*Photographs of damage developed by penetration of a ram through the pipe wall at a ram velocity of 0.51 m/s (1200 in./min) and a depth of penetration of 4.45 mm (0.175 in.).*

statistical variability of the data and was used as the standard opening for the study.

A second support plate, with a 76.2-mm (3-in.)-diameter hole, was placed in front of the specimen to permit easier removal of the probe as it retracted after puncturing the specimen.

Once the method was finalized, a designed experiment, consisting of 52 tire belt constructions, was conducted. The tire variables studied were belt cord materials, numbers of cords per inch, rubber composition, and belt thickness.

The road hazard variables were simulated by the geometry of the probe (see

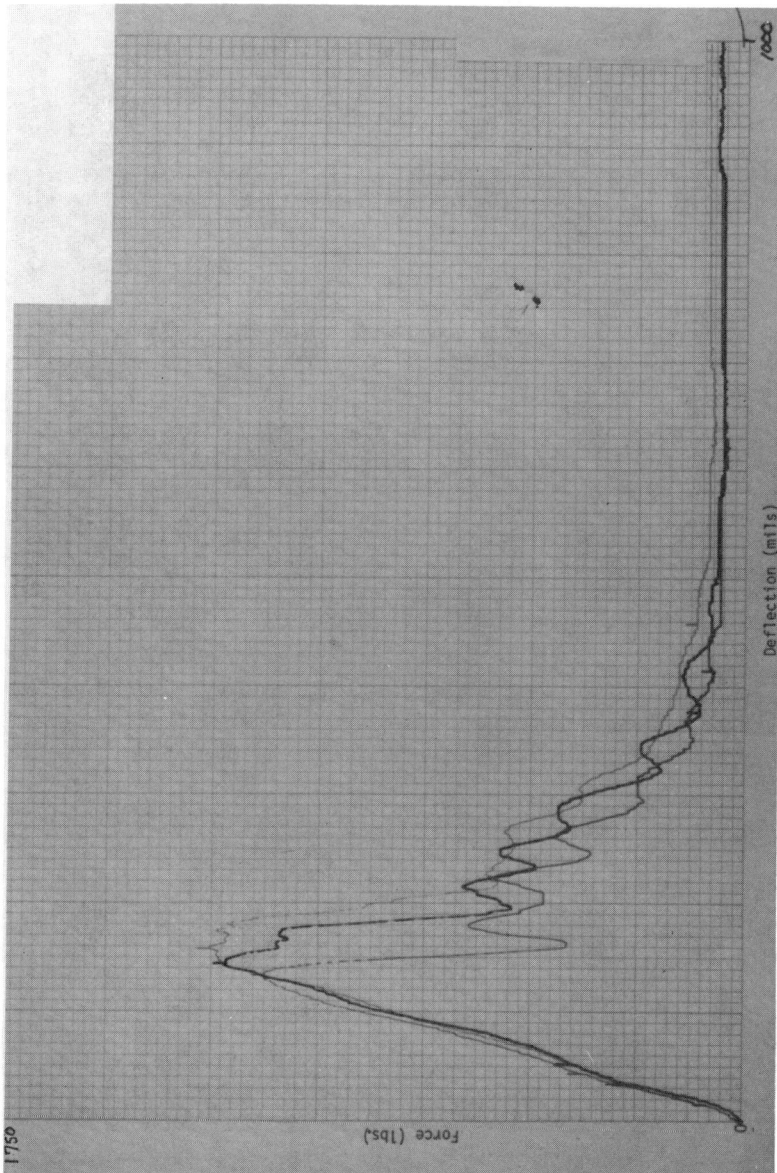

FIG. 16—*Puncture impact traces on a filament wound pipe, Specimen A, at a speed of 3.47 m/s (8200 in./min).*

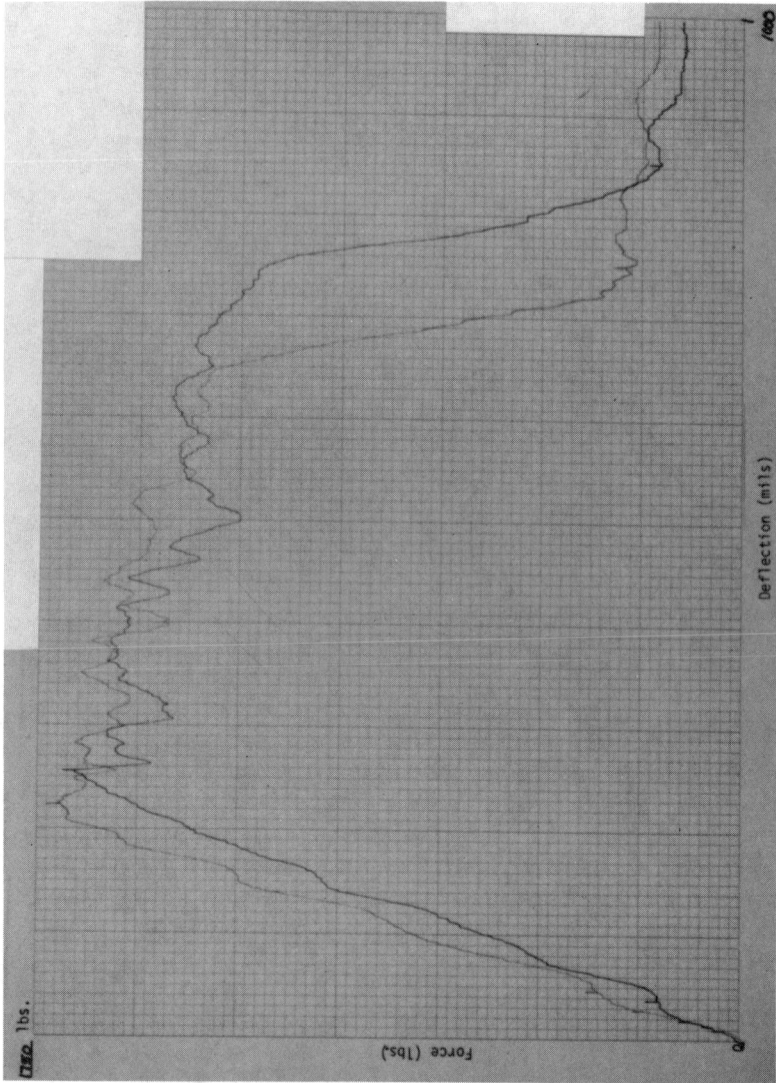

FIG. 17—*Puncture impact traces on a filament wound pipe. Specimen B. at a speed of 3.47 m/s (8200 in./min).*

TABLE 3—*Puncture impact of 57.15-mm (2.25-in.) diameter filament wound pipe as a function of velocity.*

Specimen	Velocity, m/s (in./min)	Return Point Selected, mm (in.)	Ultimate Force, N (lb)	Ultimate Energy, J (lb · in.)	Total Energy, J (lb · in.)
A	0.013	63.5	4021	7.80	17.51
	(30)	(2.50)	(904)	(69)	(155)
	3.47	. . .	5298	9.94	23.16
	(8 200)	. . .	(1191)	(88)	(205)
	8.47	. . .	6098	13.33	21.81
	(20 000)	. . .	(1371)	(118)	(193)
B	0.013	. . .	4950	14.91	23.50
	(30)	. . .	(1113)	(132)	(208)
	3.47	. . .	7557	25.40	92.20
	(8 200)	. . .	(1699)	(225)	(816)
	8.47	. . .	7477	38.60	63.38
	(20 000)	. . .	(1681)	(342)	(561)

Fig. 19). A 6.35-mm (0.25-in.)-diameter rod with a 6.35-mm (0.25-in.) hemispherical tip and a 4.57-mm (0.18-in.)-diameter pointed rod were used at speeds of 0.447 m/s (1 mph) and 6.71 m/s (15 mph) under wet and dry conditions.

A 25.4-mm-(1-in.) diameter ram was modified by drilling and tapping a hole in the end for attachment of the probe (Figs. 20 and 21). The specimens were mounted approximately 76.2 mm (3 in.) from the end of the probe, to be within the electronic sensor to capture and record the impact as the probe penetrated the tire (Fig. 22).

A sampling of the results obtained from the impact tests is given in Table 4. As expected, the 6.35-mm (0.25-in.)-diameter hemispherical probe produced significantly higher impact forces than the 4.57-mm (0.18-in.)-diameter pointed probe. Similarly, significantly higher results were obtained when the speed was increased from 0.447 m/s (1 mph) to 6.71 m/s (15 mph). The test results conducted under dry conditions were significantly higher than those results obtained under wet conditions. These conclusions parallel test data obtained from automotive road puncture trials.

The results of this study on tire construction variables show significant differences in impact resistance, and the methodology can differentiate between variations in tire constructions. The data generated from studies of this type can be used in the design of a more puncture-resistant tire. Of course, any design change can affect other tire properties, such as skid resistance and cornering, In tire designs directed at improving the handling characteristics (for example, skid resistance, cornering), this test could be used to ensure that the puncture resistance does not fall below acceptable performance levels.

FIG. 18—*Section of laboratory-prepared tire body construction, without thread* (left) *and a section of a tire with thread* (right).

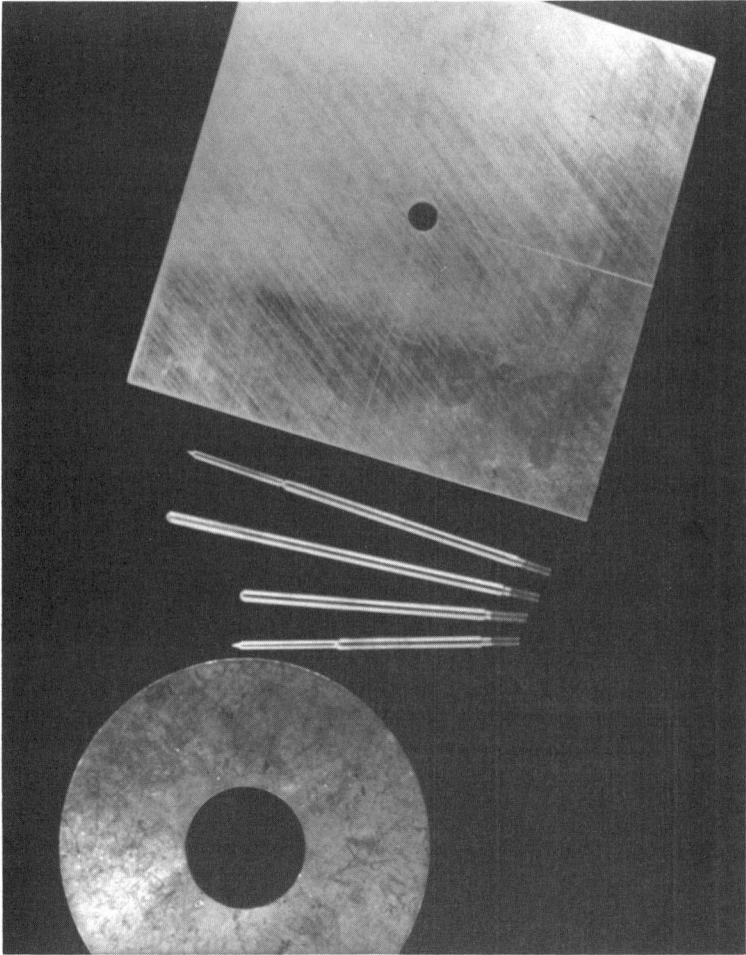

FIG. 19—*Retractor plate, probes, and support plate for a tire puncture test.*

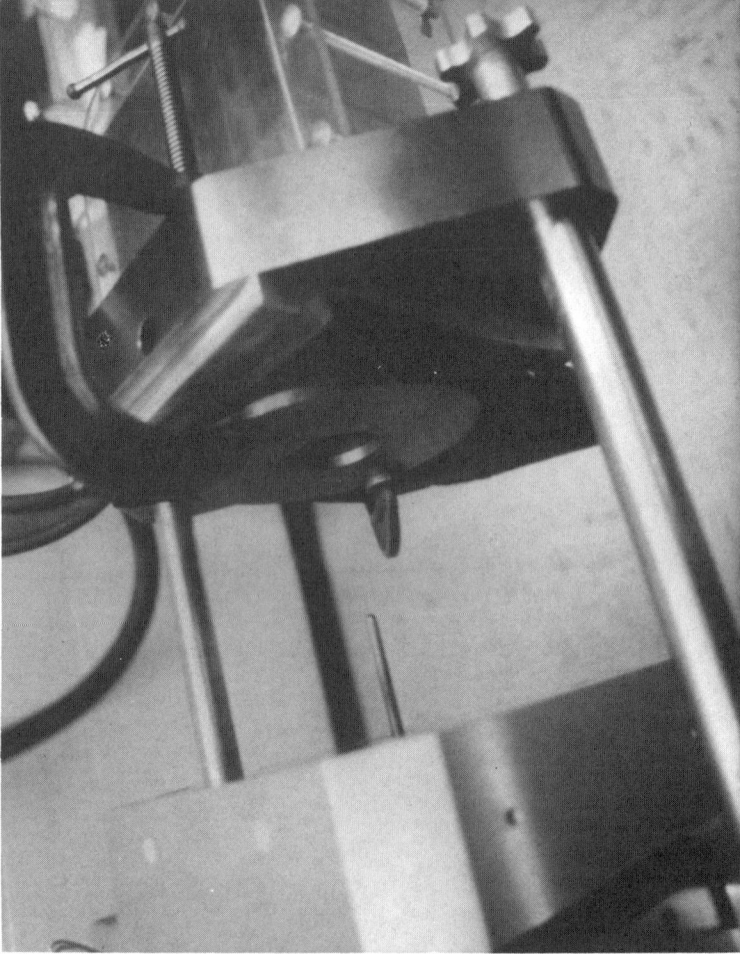

FIG. 20—*View showing a round-tip probe prior to puncturing of the tire construction.*

FIG. 21— *View showing a pointed-tip probe prior to puncturing of the tire construction.*

FIG. 22— *View showing a round-tip probe after puncturing of the tire construction.*

TABLE 4—*Puncture impact of tire constructions using a rounded and pointed-tipped probe.*

Specimen	Pooled (Average), N (lb)	0.447 m/s (1 mph), N (lb)	6.71 m/s (15 mph), N (lb)
6.35-mm (0.25-in.) Hemispherical Tip			
H	1448.49 (325.65)	. . . . . .	. . . . . .
H (dry)	1630.06 (366.60)	1386.89 (311.80)	1751.84 (393.85)
H (wet)	1266.35 (284.70)	1163.37 (261.55)	1472.00 (331.00)
4.57-mm (0.18-in.) Pointed Tip			
N	299.93 (67.43)	. . . . . .	. . . . . .
N (dry)	378.97 (85.20)	398.99 (89.70)	358.95 (80.70)
N (wet)	319.06 (71.73)	269.10 (60.50)	344.05 (77.35)

The laboratory test appears to be useful as a screening tool for the development of improved tire constructions. This test is quicker, easier, and less expensive and provides more information than other tire puncture tests. No longer is it necessary to build and road test every tire design. Laboratory-prepared specimens can be screened quickly, and the most promising designs can then be used to produce tires for road testing. The quantitative nature of the results of high-speed impact tests may help us understand and model road puncture tests more effectively. Therefore, the HRIT and road test systems will complement each other.

Conclusions

An instrumented high-rate impact tester yields more information about the impact properties of composite systems than traditional falling-weight or pendulum-type instruments, which are generally not instrumented and operate at only one velocity.

The addition of the microprocessor and electronic sensors makes impact testing with the HRIT tester easier, faster, and more sensitive, which affords a clearer understanding of the behavior of composites under impact stresses. With proper selection of the test parameters, specimen configuration, and environmental conditions, materials can be evaluated under simulated end-use conditions. This provides the scientist or engineer with a means of observing, under laboratory conditions, the impact phenomenon, which may be impossible when the product is in service.

An existing products are improved or new products developed, characterization can be accomplished rapidly at minimum cost. Screening of candidate materials can be accomplished without the time-consuming and expensive process of producing full-scale prototypes and conducting in-service tests.

Finally, as noted in this paper, the tests conducted measure only the impact properties of the composite systems. If impact occurs in combination with other stress fields, the predictability of end-use behavior may be clouded. However, the impact properties, as measured with the instrumented high-rate impact tester, will give more information and a better understanding of material behavior than was previously available.

Impact Characterization of
Selected Materials

Marilyn W. Wardle[1] and George E. Zahr[1]

Instrumented Impact Testing of Aramid-Reinforced Composite Materials

REFERENCE: Wardle, M. W. and Zahr, G. E., "**Instrumented Impact Testing of Aramid-Reinforced Composite Materials,**" *Instrumented Impact Testing of Plastics and Composite Materials, ASTM STP 936,* S. L. Kessler, G. C. Adams, S. B. Driscoll, and D. R. Ireland, Eds., American Society for Testing and Materials, Philadelphia, 1987, pp. 219–235.

ABSTRACT: Instrumented impact testing has been applied to study the effect of fiber properties on the impact damage tolerance of fiber-reinforced composites containing aramid, carbon, and glass fibers. The energy-absorbing capability of fibrous composites in impact is dependent on the tensile strain capacity (toughness) of the fibers, as well as on properties of the resin and interface. The aramid fibers are particularly efficient in energy absorption and in retention of properties after impact.

KEY WORDS: composite materials, carbon, glass, aramid, impact strength, epoxy resins, impact testing, instrumented impact test

As advanced fiber-reinforced composites find more and more applications in aircraft and aerospace hardware, concerns have been raised about the impact damage tolerance of these relatively brittle materials [1,2]. In most applications composites are replacing metals, which are highly ductile and damage tolerant. Because composites often have little or no ductility, this is a very real concern.

Among the most important factors governing the impact damage tolerance of a composite material are the properties of the reinforcing fiber. We have used the methodology of instrumented impact testing to study the relative behavior of composites reinforced with various fibers in two geometries, representative of different applications (Fig. 1): flat, fabric-reinforced panels, simulating aircraft structures, and filament-wound pressure bottles, simulat-

[1]Senior research engineer and research chemist, respectively, E. I. du Pont de Nemours and Co., Inc., Wilmington, DE 19898.

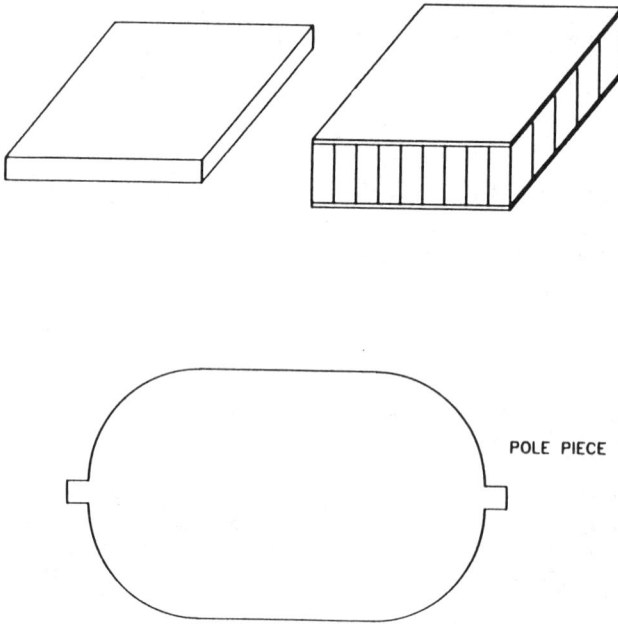

FIG. 1—*Composite forms used for impact program:* (top left) *solid panel;* (top right) *honey-comb-core panel;* (bottom) *filament-wound pressure bottle.*

ing rocket motor cases. Flat panels have been tested both as solid laminates and as honeycomb-core laminates, both structural types used in aircraft.

Experimental Procedure

Impact testing was performed using drop towers and instrumentation built by the Dynatup Division of General Research Corp. Details of the impact test conditions are given in Table 1. Pressure bottles were impacted by General Research, Santa Barbara, California; honeycomb panels were tested by Boeing Technology Services, Seattle, Washington; solid panels were tested in the authors' facilities at E. I. du Pont de Nemours and Co., in Wilmington, Delaware. A 1.27-cm (1/2-in.)-diameter hemispherical indentor was used in all cases.

The nominal properties of the fibers used, obtained from the manufacturers' literature, are listed in Table 2. The resin systems are listed in Table 3. Flat panels were fabricated from commercial prepreg fabrics woven from

TABLE 1—*Impact parameters.*

Parameter	Solid Panels	Honeycomb Panels	Pressure Bottles
Specimen size, cm (in.)	15.2 by 15.2 (6 by 6)	15.2 by 15.2 (6 by 6)	15.2 diameter (6)
Support conditions	simple	simple	rigid at poles
Support span, cm (in.)	12.7 by 12.7 (5 by 5)	12.7 by 12.7 (5 by 5)	. . .
Drop weight, kg (lb)	106 (233)	83 (184)	13 (28)
Drop height, cm (in.)	30.5 (12)	7.6 (3)	91.4 (36)
Impact energy, J (ft·lb)	312 (233)	62 (46)	≤ 114 (85)
Indentor diameter, cm (in.)	1.27 (0.5)	1.27 (0.5)	1.27 (0.5)

Kevlar 29 and Kevlar 49[2] aramid fibers and carbon fiber in resin systems typical of those specified for parts on large commercial transport airplanes. Panels varying in thickness from about 0.13 cm (0.05 in.) to 0.5 cm (0.2 in.) were fabricated. The lay-ups were from the (0,90, ±45) family, and the thickness was varied by changing the number of plies of fabric used. The fiber volume fraction of the prepregs ranged from 42 to 48%. Honeycomb-core panels were fabricated each with two plies of prepreg fabric (3K-70-PW Thornel 300 carbon, S-285 Kevlar 49 aramid, S-1581E-glass) at (0,90) and cocured on a 1.27-cm (1/2-in.), 0.048-g/cm³ (3-lb/ft³), 0.318-cm (1/8-in.) cell honeycomb made from Nomex aramid structural sheet. An aluminum face sheet (Type 6061 T6) was also included in the honeycomb comparison.

Bottles were wound by Morton Thiokol using commercial filament winding equipment. The bottles were a standard nominal 15.2-cm (6-in.)-diameter design specified by the ASTM Method for Preparation and Tension Testing of Filament-Wound Pressure Vessels [D 2585-68 (1980)] and commonly used to evaluate the effect of fiber properties on burst pressure. The bottles were rigidly supported at the poles and impacted at the equator. They were empty when impacted. Several fibers and two resin systems were used, as detailed in Table 3. New, high strain-to-failure versions of both Kevlar 49 (Type 981) and carbon fiber (IM-6) were included, along with the standard materials used in the flat panels. The material UFX 82-17 is characterized as a "soft" epoxy, having a strain-to-failure of about 50%, while UF 3298 is a "rigid" epoxy

[2]Kevlar 49, Kevlar 29, and Nomex are registered trademarks of E. I. du Pont de Nemours and Co. for its aramid fibers.

TABLE 2—Fiber properties determined by the ASTM Test for Tensile Properties of Glass Fiber Strands, Yarns, and Rovings Used in Reinforced Plastics [D 2343-67(1979)].

Fiber	Kevlar 29 Aramid	Kevlar 49 Aramid	Kevlar 49 Type 981	Thornel 300 Carbon	AS-4 Carbon	IM-6 Carbon	E Glass
Producer	Du Pont[a]	Du Pont	Du Pont	Union Carbide[b]	Hercules[c]	Hercules	Owens-Corning[d]
Strength, MPa (10^3 psi)	3620 (525)	3620 (525)	4070 (590)	3100 (450)	3590 (520)	4270 (620)	3450 (500)
Modulus, GPa (10^6 psi)	83 (12)	124 (18)	124 (18)	231 (33.5)	234 (34)	290 (42)	72 (10.5)
Density, g/cm³	1.44	1.44	1.44	1.70	1.80	1.74	2.54

[a] E. I. du Pont de Nemours and Co., Wilmington, DE.
[b] Union Carbide, Inc., Danbury, CT.
[c] Hercules, Inc., Magna, UT.
[d] Owens-Corning Fiberglas, Toledo, OH.

TABLE 3—*Resin system data (all cured at 126°C).*

Specimen Type	Resin	Producer	Type	Fibers
Solid panels	F-155	Hexcel[a]	rubber-toughened epoxy	Kevlar 49 Kevlar 29 Thornel 300
Honeycomb panels	MA 5400	McCann[b]	rubber-toughened epoxy	Kevlar 49 Thornel 300 E-glass
Pressure bottles	UF 3298	Morton Thiokol[c]	rigid epoxy	IM-6 Kevlar T-981
	UFX 82-17	Morton Thiokol	high strain-to-failure epoxy	AS-4 Kevlar 29 Kevlar 49

[a]Hexcel Corp., Dublin, CA.
[b]McCann Manufacturing Co., Oneco, CT.
[c]Morton Thiokol Corp., Brigham City, UT.

with a strain-to-failure of only 6%. Because resin properties have an important effect on the burst pressure in bottles and on the impact responses of structures, comparison of the bottle impact data needs to be done with care.

Data Interpretation

Flat Panels

Panels were impacted with sufficient energy to ensure complete penetration. Load and energy histories similar to those shown in the schematic in Fig. 2 were recorded. For the solid panels, a series of preliminary experiments in which the falling weight was stopped prior to complete penetration (limited deflection) was used to identify and explore the important feature of the load-deflection curve. These are identified in the figure as follows: i is the incipient damage point, m is the maximum load point, and t is the point of complete penetration.

Studies on similar laminates published previously have shown that the incipient damage point corresponds to the first irreversible damage in the laminate [3], in this case internal delamination or separation of fiber and resin. Very little damage is observed on superficial inspection at this point. At the maximum load point, the first visually observable fiber failure occurs, beginning with a crack on the back surface of the panel. Complete penetration occurs at t. In very brittle materials such as carbon fiber composites, the incipient and maximum load points may not be distinct—once failure initiates, it is catastrophic.

In most applications, the energy, E, absorbed by the structure is the most useful parameter. Absorbing the energy of the impacting object is the key to

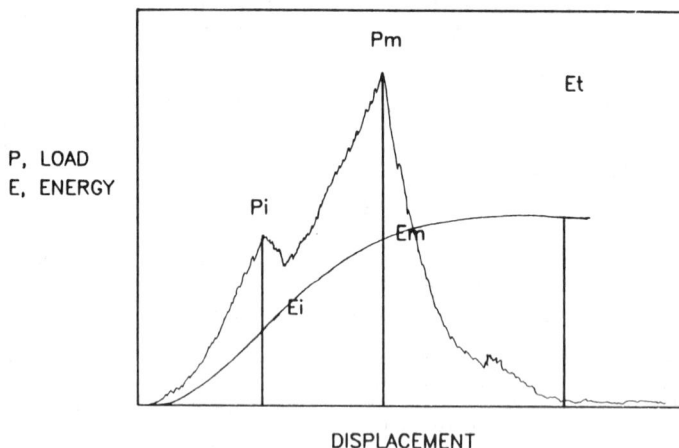

FIG. 2—*Schematic load-displacement and energy-displacement curves for impacted solid composite panels showing points of interest: incipient damage, maximum load, total penetration.*

survival. Depending on the type of structure under consideration, it is necessary to focus on one of the energies—E_i, E_m, or E_t—more than the others. One of the great advantages of instrumented impact testing is that it allows this kind of differentiation between features of the impact event.

Aircraft structure is divided into two categories: (1) the primary structure, which bears significant loads and is essential to flight, and (2) the secondary structure, which is only lightly loaded and is nonessential. In primary structure, the loss of properties—strength and stiffness—due to impact damage is important. Therefore, the energy required to start damage, the incipient energy, E_i, generally commands the most attention. In secondary structure, however, the concern is either for maintaining the integrity of the part in the face of impact damage and subsequent service or for protecting the structure underlying it. In the first case, the energy required to start fiber damage, E_m, is the most important, while in the second case, the total energy absorbed by the part as it shields the substructure, E_t, is important. On commercial aircraft today, most composite applications are in secondary structure, but increasing use of composites in primary structure is anticipated. In military aircraft, much more composite material is used in primary structure. Secondary structure is largely in the form of thin face sheets over a honeycomb core, while primary structure is mostly solid panels.

Honeycomb panels were also impacted with sufficient energy to penetrate completely. A typical load and energy trace is shown in Fig. 3. The first major peak corresponds to failure of the top face sheet, the valley corresponds to penetration of the core, and the last major peak corresponds to the interaction with the back face. For durability of honeycomb structures, it is impor-

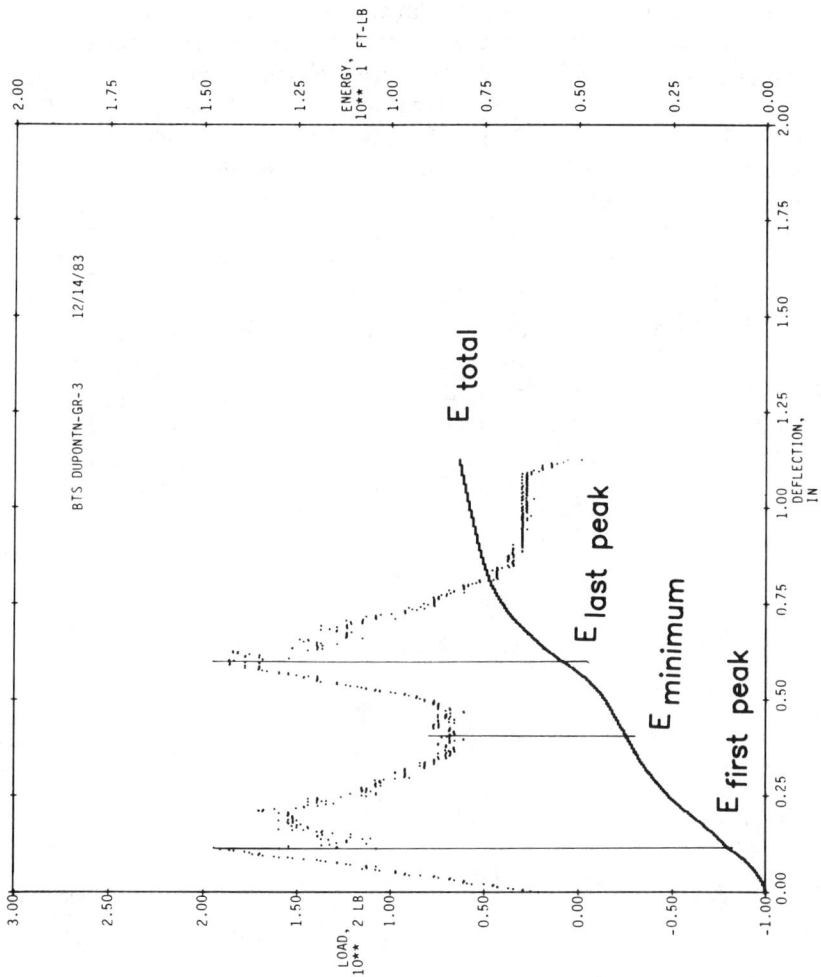

FIG. 3—Typical impact trace from honeycomb-core panel showing points of interest.

tant that the front face remain intact—otherwise moisture penetrates into the core and the structure becomes waterlogged. Therefore, it is common to focus on the interaction with the first face. This interaction is complex—there may be more than one distinct peak in this part of the load trace, as is shown. In the aluminum face sheet there is only one peak, although there is evidence of yielding on the leading edge of the load trace.

In honeycomb-core composites, the nature of the damage at the load peaks has not been established, so we have chosen not to use the previous letter subscript. $E_{first\ peak}$ is probably analogous to E_i in solid panels, although in very thin face sheets there may not be a true incipient damage point. The energy, $E_{minimum}$, at the valley represents the total energy of penetration of the front face. However, it also includes some contribution from crushing of the core beneath. The latter is small and is assumed to be approximately the same for each of the face sheets.

Pressure Bottles

By definition, a pressure bottle is primary structure—it is designed to contain gases under high pressure and hence is heavily loaded in service. Its retained properties, in this case burst strength, after impact, are as important as the energy required to cause damage. Impact traces from bottles are qualitatively similar to those from solid panels (Fig. 4) in that there is an incipient, maximum load, and full penetration point, although, in carbon, the incipient

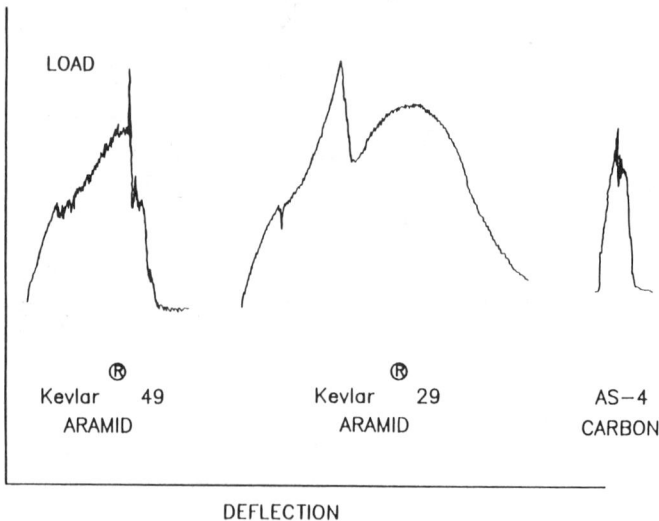

FIG. 4—*Typical impact traces from filament wound pressure bottles reinforced with Kevlar aramid fibers and carbon fiber.*

and maximum load points are identical. These parameters were determined on a single bottle, and then subsequent virgin bottles were impacted at various percentages of the failure energy, E_t. The impact energy was controlled by varying the drop height. The damaged bottles were then returned to Morton Thiokol for burst testing, and the residual strength was determined as a function of impact energy. Raw data for impact of the pressure bottles are given in Ref 4.

Results and Discussion

For solid panels, the three energy parameters—E_i, E_m, and E_t—are shown in Figs. 5, 6, and 7, plotted against the areal density of the panel (raw data are presented in Table 4). This presentation is relevant to aircraft because advanced composites are being used primarily to save weight over metals or glass-reinforced composites. For clarity, about one third of the data are not included here, but the trends are clearly established. In all three cases, the panels reinforced with Kevlar 29 aramid fibers show the highest energy absorption at a given weight. In incipient energy (Fig. 5), there is little difference between panels reinforced with Kevlar 49 and those reinforced with carbon fiber. Incipient damage is heavily dependent on the nature of the resin and the fiber-resin interface [3], so it is not apparent why Kevlar 29 is superior in this case.

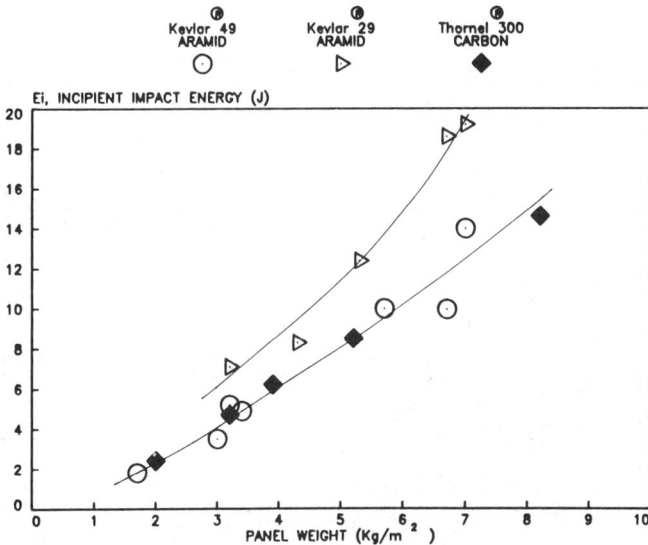

FIG. 5—*Energy absorbed by solid panels up to the incipient damage point versus areal density of the panel.*

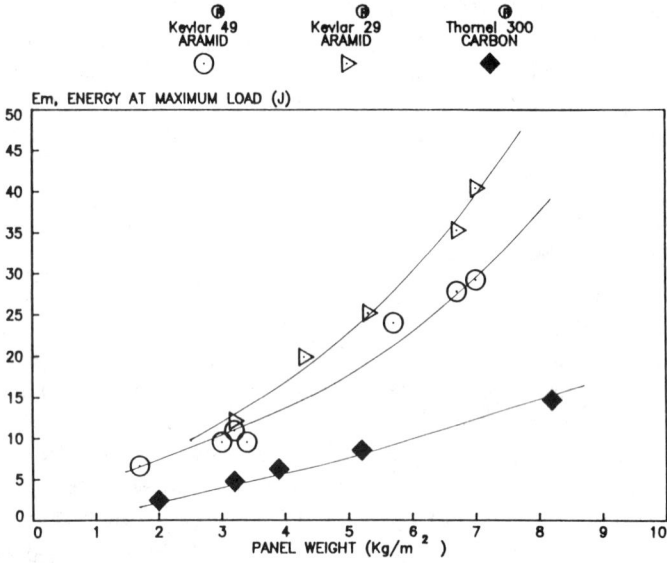

FIG. 6—*Energy absorbed by solid panels up to the maximum load point (first fiber failure) versus areal density of the panel.*

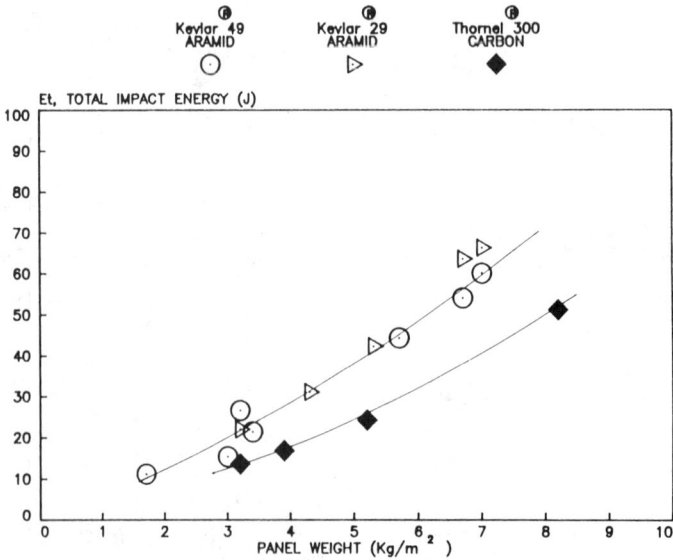

FIG. 7—*Energy absorbed by solid panels in complete penetration versus areal density of the panel.*

TABLE 4—Solid panel impact data: F-155 prepreg.

Specimen Identification	Fabric	Lay-Up	Thickness, mm	Weight, kg/m²	P_i, N	E_i, J	P_m, N	E_m, J	E_t, J
		THORNEL 300 CARBON							
C-17	3K-70-PW	$(+45/0/90)_s$	1.30	1.95	1121	2.4	1121	2.4	...
C-2A	3K-70-PW	$(0_2/-45/90_2)_s$	2.18	3.22	2006	4.7	2006	4.7	13.8
C-19	3K-70-PW	$(+45/0_2/-45/90_2)_s$	2.57	3.86	2669	6.0	2669	6.0	17.0
C-11	3K-70-PW	$(+45/0_2/-45/90_2)_s$	2.62	3.91	2406	6.3	2406	6.3	17.1
C-29	3K-70-PW	$(0/90/45)_{2s}$	2.57	3.91	2486	5.5	2486	5.5	17.2
C-27	3K-70-PW	$(\pm45/0_2/\pm45/90_2)_s$	3.40	5.17	4172	8.5	4172	8.5	24.7
C-97A	3K-70-PW	$(\pm45/0_2/\pm45/90_2)_s$	3.45	5.22	3977	8.1	3977	8.1	22.2
C-18	3K-70-PW	$(+45/0_2/-45/90_2)_{2s}$	5.36	8.15	7922	14.6	7922	14.6	51.9
		KEVLAR 49 ARAMID							
C-20	S-285	$(+45/0/90/0/-45)_t$	1.27	1.66	774	1.9	1027	6.6	11.3
C-1	S-285	$(+45/0_2/-45/90)_s$	2.59	3.32	1895	5.3	2366	10.9	23.9
C-30	S-285	$(0/90/+45/0/90)_s$	2.62	3.42	1913	5.7	2251	9.5	21.7
C-21	S-285	$(+45/0_2/-45/90)_s$	2.59	3.42	1850	4.9	2429	11.0	23.2
C-33	S-285	$(+45/0_2/\pm45/0/90_2)_s$	4.01	5.71	3892	10.2	4510	24.0	45.0
C-22	S-120	$(\pm45/0_2/-45/90_2)_s$	5.08	6.98	5444	14.2	5925	29.2	61.0
C-96	S-120	$(\pm45/0_4/\pm45/90_4)_s$	2.08	2.59	...	...	1810	7.5	14.6
C-31	S-120	$(0/90/+45)_{4s}$	2.18	2.98	1477	3.8	1855	8.3	15.6
C-8	S-120	$(\pm45/0_4/\pm45/90_2)_s$	2.34	2.98	1459	3.5	1926	9.5	...
C-26	S-120	$(\pm45/0_4/\pm45/90_4)_{2s}$	4.27	5.96	4386	11.8	4404	20.3	41.6
C-10	S-281	$(+45/0_2/-45/90)_s$	2.34	3.22	1672	4.0	2309	8.1	21.0
C-32	S-281	$(45/0_2/-45/90)_{2s}$	4.70	6.64	4221	10.0	5742	27.8	54.9
		KEVLAR 29 ARAMID							
C-50-2	S-728	$(+45/0/-45/90)_s$	2.46	3.22	2326	7.2	2531	12.1	22.4
C-52-1	S-728	$(+45/0/90)_{2s}$	3.94	5.32	3870	12.5	4123	20.9	37.8
C-51-1	S-728	$(+45/0/-45/90)_{2s}$	5.21	6.83	5560	18.8	6774	35.3	64.5
C-51-2	S-728	$(+45/0/-45/90)_{2s}$	5.13	6.98	5862	19.5	6757	40.4	67.3
C-52-2	S-728	$(\pm45/0/90)_{2s}$	4.06	5.33	3892	12.6	4444	25.2	43.0
C-46-2	S-740	$(\pm45/0_4/\pm45/90_4)_s$	2.51	3.22	2019	6.0	2602	12.6	20.5
C-49-2	S-740	$(\pm45/0_2/\pm45/90_2)_{2s}$	3.30	4.30	2962	8.4	3759	19.9	31.6
C-48-2	S-740	$(\pm45/0_4/\pm45/90_4)_{2s}$	5.03	6.93	5609	17.0	6023	36.7	60.5

In the energy absorbed before the first fiber failure (Fig. 6), Kevlar 49 is much closer to Kevlar 29 than to carbon fiber. It is important to note that, in these panels, the carbon fiber did not show distinct incipient and maximum load points, while the Kevlar fibers did (Fig. 8). Thus, the same values are plotted here for the maximum load point of carbon as were plotted in Fig. 5 for the incipient energy. In total energy absorbed (Fig. 7), again, the two aramid fibers are grouped together, showing higher energy absorption than the carbon fiber. Note that the curves in Figs. 5, 6, and 7 tend to converge as the panel weight (thickness) becomes small.

An attempt was made to rationalize the relative performance of the three fibers by examining the tensile toughness—the strain energy required for each to fail in tension (Fig. 9). This parameter is proportional to the area under the tensile stress-strain curve. The values shown here were obtained from the data in Table 1. The three fibers used in solid panels fall in the same order as the impact energies—Kevlar 29 is highest, followed by Kevlar 49 and Thornel 300 carbon. To show that this parameter dominates the impact energy up to the point of first fiber failure, the data from Fig. 6 were normalized to the panel weight and averaged for each of the fibers and then plotted against the specific toughness (data in Fig. 9 divided by density), with the result shown in Fig. 10. The impact energy, E_m, is determined by the fiber's strain energy to failure in tension. (In the test geometry used, the stress state

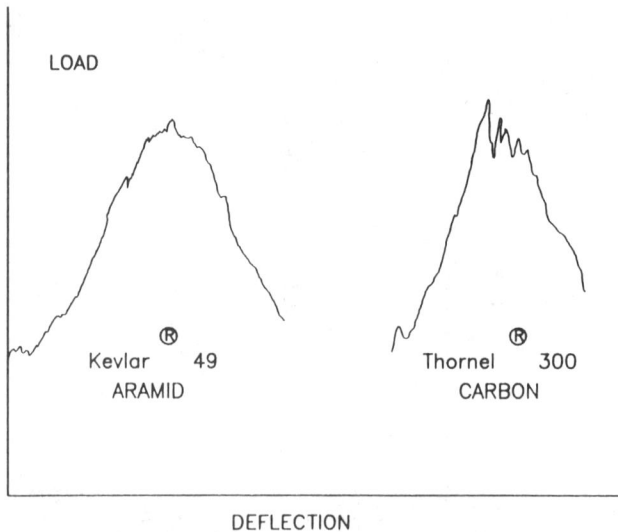

FIG. 8—*Typical impact traces from solid panels reinforced with fabrics of Kevlar 49 aramid fiber and Thornel 300 carbon fiber.*

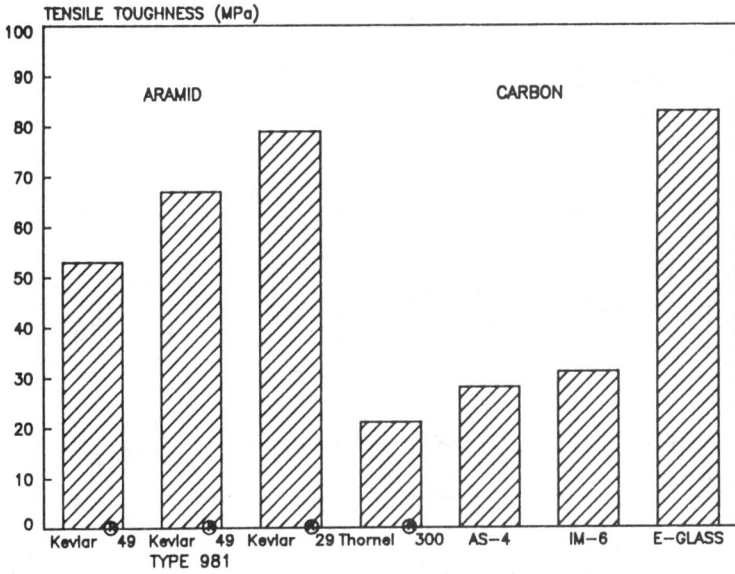

FIG. 9—*Tensile toughness (area under the tensile stress-strain curve) for various reinforcement fibers.*

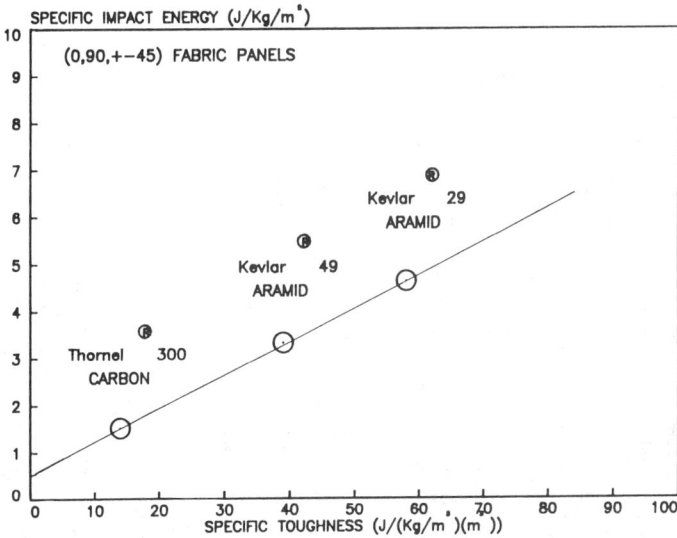

FIG. 10—*Specific impact energy (E$_m$ divided by areal density of the panel) versus specific tensile toughness (toughness divided by panel density).*

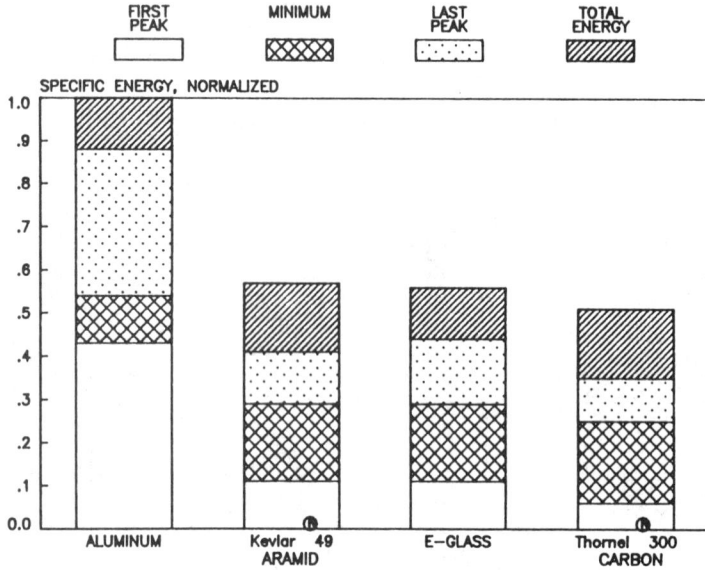

FIG. 11—*Specific impact energy (E divided by composite weight), normalized to aluminum, for various reinforcement fibers in honeycomb-core composites.*

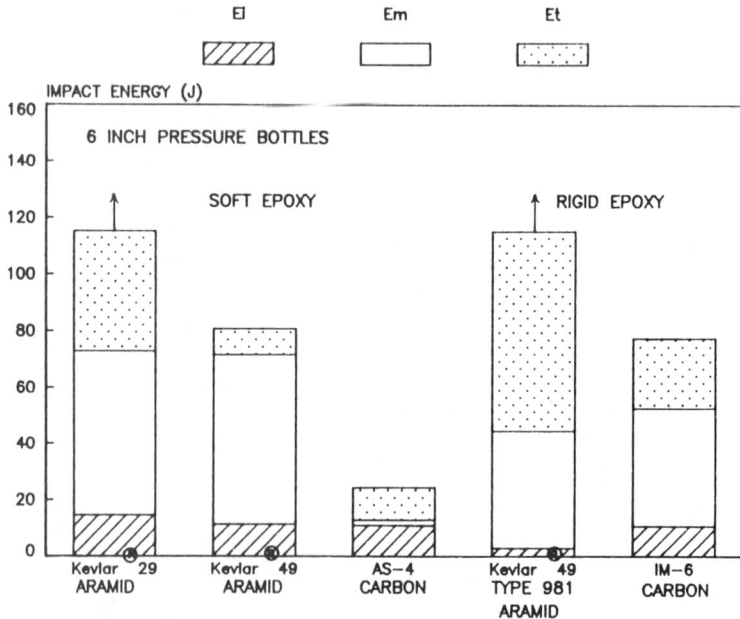

FIG. 12—*Impact energy for filament-wound pressure bottles reinforced with various fibers. The arrow indicates that the bottle was not fully penetrated at the highest impact energy.*

is predominantly tensile or membrane stress because of the high span-to-depth ratio of the panels. In other geometries, other factors such as shear stress will be more important.)

In the honeycomb panels, only one face-sheet thickness (two plies, or 0.05 cm) was tested for each fiber. The energy parameters, divided by the composite weight (core plus two face sheets) and normalized to the value for aluminum are summarized in Fig. 11. The composites still have some way to go to match the impact performance of aluminum. Among the composites, Kevlar 49 shows the highest energy absorption per unit of weight (Kevlar 29 was not included in this test), followed by glass and carbon. The differences among the fibers are not as large, proportionately, as was observed for the solid panels. This is partly due to the underlying contribution of the honeycomb core, which is present in each case, and to the face-sheet thickness.

For the pressure bottles, similar observations can be made (Fig. 12). (These data are not weight normalized.) Among the standard fibers in the so-called soft epoxy resin, Kevlar 29 has the highest energy parameters, followed by Kevlar 49 and AS-4 carbon. Even at the highest impact energy, 114 J (85 ft·lb), the bottle wound with Kevlar 29 was not fully penetrated. For the "improved" fibers, additional complications enter. First, these fibers were wound in a different resin, a rigid epoxy. In comparing these to the standard materials, one needs to bear in mind that generally soft epoxies give both higher burst pressure and better impact performance, particularly E_i, than rigid epoxies. Second, the improved Kevlar fiber, Type 981, embodies not only a fiber with higher tensile strength but also a new, proprietary surface treatment, which improves the translation of strength from the fiber to the bottle. This treatment tends to promote early separation between the fiber and the resin—hence, the low E_i of this material. However, E_m and E_t of this fiber are very high—complete penetration was not achieved at the maximum impact energy. (Delamination does not necessarily result in loss of performance in pressure bottles because that performance is dominated by the tensile properties of the fibers. Major losses in burst pressure will probably not be seen until the first fiber damage occurs.)

The residual burst pressure of damaged bottles is shown versus the impact energy in Fig. 13. The pressure is normalized to the average burst pressure of virgin bottles. Looking first at the standard fibers in soft epoxy (open symbols), AS-4 carbon loses significant strength at very low impact energy, while bottles with Kevlar fibers hold up much longer. Kevlar 29 retains half of its initial strength at the highest energy tested.

Of the two "improved" fibers in rigid epoxy (filled symbols), IM-6 behaves about like AS-4, while Type 981 Kevlar shows very little degradation in strength out to the highest impact energy. The high tensile toughness and the surface treatment on this fiber combine to make a structure that is extremely damage tolerant.

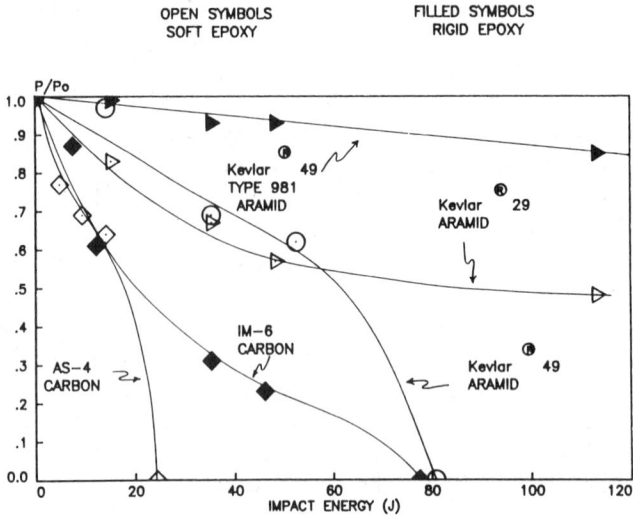

FIG. 13—*Residual burst pressure versus impact energy for filament-wound pressure bottles reinforced with various fibers.*

Conclusions

The methodology of instrumented drop-weight impact testing has been applied to studying the role of fiber properties in impact damage tolerance of structures, simulating both types of aircraft structures—primary and secondary—and filament-wound rocket-motor cases. The following conclusions have been reached:

1. The aramid fibers, and particularly Kevlar 29, can absorb large amounts of energy prior to failure, largely because of the high-tensile-strain energy or inherent toughness of the fibers.

2. Carbon fibers with lower toughness tend to be less efficient energy absorbers and to sustain more strength loss due to impact in pressure bottles.

3. Aluminum as a face sheet on honeycomb is an efficient energy absorber, and composite materials have a way to go before they can match the impact damage tolerance of aluminum even on a specific basis.

4. In pressure bottles, the combination of high strength and surface modification represented by Type 981 Kevlar produces a uniquely damage-tolerant structure.

References

[*1*] McCarty, J. E., "Damage Tolerance of Composites," NASA Conference Publication No. 2322, National Aeronautics and Space Administration, Washington, DC, August 1984.

[2] Davies, G. A. O., Ed., *Structural Impact and Crashworthiness*, Elsevier Applied Science Publishers, London, 1984.
[3] Miller, A. G., Hertzberg, P. E., and Rantula, V. M., *Society for the Advancement of Materials and Process Engineering Quarterly*, Vol. 12, No. 2, Azusa, CA, 1981, pp. 36–42.
[4] Adler, W. F., Carlyle, J. P., and Dorsey, J. J., "Damage Tolerance Assessment of Six-Inch Composite Pressure Vessels," CR-84-1263, Advanced Technologies Division, General Research Co., Santa Barbara, CA, 1984 (available from E. I. du Pont de Nemours and Co., Wilmington, DE).

Satish K. Gaggar[1]

Effects of Test Rate and Temperature on Fracture Behavior of Some Rubber-Modified Polymers

REFERENCE: Gaggar, S. K., **"Effects of Test Rate and Temperature on Fracture Behavior of Some Rubber-Modified Polymers,"** *Instrumented Impact Testing of Plastics and Composite Materials, ASTM STP 936,* S. L. Kessler, G. C. Adams, S. B. Driscoll, and D. R. Ireland, Eds., American Society for Testing and Materials, Philadelphia, 1987, pp. 236–247.

ABSTRACT: The fracture behavior of acrylonitrile-butadiene-styrene (ABS) and rubber-modified polyvinyl chloride (PVC) at various test temperatures and test rates was studied. The ductile-brittle transitions, in terms of test rate and test temperature, were established in each case, and the effects of material structural parameters on such transitions were studied. In modified PVC it is shown that the presence of the rubber modifier causes a shift of the ductile-brittle transition of unmodified PVC to a higher test rate at a given test temperature. The presence of an elastomeric phase in ABS induces a ductile failure mode in an otherwise brittle matrix over a broad range of test rates and temperatures. The ductile-brittle transition in ABS shifts to lower test temperatures as the rubber level is increased.

KEY WORDS: acrylonitrile-butadiene-styrene, polyvinyl chloride, ductile-brittle transition, fracture energy, crazes, impact strength, notch sensitivity, apparent activation energy, orientation, impact testing

Toughened polymers are being used in many load-bearing applications in which material toughness behavior over a wide range of temperatures and strain rates is critical. Rubber-modified polymers such as high-impact polystyrene (HIPS) and acrylonitrile-butadiene-styrene (ABS) contain a dispersed rubber phase in a relatively brittle matrix to achieve a desired level of impact strength. The presence of rubber modifier in polyvinyl chloride (PVC) reduces the notch sensitivity of an otherwise ductile matrix phase. The glass transition temperature of the rubbery phase can be as low as $-85°C$, so that

---

[1]Staff scientist, Borg-Warner Chemicals, Inc., Technical Centre, Washington, WV 26181.

a ductile failure mode with good impact strength is usually maintained much below room temperature. Izod impact testing on various rubber-modified polymers has indicated signs of ductility at test temperatures as low as $-70°C$ [1]. The toughening mechanisms in rubber-modified polymers have been studied by a number of investigators, and a detailed description can be found in Ref 2.

It is generally agreed that the role of rubber particles in HIPS and ABS is to promote multiple crazing in the matrix phase and to control the growth of crazes and the formation of unstable cracks during impact-loading conditions. A large amount of deformation in the matrix phase can take place, resulting in high toughness values. It is expected that the amount and size of the dispersed phase, the matrix mechanical behavior, and the rubber-matrix interface will have significant influences on the toughness of the modified polymers. In HIPS and ABS, the rubber particles are grafted with a polymer whose composition is the same as that of the matrix phase to achieve a good interfacial bond. The size of rubber particles needed to achieve optimum impact strength depends strongly on the nature of the matrix phase.

This study was undertaken to determine the influence of the rubber level and matrix molecular weight on the ductile-brittle (DB) transition in ABS; the rubber particle size and graft structure were not varied. The DB transition in rubber-modified PVC was established at various levels of a commercial impact modifier.

Experimental Procedure

A commercial-grade high-impact ABS was extruded into a 3-mm-thick sheet. Test specimens were prepared parallel to and perpendicular to the extrusion direction to study the effect of orientation on the DB transition. Model ABS specimens with a controlled rubber level and matrix molecular weight were prepared by compounding grafted rubber with polystyrene acrylonitrile (PSAN) in a Banbury intensive mixer followed by milling, pelletizing, and injection molding in the form of standard tensile bars of 3-mm thickness. Modified PVC specimens were prepared by extrusion compounding a molding grade of PVC with a commercial rubber modifier and then injection molding tensile bars of 3-mm thickness. Standard Izod impact specimens [ASTM Test for Impact Resistance of Plastics and Electrical Insulating Materials (D 256-84)] were prepared in each case and were tested in a three-point bend mode (50-mm span) on an MTS high-speed tester and a Plastechon high-rate tester, each fitted with an environmental chamber. The test rates and temperatures were varied as required, and the load-time (deflection) curve for each test was recorded on a Nicolet digital oscilloscope. The area under the load-deflection curve was measured by using a planimeter, and the energy to fracture was normalized with respect to the crack surface area as follows

$$\text{Fracture energy (FE)} = \frac{W}{2(w - a)t}$$

where W is the area under the load deflection curve (in joules), w is the depth of the specimen, a is the initial crack length, and t is the thickness of the specimen. The fracture energy values were averaged for three specimens in most cases, and the maximum variation from the mean FE value was less than 10% in each set of data.

Results and Discussion

Effect of Orientation on the DB Transition in ABS

Although the level of orientation in an extruded sheet caused by stretching during the takeoff stage is relatively low, its influence on the impact strength can be significant. Figures 1 and 2 show the FE data at various test temperatures and deformation rates for the specimens cut parallel to the extrusion direction (crack propagation perpendicular to orientation) and in the transverse direction (crack propagation along orientation direction), respectively. The residual orientation, measured by an oven shrinking test (2 h at 175°C), was about 5%. At each test temperature, the residual orientation in parallel specimens results in higher FE values over the entire range of deformation

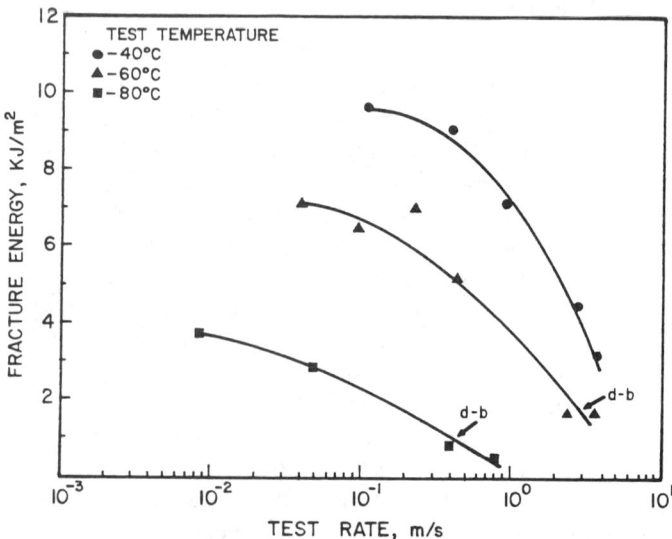

FIG. 1—*Fracture energy data for parallel specimens (initial notch perpendicular to the extrusion direction).*

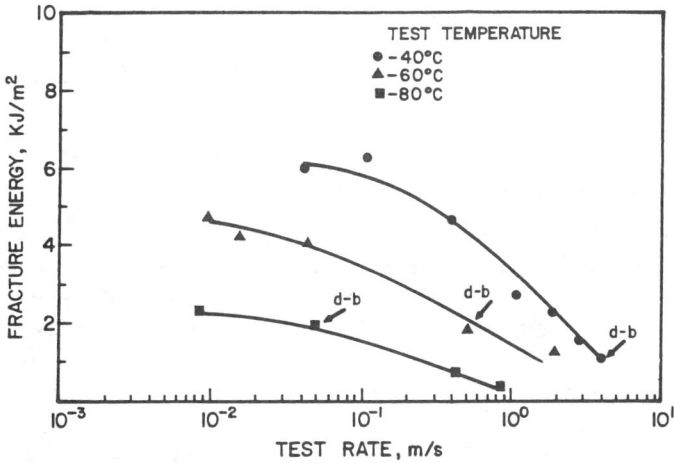

FIG. 2—*Fracture energy data for transverse specimens (initial notch along the extrusion direction).*

rate. The DB transition for parallel specimens occurs at higher test rates than that for transverse specimens at each test temperature. The transition from ductile to brittle was observed to be gradual in each case.

In the ductile failure region, the fracture surface is fully craze whitened, and the whitening extends some distance along the length of the specimen on both sides of the fracture plane. The total volume of the craze-whitened material appears to decrease as the test rate increases and the test temperature decreases. The DB transition refers to the test speed at which the whitening did not cover the entire fracture surface and was confined to a distance of about 2 mm ahead of the initial notch tip. The DB transition for parallel specimens is about an order of magnitude higher in the test speed than that for transverse specimens.

In the direction of extrusion, the matrix polymer chains are aligned to some extent and the rubber particles are also known to elongate. The stress necessary to initiate crazes at the equator of the rubber particles is expected to be higher because of the matrix orientation and reduced stress concentration [3]. Since the yield strength and elongation to breaking also increase with orientation, a greater amount of energy is absorbed during the fracture of the parallel specimens. The amount of stress whitening along the fracture plane in an oriented specimen was observed to be significantly greater, indicating an increased amount of microcrazing that resulted in the DB transition being shifted to higher test rates. The shift of DB transition with test rate and test temperature in Figs. 1 and 2 demonstrate that the DB transition in ABS and other polymers can be adequately defined only when both test rate and test temperature are taken into account. One can define DB transition in terms of a test rate at a given test temperature or in terms of a test temperature at a given test rate. From results shown in Figs. 1 and 2, an order of magnitude change in the test rate corresponds to about a 20-degree-Celsius shift in the DB transition.

Effect of Rubber Amount on DB Transition

ABS specimens were prepared with rubber levels of 20, 25, 30, and 35% by weight. The notch-bend test (NBT) specimens were prepared from the injection-molded tensile bars. These were tested at a constant deformation rate of about 2 m/s at various test temperatures. The test temperature was lowered in steps of 10 degrees Celsius, starting with an initial test temperature of 0°C. At each test temperature the specimens were conditioned for 20 min before fracturing. The range of test temperatures over which the specimen changed from a completely ductile failure to an almost brittle failure (trace of whitening at the initial notch tip) was determined in each case. At any test temperature, if fracture was partially brittle, it was taken as the DB transition.

Table 1 lists the data from the preceding tests which clearly indicate that increasing the rubber level has a significant effect on DB transition in ABS. A shift of almost 40 to 50 degrees Celsius in the DB transition occurs as the rubber level increases from 20 to 35%. As the rubber amount is increased, enhanced crazing and more efficient craze termination are possible, thus resulting in a lower DB transition temperature.

Effect of Matrix Molecular Weight on DB Transition in ABS

Matrix PSAN materials (30% acrylonitrile) were prepared with molecular weights in the range of $M_n = 24\,000$ to $108\,000$, where M_n is the number average molecular weight, as determined by gel permeation chromatography. Compression-molded specimens (1 mm thick) were prepared for mechanical characterization of rigid PSAN specimens. The tensile strength was established for each specimen. The craze initiation stress was measured by using

TABLE 1—*Effect of rubber amount on ductile-brittle transition in ABS.*

Material	Ductile Brittle Transition Temperature (Test Rate 2 m/s), °C
20% rubber	-10
25% rubber	-20 to -30
30% rubber	-30 to -40
35% rubber	-50 to -60

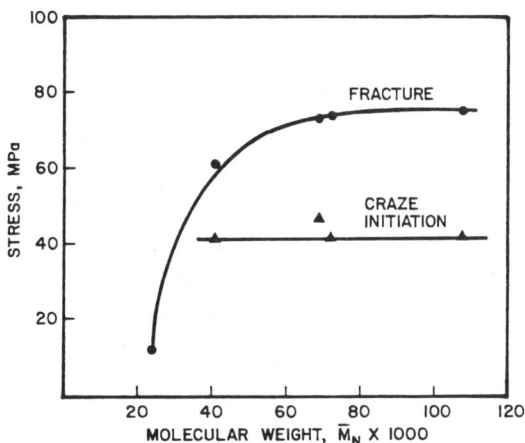

FIG. 3—*Effect of molecular weight of PSAN on fracture strength and craze initiation stress values.*

an optical arrangement discussed in Ref *4*. Figure 3 shows the tensile strength and craze initiation stress data for PSAN specimens at five molecular weight values. All the specimens fractured in a brittle manner. The tensile strength increased as the M_n value changed from 24 000 to 41 000 and then gradually leveled off to a constant value beyond $M_n = 69 000$. The craze initiation stress is virtually independent of molecular weight. These results are consistent with the similar results reported for polystyrene [*4*]. The lowest-molecular-weight specimen ($M_n = 24 000$) failed in a glasslike manner without craze formation, and thus it can be anticipated that modifying this PSAN specimen with grafted rubber will not result in improved toughness.

Figure 4 shows the NBT data for three ABS specimens prepared from low, medium, and high-molecular-weight PSAN and grafted polybutadiene rubber. The rubber level in each case was maintained at 20%. The data indicate that except for the low-molecular-weight specimen that remained brittle, the DB transition for medium and high-molecular-weight specimens is not strongly dependent on the PSAN molecular weight. The energy absorbed during fracture is significantly larger for the higher-molecular-weight material. This should be expected because of the higher tensile strength of high-molecular-weight matrix phase, which is an indication of the strength of the crazes formed. The DB transition may be related to the craze initiation stress, which has been shown to be independent of PSAN molecular weight. A slight shift of the DB transition to a lower test rate, shown in Fig. 4 for a medium-molecular-weight specimen, may be due to the lower stability of the crazes that are formed. The author thus concludes that except for very low molecular weight ABS, the DB transition remains relatively constant with respect to matrix mo-

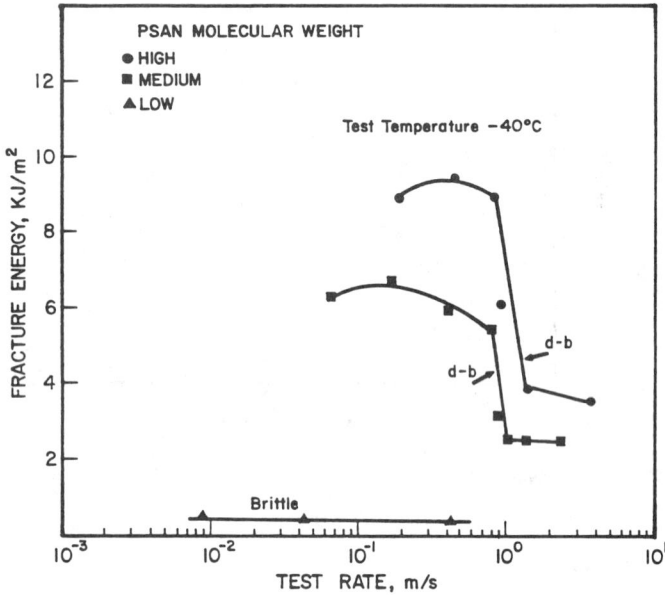

FIG. 4—*Effect of molecular weight of PSAN on DB transition in ABS (rubber level 20% by weight).*

lecular weight, and the rubber content has the major influence on the DB transition phenomenon.

Fracture Behavior of Rubber Modified PVC

Rigid PVC is known to be very notch sensitive and fails in a relatively brittle manner in the Izod impact test. Unnotched PVC specimens show considerable ductility in tension tests, and the primary mode of deformation is shear yielding and neck formation. The mechanical behavior of the ductile matrix in toughened PVC is thus quite different from that of the brittle matrix in ABS. The addition of a rubber modifier to PVC significantly reduces the notch sensitivity at impact test rates, and the fracture behavior is expected to be dependent on the level of the rubber modifier.

Figures 5 through 8 show the fracture energy data at various test temperatures and rates for unmodified PVC and toughened PVC at three levels of a rubber modifier. It should be noted in Fig. 5 that notched unmodified PVC fails in a completely ductile manner up to a test rate of 0.08 m/s and then undergoes a sharp DB transition. In the ductile failure mode, the material around the fracture plane was drawn considerably, and a significant reduction in the thickness of the specimen occurred on both sides of the fracture

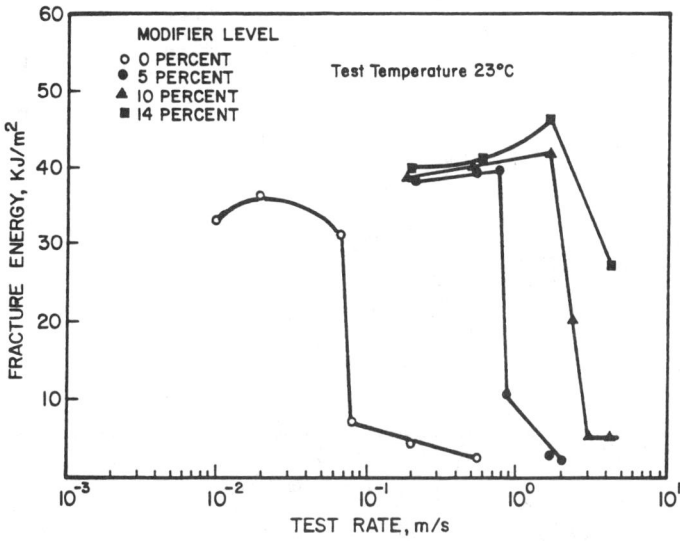

FIG. 5—*Fracture energy data for PVC and modified PVC at room temperature.*

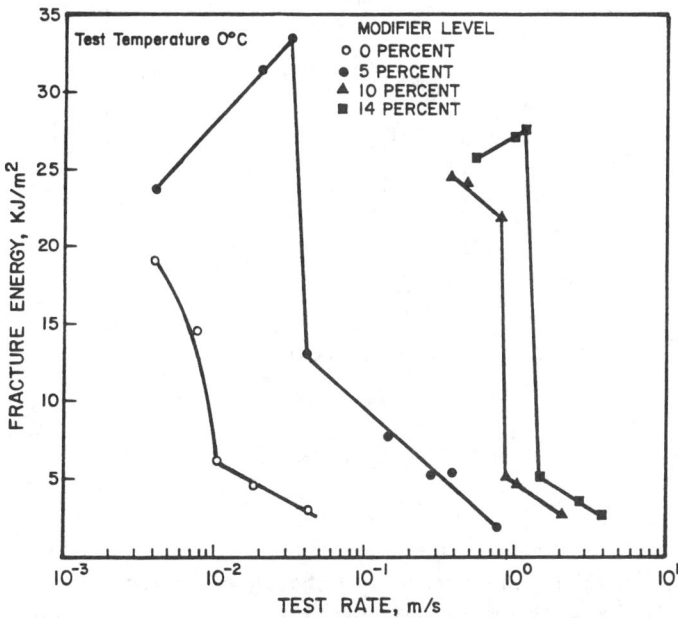

FIG. 6—*Fracture energy data for PVC and modified PVC at 0°C.*

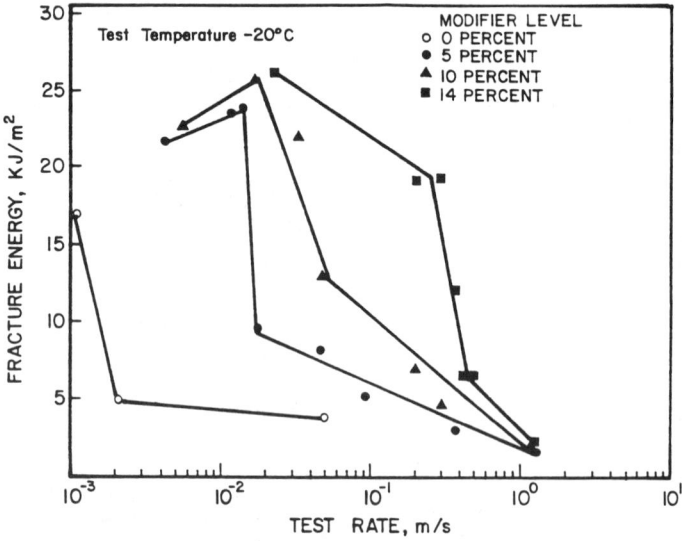

FIG. 7—*Fracture energy data for PVC and modified PVC at −20°C.*

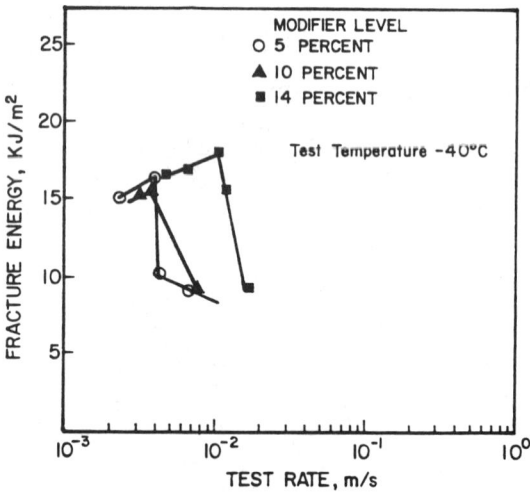

FIG. 8—*Fracture energy data for modified PVC at −40°C.*

plane. The fracture energy values are quite high compared with that for ABS (Fig. 1). The DB transition occurs very sharply over a very narrow range of test rate. In the brittle region where the FE value is low, the fracture surface was flat, and the fracture occurred without any macroscopic yielding. It is thus obvious that the notch sensitivity of PVC is strongly rate dependent. The effect of the addition of a modifier, as shown in Fig. 5, is to shift the DB transition to higher test speeds. The amount of shift in the DB transition rate is shown to be dependent on the modifier content. A similar behavior is shown at lower test temperatures in Figs. 6, 7, and 8.

It is interesting to note that the presence of a modifier does not result in any significant increase of FE of PVC in the ductile failure region. The DB transition of all modified PVC specimens was observed to be very sharp, like that of unmodified PVC. This is in contrast with the behavior of ABS, as discussed earlier. In other words, the deformation mode in PVC changes sharply from yielding at the notch tip to a completely brittle behavior, in contrast with ABS in which crazing deformation dominates and the amount of crazing changes gradually with the test rate and test temperature.

Various investigators have explored the possibility of relating DB failure transition in PVC and other polymers to the low-temperature relaxation phenomenon [5,6]. It is generally agreed that the existence of a low-temperature β transition may be a necessary but not a sufficient condition for the occurrence of a DB transition in most polymers. Radon [7] measured the fracture toughness of PVC by three-point bend testing and observed that the peak in fracture toughness occurred at the same test temperature where the β relaxation peak occurred when measured at a frequency corresponding to the strain rate of the fracture test. The DB transition test rate and test temperature data from Figs. 5 through 8 can be fitted to an Arrhenius rate equation to obtain the apparent activation energy, Ea, from the following

$$v = Ke^{-Ea/RT}$$

where v is the test speed, K is a constant, R is the gas constant, and T is the absolute temperature.

A plot of $\ell n(v)$ versus $1/T$ is shown in Fig. 9 for PVC and modified PVC. An apparent activation energy value of 14 kcal/mole is obtained for PVC, which is in excellent agreement with the reported value of apparent activation energy for the β relaxation process in PVC [8]. For modified PVC, the apparent activation energy values are calculated to be in the range of 13 to 15 kcal/mole. The addition of a modifier to PVC did not change the rate–temperature sensitivity of the DB transition. Based on the apparent activation energy values, it can be stated that there is a close correspondence between the DB transition and the β relaxation process in PVC.

Conclusions

The DB transition in ABS is shown to be affected by residual orientation and the rubber level in the specimens. It has been shown that the matrix molecular weight above a critical value does not significantly influence the DB transition. The craze initiation stress value is shown to be independent of

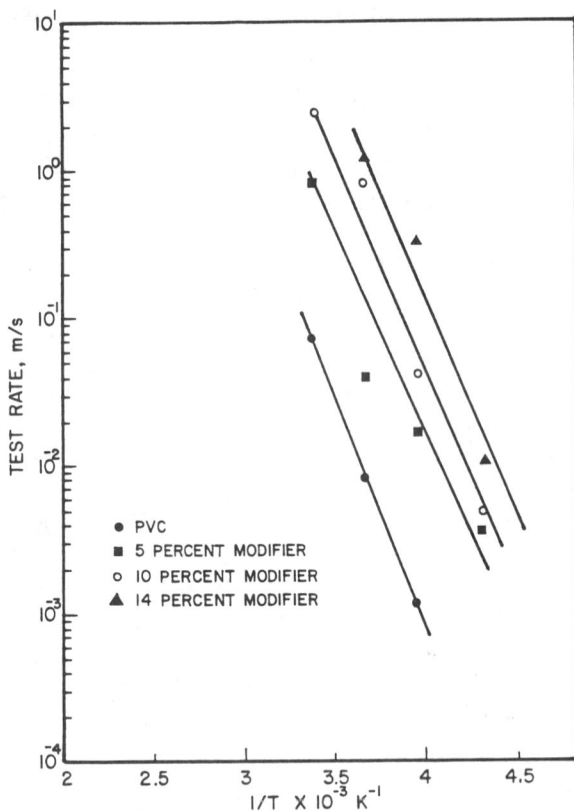

FIG. 9—*Arrhenius plot for PVC and modified PVC.*

PSAN molecular weight except at very low molecular weights ($M_n < 30\,000$) where crazing does not seem to occur. Addition of grafted rubber to PSAN when the crazing mechanism is absent does not result in impact modification. The DB transition in ABS is not sharp at various test temperatures, in contrast with that of modified PVC. The rubber modification of PVC results in the shift of the DB transition to higher test rates compared with that for matrix PVC. The apparent activation energy value for PVC and modified PVC calculated from the fracture energy data compares closely with the reported values of activation energy for the β relaxation, indicating a significant role of low-temperature molecular relaxation in the fracture process.

References

[1] Matsuo, M., Veda, A., and Kondo, Y., *Polymer Engineering and Science*, Vol. 10, No. 5, September 1970, pp. 253-260.

[2] Bucknall, C. B., *Toughened Plastics*, Applied Science Publishers, London, 1977, p. 182.

[3] Yokouchi, M., Yokota, A., and Kobayashi, Y., *Journal of Applied Polymer Science*, Vol. 27, 1982, pp. 3007–3018.

[4] Fellers, J. F. and Kee, B. F., *Journal of Applied Polymer Science*, Vol. 18, No. 8, August 1974, pp. 2355–2365.

[5] Boyer, R. F., *Polymer Engineering and Science*, Vol. 8, No. 3, July 1968, pp. 161–185.

[6] Retting, W., *European Polymer Journal*, Vol. 6, No. 6, June 1970, pp. 853–863.

[7] Radon, J. C., *Polymer Engineering and Science*, Vol. 12, No. 6, November 1972, pp. 425–431.

[8] McCrum, N. G., Read, B. E., and Williams, G., *Anelastic and Dielectric Effects in Polymeric Solids*, Wiley, New York, 1967, p. 434.

Lee W. Gause[1] and Leonard J. Buckley[1]

Impact Characterization of New Composite Materials

REFERENCE: Gause, L. W. and Buckley, L. J., "Impact Characterization of New Composite Materials," *Instrumented Impact Testing of Plastics and Composite Materials,* *ASTM STP 936,* S. L. Kessler, G. C. Adams, S. B. Driscoll, and D. R. Ireland, Eds., American Society for Testing and Materials, Philadelphia, 1987, pp. 248-261.

ABSTRACT: Drop-weight instrumented impact tests were used in conjunction with ultrasonic C-scan inspection to characterize the impact responses of several new graphite-fiber composite material systems. The AS4/Hercules 3501-6 graphite/epoxy was compared with the newer AS4/Hercules 2220-1, Celion high-strain/Narmco 5245 and IM6/Narmco 5245C systems in the 394 K (250°F) service category. The materials tested in the 450 K (350°F) service category were T300/Avco 130B, T300/Hexcel 81-5, T300/U.S. Polymeric V378A, XAS/Hysol 9101-3, and HX/Hexcel 1516 graphite/bismaleimides. A material ranking is given, along with measured impact parameters and impact energy versus damage relationships. The impact behavior of all four 394 K (250°F) systems was similar, except that IM6/5245C had the highest impact resistance. The bismaleimides had incipient damage levels similar to those of the epoxies but were more easily punctured. The impact resistance of the five 450 K (350°F) systems was similar except that T300/130B had the lowest impact resistance.

KEY WORDS: composite materials, impact testing, instrumented impact tests, material damage, graphite fiber, epoxy, bismaleimide

Graphite-fiber reinforced resin-matrix composites are firmly established as a major aerospace material expected to comprise more than half the structural weight of near-future aircraft. The widespread use and acceptance of graphite/epoxy composites in components of such advanced aircraft as the F-18 and AV-8B results from the structural efficiency, extensive characterization, and manufacturability of the current, mature systems, such as AS/

The opinions and assertions expressed in this paper are the authors' private opinions and are not to be construed as official or reflecting the views of the U.S. Department of the Navy or the Naval Services at large.

[1]Aerospace engineer and materials engineer, respectively, Aircraft and Crew Systems Technology Directorate, Naval Air Development Center, Warminster, PA 18974.

3501-6 and T300/5208. The epoxy matrixes of these materials, however, have prevented the structural engineer from taking full advantage of the performance improvements possible through the use of graphite-fiber composites. While it is the high tensile strength and modulus of the fiber that is responsible for the strength and stiffness of a composite structure, the matrix is an essential element in maintaining fiber alignment, stabilizing the fibers against buckling, and providing for load transfer between fibers. The current epoxy resins are degraded by environmental moisture, drastically reducing their strength at elevated temperatures and limiting their continuous-service capabilities to temperatures below 394 K (250°F). They are brittle and easily damaged by low-velocity impact, in some circumstances incurring substantial internal damage while showing no visible signs of being struck. The designer is thus forced to restrict these composites to load levels far below the capabilities of the fibers in order to compensate for environmental effects and possible impact damage. Bismaleimide resin systems have provided improved thermal resistance over those of epoxies but possess the same limitations. The material suppliers have undertaken to address these limitations by formulating new resin systems to provide better impact resistance, higher strain-to-failure values, and improved hot-wet strength. The impact characterization described in this paper was one part of a larger overall program fully characterizing several new prepreg systems with respect to their physical and mechanical properties [1].

The materials being evaluated were divided into two classes based on the operational service temperature. The AS4/Hercules 3501-6 graphite/epoxy was compared with the newer AS4/Hercules 2220-1, Celion high-strain/Narmco 5245, and IM6/Narmco 5245C toughened epoxy systems in the 394 K (250°F) service category. Materials tested in the 450 K (350°F) service category were T300/Avco 130B, T300/Hexcel 81-5, T300/U.S. Polymeric V378A, XAS/Hysol 9101-3, and HX/Hexcel 1516 graphite/bismaleimides.

Procedure

Equipment

A Dynatup Model 8200 drop tower with a Dynatup Model 371 instrumented impact system was used for the impact tests (Fig. 1). The crosshead weight can be varied from 3.2 to 14.5 kg (7.0 to 32 lb) and impact velocities up to 7.6 m/s (25 ft/s) can be achieved. This tower can impose impact energies in the range from 1.4 to 434 J (1 to 320 ft · lb), so that the complete spectrum of composite failure mechanisms from incipient damage up to through-penetration can be studied. Impact-force versus time and velocometer output data from the instrumented impact system are captured on a Nicolett Explorer III model 206-2 digital oscilloscope. The curser trigger feature of the digital oscilloscope simplifies testing as the force-time analog output itself is used to

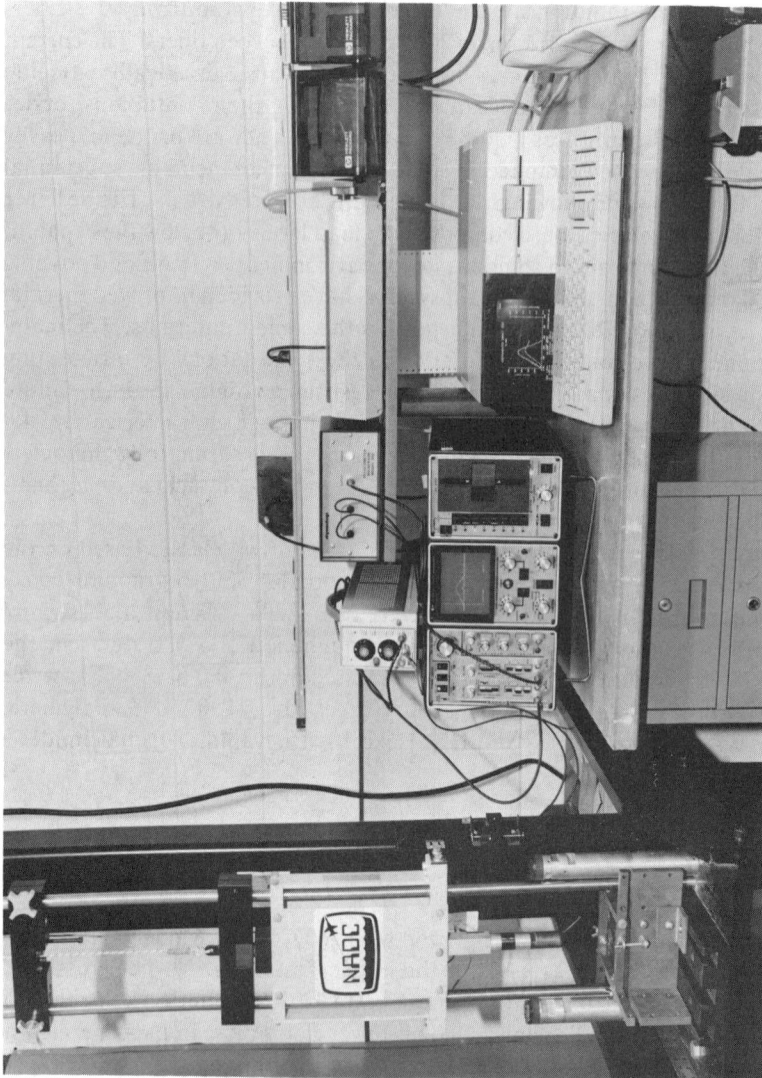

FIG. 1—*Naval Air Development Center (NADC) Instrumented drop weight system.*

trigger signal capture. Further, the digitized wave form then stored by the oscilloscope is directly output to a Hewlett-Packard HP9826 desktop computer for analysis and data presentation.

Specimen Preparation

Quasi-isotropic, 16-ply laminates with a $[\pm 45_2/(0/90)_2]_s$ stacking sequence were fabricated of each material system from which individual, 152-mm-(6-in.) square impact test specimens were cut. The nominal specimen thickness was 3 mm ($1/8$ in.). All the specimens where fabricated from prepreg tape except the T300/V378A specimens, which were made from balanced plain weave cloth.

Procedure

Each plate impact specimen was clamped in the drop tower along its edges between two steel frames, leaving a 127 by 127-mm-(5 by 5-in.) square test section. A 12.7-mm ($1/2$-in.)-radius hemispherical steel indenter was attached to the crosshead, and each specimen was struck once at its center normal to its surface. The crosshead was caught after rebound to prevent multiple impacts. All the specimens were inspected by ultrasonic C-scan before and after each test. All testing was performed at room temperature in a laboratory environment. The impact energy was controlled by adjusting the crosshead weight and drop height. The critical parameters determined for comparing the impact response of each type of material were the following:

(a) load at incipient damage, P_{inc},
(b) energy absorbed at incipient damage, E_{inc},
(c) maximum load, P_{max},
(d) energy absorbed to maximum load, E_{max}, and
(e) total absorbed energy for through-penetration, E_{tot}.

A typical instrumented-impact output for a through-penetration test of an AS4/3501-6 specimen is shown in Fig. 2, which identifies the various critical loads and energies. While the load and time response are directly measured, the absorbed energy and displacement values are computed incrementally from the measured initial velocity, the crosshead mass, and the load-time history using the methods described in Ref 2.

Three impact energy levels were studied:

(a) through-penetration (puncture),
(b) maximum load impact energy, E_{max}, and
(c) incipient damage.

In this manner, a damage gradient was obtained for each prepreg system indicating its response over the entire range of damage mechanisms, from in-

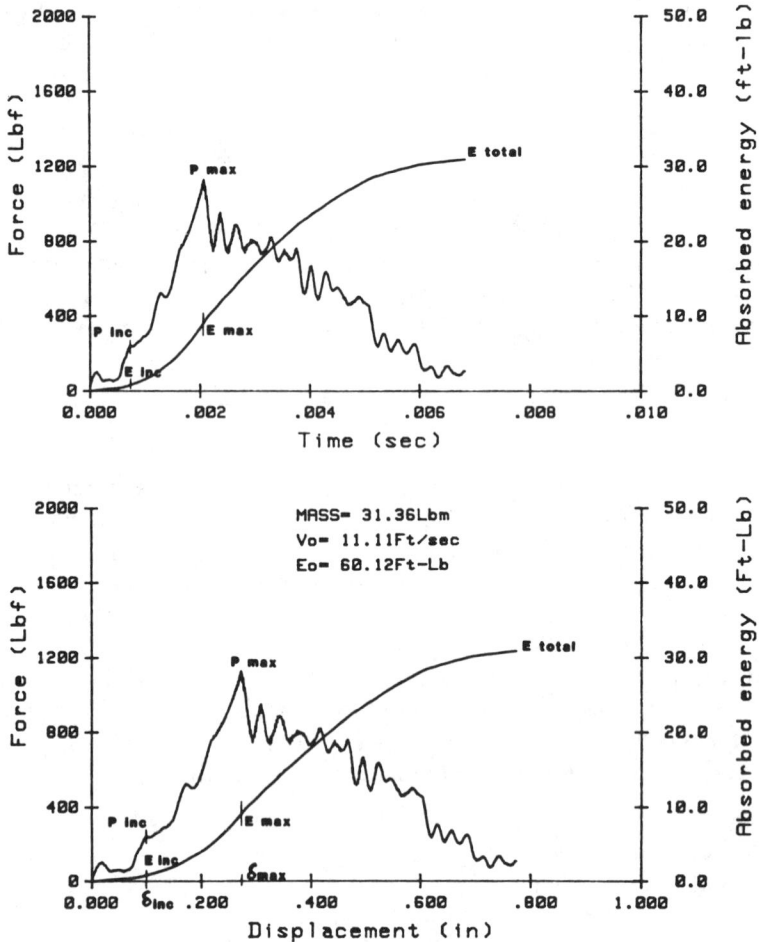

FIG. 2—*Instrumented impact test output.*

cipient damage to total puncture. Only one test was performed per prepreg system per energy level. Through-penetration instrumented-impact test results were used to establish the peak load energy level. It was not always obvious, from the through-penetration test results, when incipient damage had occurred. In most cases, incipient-damage impact levels could be determined from the initial load drop (P_{inc} in Fig. 2) during the peak-load impact energy level test or a lower-impact energy level test. In those cases in which none of the instrumented impact test traces clearly established incipient damage, it was determined by testing additional specimens, reducing the impact level until no damage was detectable by ultrasonic C-scan.

Results

Table 1 presents a summary of the critical impact parameters measured during these tests for both the 394 K (250°F) and 450 K (350°F) service systems. Detailed data sheets for each test specimen can be found in Ref *3*.

394 K (250°F) Service Systems

The impact force versus displacement response of the four 394 K (250°F) systems are plotted in Fig. 3. Figures 4 and 5 compare the energy and force results, respectively, and Fig. 6 plots the C-scan damage area versus impact energy (crosshead kinetic energy at impact) results. The IM6/Narmco 5245C demonstrated the best impact resistance, requiring nearly twice the energy to cause incipient damage and a third more energy to penetrate than the other systems. There was little difference between the responses of the other three materials. In the C-scan damage area versus impact energy, the AS4/Hercules 3501-6 sustained the greatest damage at the 13.6-J (10-ft · lb) impact level, but at higher energy levels the four systems were similar.

Our ranking of the impact resistance of the 394 K (250°F) service systems is as follows:

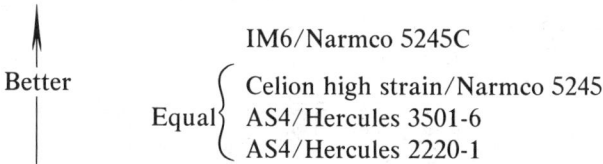

Better

IM6/Narmco 5245C

Equal⟨ Celion high strain/Narmco 5245
AS4/Hercules 3501-6
AS4/Hercules 2220-1

450 K (350°F) Service Systems

The impact force versus displacement response of the five bismaleimide systems is shown in Fig. 7. Figures 8 and 9 present the energy and force result comparisons, and Fig. 10 plots the C-scan damage area versus impact energy. Since the T300/U.S. Polymeric V378A test specimens were made from plane-weave cloth, the impact response of this material cannot be directly compared with those of the other, tape-layup systems. The use of woven cloth generally results in smaller damage areas, as the delamination is inhibited by the direct mechanical reinforcement of the interlocking fabric yarns. Of the tape systems, all had a similar impact response, except for the T300/Avco 130B. This system experienced incipient damage at a third the energy of the other systems and required only half the energy to be punctured. It also experienced the largest damage areas of all the systems tested.

Although the incipient damage levels of the better bismaleimides were equal to those of the lower-temperature materials, the bismaleimides were

TABLE 1—Summary of impact results.

Service Temperature, K (°F)	Material	Thickness, mm (in.)	At Incipient Damage		At Peak Load		Total Energy for Puncture, E_{tot}, J (ft · lb)
			P_{inc}, N (lb)	E_{inc}, J (ft · lb)	P_{max}, N (lb)	E_{max}, J (ft · lb)	
394 (250)	AS4/3501-6	2.06 (0.081)	1222 (274.7)	161 (1.19)	5008 (1126)	12.30 (9.07)	43.25 (31.90)
	AS4/2220-1	2.49 (0.098)	<1340 (<301.3)[a]	<1.76 (<1.30)[a]	4960 (1115)	12.16 (8.97)	51.34 (37.79)
	Cel/5245	2.16 (0.085)	1344 (302.1)[a]	1.91 (1.41)[a]	5504 (1237)	16.56 (12.21)	51.32 (37.85)
	IM6/5245C	2.16 (0.085)	2227 (500.6)	3.67 (2.71)	5824 (1309)	15.10 (11.14)	69.23 (51.06)
450 (350)	T300/130B	2.18 (0.086)	769 (172.9)	0.62 (0.46)	1312 (294.9)	2.69 (1.98)	13.14 (9.69)
	T300/81-5	2.46 (0.097)	1585 (356.3)	2.40 (1.77)	2864 (643.9)	11.70 (8.63)	27.90 (20.58)
	XAS/9101-3	1.93 (0.076)	<1222 (<274.7)[a]	<1.71 (<1.26)[a]	3264 (733.8)	7.74 (5.71)	30.51 (22.50)
	HX/1516	2.13 (0.084)	1596 (358.8)	2.24 (1.65)	3472 (780.5)	8.49 (6.26)	30.55 (22.53)
	T300/V378A (cloth)	2.76 (0.110)	2243 (504.2)	3.12 (2.30)	3952 (888.4)	10.93 (8.06)	39.32 (29.00)

[a]Estimated from C-scan results.

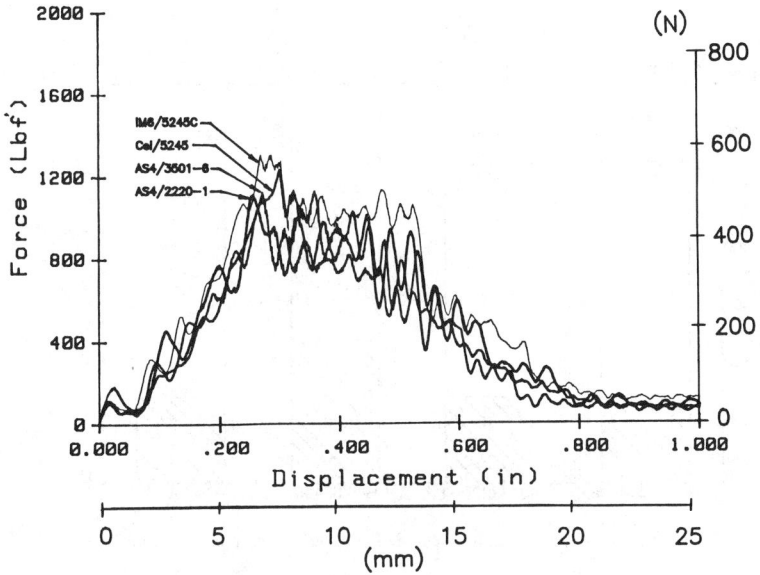

FIG. 3—*Impact force versus displacement for the 394 K (250°F) service materials.*

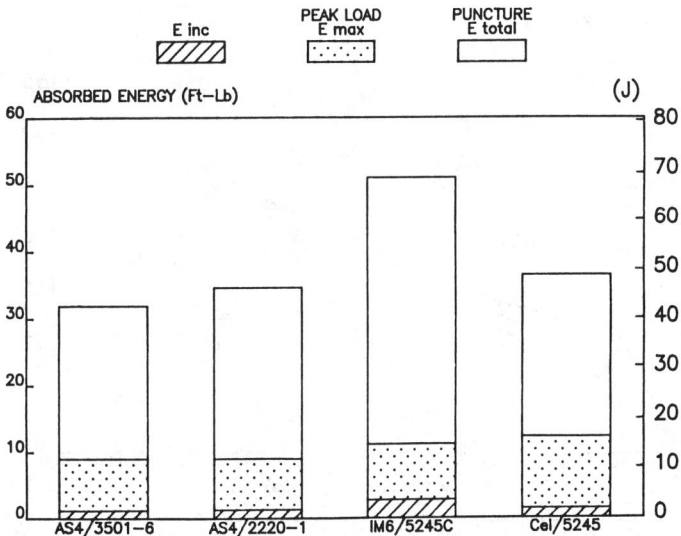

FIG. 4—*Energy results for the 394 K (250°F) service materials.*

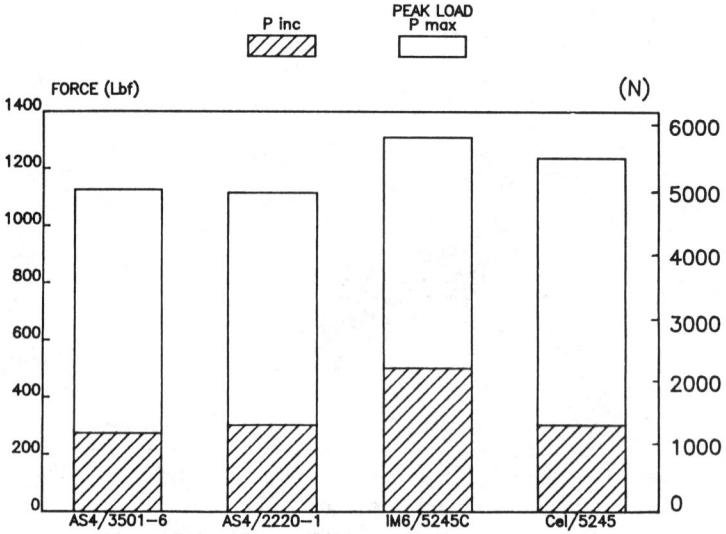

FIG. 5—*Force results for the 394 K (250°F) service materials.*

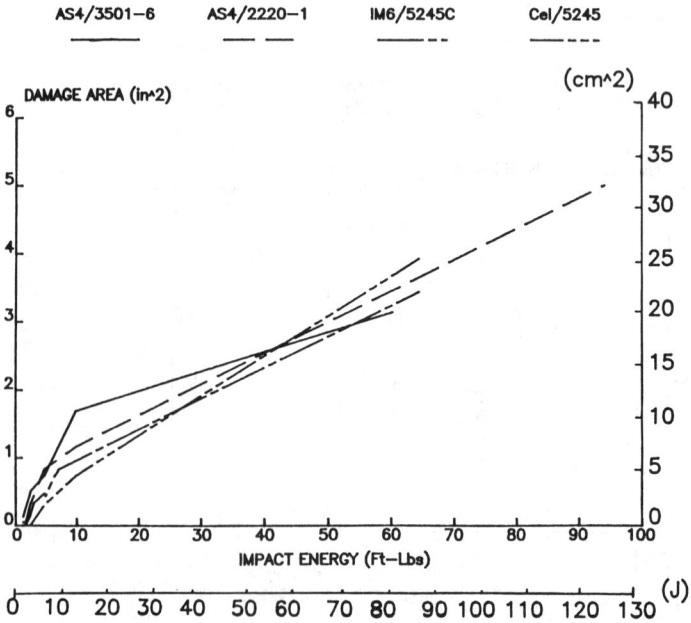

FIG. 6—*C-scan results for the 394 K (250°F) service materials.*

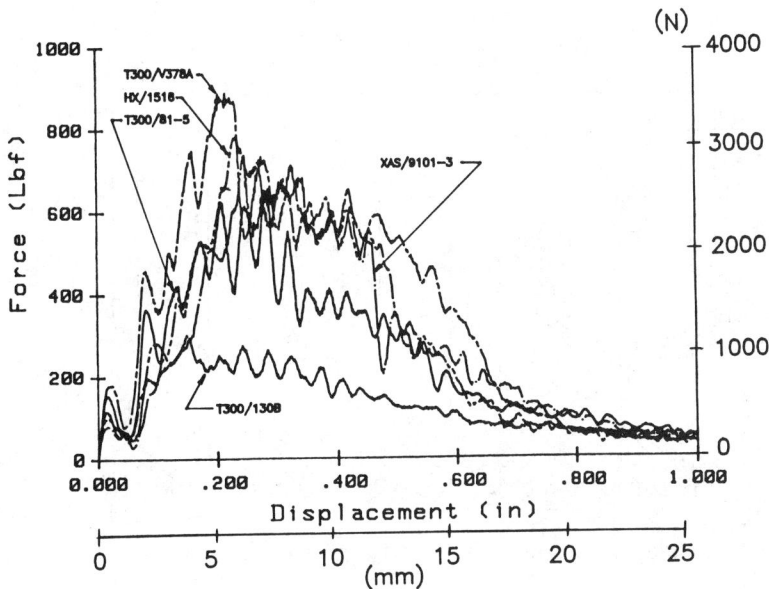

FIG. 7—*Impact force versus displacement for the 450 K (350°F) service materials.*

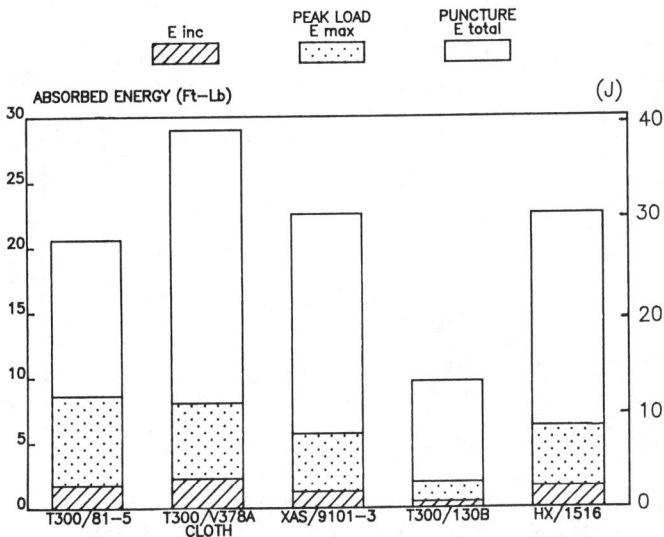

FIG. 8—*Energy results for the 450 K (350°F) service materials.*

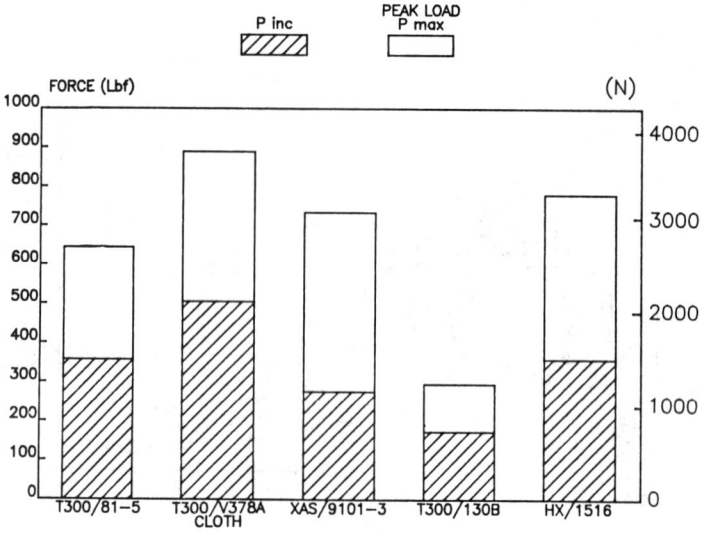

FIG. 9—*Force results for the 450 K (350°F) service materials.*

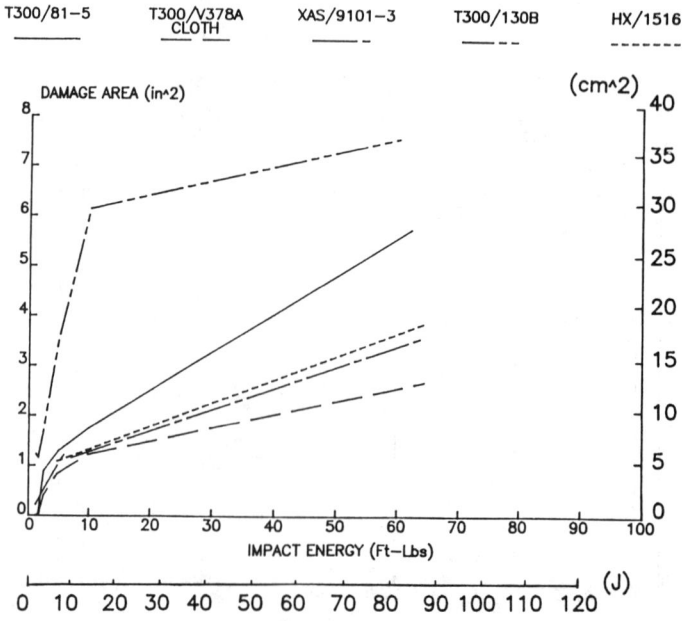

FIG. 10—*C-scan results for the 450 K (350°F) service materials.*

more easily punctured than the latter. Our ranking of the impact resistance of the bismaleimide systems is as follows:

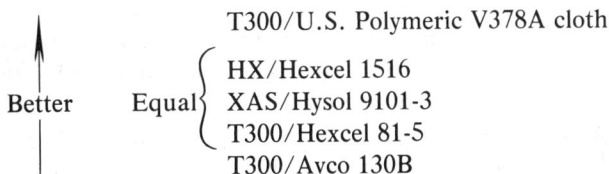

$$
\text{Better} \qquad \text{Equal}
\begin{cases}
& \text{T300/U.S. Polymeric V378A cloth} \\
& \\
& \text{HX/Hexcel 1516} \\
& \text{XAS/Hysol 9101-3} \\
& \text{T300/Hexcel 81-5} \\
& \text{T300/Avco 130B}
\end{cases}
$$

Discussion

It is normally assumed that the initiation of damage in the composite-plate impact specimen causes a reduction in stiffness, which is indicated as a dip in the load-time trace. This point could not always be established from the impact test output because of the presence of higher-frequency oscillations on the load analog signal, which are the result of dynamic interactions between the target plate and crosshead (Fig. 2). Further, the initial damage experienced in the matrix of the composite plate may be so slight as to have a negligible effect on the bending stiffness, and so not appear on the impact test trace. Thus, ultrasonic C-scan inspection was essential to establish the presence of damage and aid in determining the incipient-damage impact level in three of the nine systems tested.

Because of the amount of material needed to fabricate the relatively large test specimens required for drop tower testing, only one impact test was performed per system per energy level. The 152 by 152 by 3-mm (6 by 6 by 1/8-in.) specimen size serves both to simulate typical service support conditions and to suppress the higher-frequency oscillations present on the force analog output by reducing the effective stiffness of the target plate relative to the crosshead. Since the same large amount of scatter can be expected in the impact test results as is typically encountered in composite static testing, this is a serious disadvantage.

In ranking the impact resistance of different materials, high resistance to incipient damage is clearly a desirable trait; so, too, is high maximum load. The maximum load can be interpreted as corresponding to the fiber failure, while the incipient damage load corresponds to matrix failure. How the composite should be ranked based on the total absorbed energy, on the other hand, is not so clear-cut. The total absorbed energy includes the energy absorbed by the composite in the creation of damage. It also includes the energy lost through the various dissipative mechanisms during the impact event. Such examples are damping in the crosshead and within the specimen and the frictional drag between the tup shaft and specimen at the edge of the hole created in the specimen during the puncture test. The authors chose to ignore all these effects except for the creation of damage in the specimen. Here, again, the use of the ultrasonic C-scan inspection was essential in interpreting the impact test results.

Materials that incur large areas of delamination, which could reduce a structure's compressive strength, are less desirable than those materials that confine the damage to a relatively small area. Materials that absorb little total energy but produce a small, clean hole may be desirable in some application, providing they also have high initial damage and peak load values. If the structure is such that holes cannot be tolerated, then having materials that absorb large amounts of energy in the creation of delamination surfaces may be preferable to having the threat of a low-velocity impact penetrating the component. Generally, the best materials are those that absorb the greatest amount of energy for the least amount of damage. Thus, the total absorbed energy is not in itself a useful parameter in ranking material impact responses. The extent of damage must also be known along with the type of damage that the intended structure can best tolerate.

The incipient damage energy level for the AS4/Hercules 3501-6 is 1.6 J (1.2 ft · lb). This represents the baseline level from which to assess the improvements achieved by the newer materials. The IM6/Narmco 5245C, which had the highest incipient damage level of the materials tested here, more than doubles this value with an incipient damage energy value of 3.7 J (2.7 ft · lb). In practical terms, however, a twofold or threefold improvement over the 1.6-J (1.2-ft · lb) level still results in an easily damaged material. The designer must still allow for the possibility of subvisual damage occurring in the structure. Thus, residual strength and fatigue testing of damaged composite specimens is required to assess fully the effects of impact on the structural performance of a composite material system.

Conclusions

The purpose of the instrumented-impact testing performed here was to make direct comparisons between the impact damage resistance of different composite material systems using geometrically identical test specimens. The through-penetration or puncture test, usually the single-impact level investigated with drop-weight impact towers, provides the majority of impact response data but by itself is insufficient to describe the range of composite damage levels likely to be experienced in service. Nor can it be relied on to provide incipient-damage impact data consistently. A number of impact energy levels need to be imposed in concert with an adjunct damage measurement method, such as ultrasonic C-scan, to describe the impact response fully.

References

[1] Poveromo, L., "Composite Matrix Resin System Study for Advanced Aircraft," Report No. NADC-84164-60, Naval Air Development Center, Warminster, PA, April 1984.

[2] Wogulis, S. G., Whitney, B. W., and Ireland, D. R., "Automated Data Acquisition and Analysis for the Instrumented Impact Test," paper presented at ASTM Symposium on Computer Automation of Materials Testing, Philadelphia, PA, 8 Nov. 1978.
[3] Gause, L. W. and Buckley, L. J., "Impact Characterization of New Composite Materials," Report No. NADC-85023-60, Naval Air Development Center, Warminster, PA, March 1985.

Stan Turner[1] *and Peter E. Reed*[1]

Flexed-Plate Impact Testing of Polyether Sulphone: Development of a Notched Specimen

REFERENCE: Turner, S. and Reed, P. E., **"Flexed-Plate Impact Testing of Polyether Sulphone: Development of a Notched Specimen,"** *Instrumented Impact Testing of Plastics and Composite Materials, ASTM STP 936,* S. L. Kessler, G. C. Adams, S. B. Driscoll, and D. R. Ireland, Eds., American Society for Testing and Materials, Philadelphia, 1987, pp. 262–277.

ABSTRACT: In flexed-plate impact tests at room temperature, injection-molded disks of polyether sulphone could not be broken or penetrated by an impactor with an incident energy of 238 J. A central hole 1 mm in diameter was sufficient to embrittle a high proportion of the specimens subjected to the same impact. The failures could be segregated into three classes. A few subsidiary tests on thinned specimens with no hole suggest that in this particular molded form, and under the specific test conditions used, polyether sulphone has a tough-brittle transition within the thickness range of 1.4 to 2.0 mm.

KEY WORDS: flexed-plate impact test, impact testing, polyether sulphone, notched specimens, tough-brittle transition

In the parlance useful for broad generalizations, polyether sulphone (PES) is classifiable as tough but brittle when sharply notched. To be slightly more precise, one might say that, under impact at room temperature, flat molded plaques fail in a ductile manner and, in so doing, absorb a high energy commensurate with the high-tensile-yield stress [manufacturer's published value of 84 MN/m^2 by the ASTM Test for Tensile Properties of Plastics (D 638-82a)]. The sensitivity to notches, though widely recognized, is not well documented: an unnotched Charpy bar does not break under impact flexure

[1]Senior research assistant and senior lecturer, respectively, Department of Materials, Queen Mary College, London, E1 4NS, England.

at 23°C, but the same bar with a notch with a tip radius of 2 mm has an energy to break of about 50 KJ/m^2, and one with a notch with a tip radius of 0.25 mm has an energy to break of less than 5 KJ/m^2—that is, it may be described loosely as brittle. The numerical values depend on the grade of polymer, the type of stock from which the specimen is taken, the nature of the notch (machined or molded), the water content, and the thermal history of 'the specimen. Polysulfone and polycarbonate behave similarly.

The results presented in this paper are for some notched specimens tested by the flexed-plate (or falling weight) method using an instrumented falling weight apparatus (CEAST advanced fractoscope system Mk3) according to International Organization for Standardization (ISO)/Discussion Document (DIS)6603/1. They have been obtained as part of a major program directed toward the exploitation of that method for the meaningful evaluation of the impact resistance of polymeric materials and artifacts made from them. PES is only one of several materials under study and, in fact, has only recently been embraced by the program. The current phase of that program, in which these particular results have been generated, is actually less a study of the notch sensitivity of PES than an exploration of possibilities for the development of a notched version of the test. The lack of such a notched version has hitherto put the flexed-plate impact method at a disadvantage compared with the flexed beam and tensile impact methods whenever some measure of notch sensitivity has been deemed important. The deficiency is a serious drawback to the flexed-plate method, which otherwise provides data more directly relevant to the impact resistance of end products in service than those generated by any other methods, and it provided the incentive for the presently described investigation. Descriptions of the development and application of unnotched flexed-plate testing are increasing in the literature [1-4].

It is neither appropriate nor necessary for the earlier experiments in the provision of a suitable notched specimen to be described here. It suffices to note that ease of preparation of the specimen would be important if the method were to have any chance of becoming widely acceptable and that retention of the cylindrical symmetry of the test geometry is highly desirable, in order to preserve the basic nature of the test method. If one replaces "notch" with "stress concentrating feature," simplicity and cylindrical symmetry can be ensured by a small hole through the specimen. A hole at the center of the specimen was shown to be an effective embrittling agent for high-density polyethylene moldings and similarly, later, for disks molded from a propylene-ethylene copolymer [5]. For those materials, the effects of the hole are very much in line with what would be expected, but the situation is rather less straightforward for PES moldings, although the hole undoubtedly promotes failure. The authors report the results here, even though the work is incomplete, partly because of their intrinsic interest and partly because of the implications that they have for the processes of interpretation of the data from instrumented falling weight impact tests.

Experimental Details

The specimens were injection-molded, edge-gated disks 114 mm (4½ in.) in diameter and nominally 3.2 mm (⅛ in.) thick. The material was an easy flow grade of PES (ICI Victrex PES 3600G). The specimens were freely supported on an annulus of 40-mm internal diameter and 60-mm external diameter and the impactor had a hemispherical tip of 20-mm diameter. Those dimensions conform to the specification in ISO/DIS 6603/1 and are used widely in Continental Western Europe. The apparatus was checked for alignment to ensure that the striker fell centrally inside the specimen support ring. The specimen support system was also fitted with a V-block locating device, each new specimen being placed against this locating block, thus ensuring reproducible and identical relative location for all specimens. This technique served to minimize variation due to misalignment. Under the ISO/DIS 6603/1 specification, the specimens should be either disks of 60-mm diameter or 60-mm squares, but generally in our experiments complete molded disks were tested, although a few small squares were impacted as a check that there were no serious effects attributable to the magnitude of the lateral dimensions. The deviation from recommended practice can be excused, though not strictly justified, on the grounds that it is the usual practice in the United Kingdom at present, although such expediency is slowly giving way to enthusiasm for, and compliance with, international standards. In the context of these particular experiments, it is immaterial whether the specimen was an entire disk or a piece cut from it.

The velocity of the impactor was varied between 1 and 5 m/s during exploratory tests, but most of the experiments were carried out at 5 m/s. At room temperature, an unblemished disk 3.2 mm (⅛ in.) thick cannot be broken or penetrated by the impactor at that speed, which corresponds in this apparatus to an incident energy of 238 J, although, of course, it is severely indented. The force experienced by the impactor exceeds 9500 N, which is the safe working limit of the sensors. If, on the other hand, the disk has a blemish on the tension face directly opposite the point of impact, or nearly so, the behavior can be very different, depending on the degree of stress concentration that has been introduced. A sharp scratch, for instance, induces brittle failure and, in some cases, so does a hole bored through the specimen. Specimens through which a hole has been bored are the focus of attention in this paper. Holes of various diameters between 0.5 and 4 mm were bored on a lathe. Backing plates were used to ensure clean exit of the drill bit, and, in the subsequent tests, the face into which the bit had entered was the one that was tensioned; that face was always the same one in relation to the mold cavity.

Scouting experiments failed to highlight special merit for any particular diameter of the hole, and, in fact, the apparent trends with diameter were confusing. Therefore, on the grounds that the larger holes would tend to reduce the stiffness of the specimen and the smallest hole would be of dubious quality because of flexing in the drill during the boring operation, the main investigation was developed for holes 1 mm in diameter. The overall behavior of specimens with holes of other diameters was very similar, and the results quoted here for the one thickness are of general validity.

The impacting of hard stiff specimens inevitably sets up vibrations in the apparatus and raises the question of whether the electrical signals should be filtered. As a general policy, the authors prefer to work with unfiltered data,

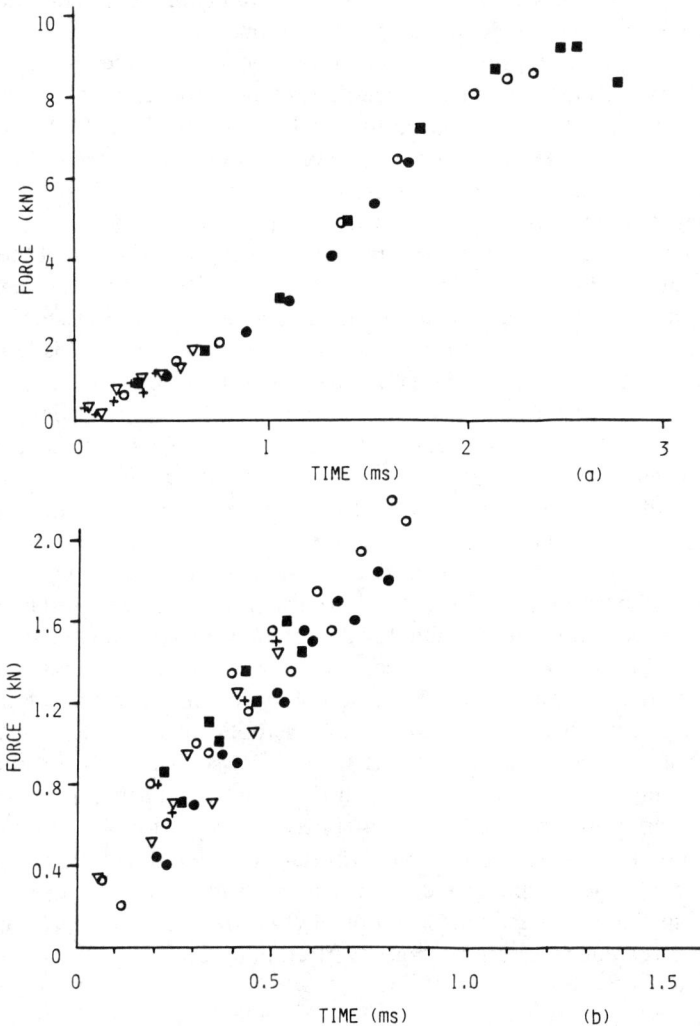

FIG. 1—(a) *Superimposed force-time data for five impacted specimens having a 1-mm-diameter central hole. The data are almost to peak value in each case; there are two brittle and three ductile specimens.* (b) *Early portion of data of Part a replotted on an enlarged scale. The datum points are for maxima and minima of the oscillating force-time signal; the specimen identification symbols are the same in both figures.*

provided the noise is not overwhelming, and most of the results discussed in this paper were unfiltered data. The authors have clear evidence of signal distortion due to injudicious filtering with some other materials. Thus, the force-time curves have a superimposed vibration of a frequency of approximately 9 kHz, which is characteristic of the apparatus. The degree to which the noise intrudes depends on the nature of the fracture; for the tests discussed here, it was never a serious source of confusion in the interpretation of the results, although it contributed to the uncertainty in the measurement of the gradient of the force-deflection and force-time curves.

In this particular program the gradient is of no special concern, except insofar as it might shed light on any variations in the early stages of the fracture process, but a few results are given in Fig. 1a and b to show, first, that the derived modulus agrees with the widely accepted values and, second, that the noise presents only a minor difficulty. In both parts of this figure, force is plotted against time rather than deflection for precision, since such data have been through no numerical processing stage. Any error arising out of the substitution of time for displacement is small over most of the curve and negligible near the origin in these experiments. The data relate to five specimens, each with a central hole 1 mm in diameter: three failed in what is later referred to as a ductile manner, and the other two were brittle. Figure 1a shows the force-time curves up to the peak force, and Figs. 1b shows the first part of the same curves, in greater detail, with the datum points having been extracted from the original signals at successive peaks and troughs.

The broad conclusion to be drawn is that the individual curves superimpose remarkably well, and there is no detectable difference between the curves for the two classes of failure (brittle or ductile) apart from the early termination of those for the brittle specimens. The initial slope of the force-time curve can be used to derive a modulus value for the material by assuming (1) that the impactor velocity remains virtually constant during the early stages of the deformation and (2) that standard elastic theory for the deflection of simply supported, centrally loaded circular plates is applicable. The slope near the origin thus gives a modulus of 2.8 GN/m^2 if Poisson's ratio is assumed to be 0.35, which agrees well with the manufacturer's published value of 2.6 GN/m^2, based on the ASTM Test for Flexural Properties of Unreinforced and Reinforced Plastics and Electrical Insulating Materials (D 790-81), when allowance is made for the difference in straining rate. The modulus is calculated using standard elastic deformation circular flat-plate theory [6], fitting the modulus value to match the force-deflection data.

Results and Discussion

Of 14 specimens, each with a hole 1 mm in diameter, 5 failed in an ostensibly ductile manner, and the rest were brittle. Figure 2a shows a force-time record typical of the ductile failures. The brittle specimens could be subdi-

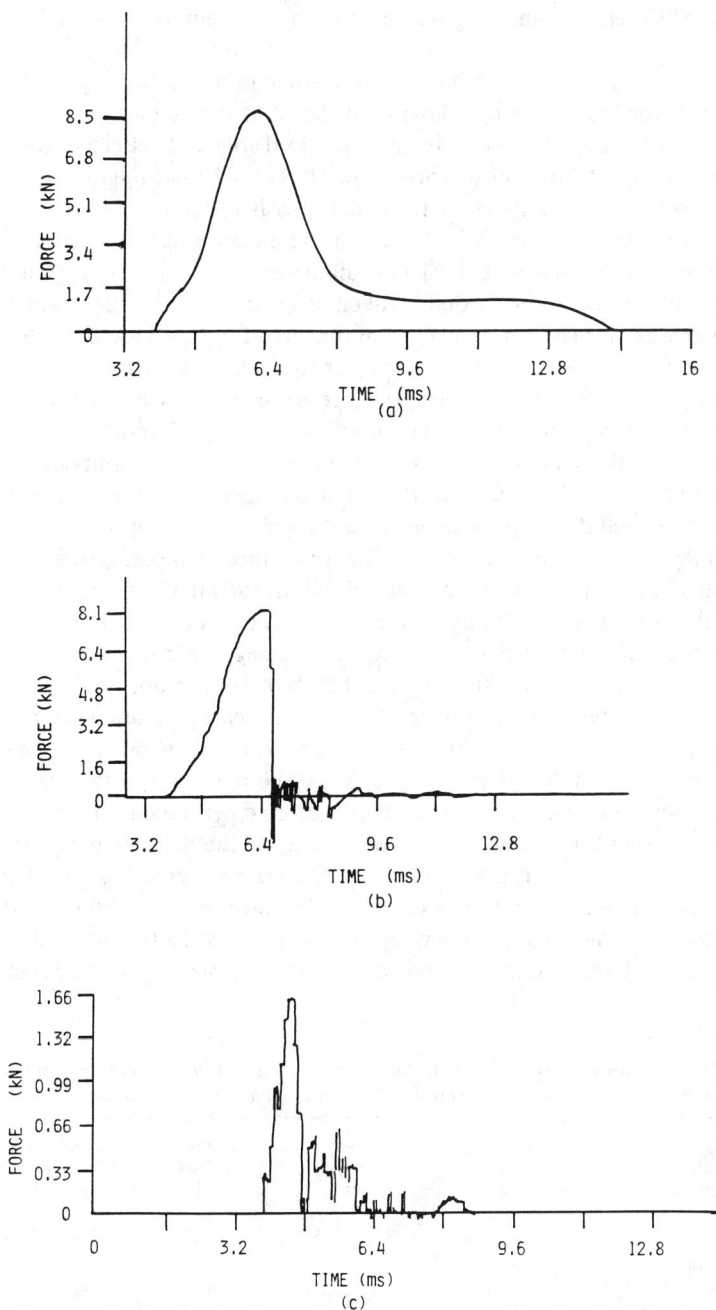

FIG. 2—*Experimental force-time records for* (a) *ductile*, (b) *high-energy brittle, and* (c) *Low-energy brittle failures.*

vided into two classes on the grounds of both the magnitude of the impact energy and the appearance, depending on whether the destructive cracks developed from the edge of the hole or from some other point. Typical force-time curves for the two brittle classes are shown as Fig. 2b and c. Pertinent impact data for the three subsets are given in Table 1. It is clear from the table that the two brittle subsets are very different, whereas the brittle subset characterized by high impact energy is not grossly different from the ductile subset except in the appearance of the impacted specimens. Thus, a statement about the efficiency of the 1-mm-diameter hole as an embrittling feature depends on the criterion that is taken as indicative of brittleness: on the basis of energy absorbed and the form of the force-time (and force-deflection) curves, only 5 of the 14 specimens were brittle, whereas on the basis of appearance, at least 9 were, and even the 5 regarded initially as ductile should possibly also be regarded as brittle, as will now be considered.

In the ostensibly ductile failures, three or sometimes four splits or cracks grew radially from the hole, and the impactor penetrated the specimen by forcing the triangular flaps to bend as cantilevers. The penetrated specimen gripped the impactor very tightly, and that resistance to the progression of the hemispherical tip would have contributed significantly to the energy absorbed during the impact event. The appearance of the faces of the cracks show them to have been brittle (see Fig. 3a), and Fig. 3b suggests that they ceased to grow beyond a certain length more from lack of an appropriate stress field than because of plasticity at the crack tip. There is supporting evidence for the contention that these failures started as brittle ones. At lower impact speeds (and associated lower energies, since the mass of the impactor was held constant), the specimens survived but suffered permanent distortion and other damage in the contact zone. The severity of the damage depended on the incident energy; the lowest energies caused mild creasing (or shallow blunt grooves) along radial lines in the tension face of the specimen, higher energies caused the creases to develop into surface cracks (again on the tension face), and still higher energies caused the surface cracks to penetrate

TABLE 1—*The impact resistance of injection-molded edge-gated disks of polyether sulphone specimens 3.34 mm thick with a central hole 1 mm in diameter (impactor speed 5 m/s).*

Type of Failure	No. of Specimens	Temperature, °C	Peak Force, N[a]	Energy to Peak, J[a]	Total Failure Energy, J[a]
Ductile	5	17.3	8790 (325)	47.2 (2.6)	88.6 (5.2)
High-energy brittle	5	15.8	8290 (1021)	40.1 (11.8)	50.9 (22.0)
Low-energy brittle	4	16.9	1690 (768)	3.1 (2.3)	4.5 (1.8)

[a]Standard deviations are in parentheses.

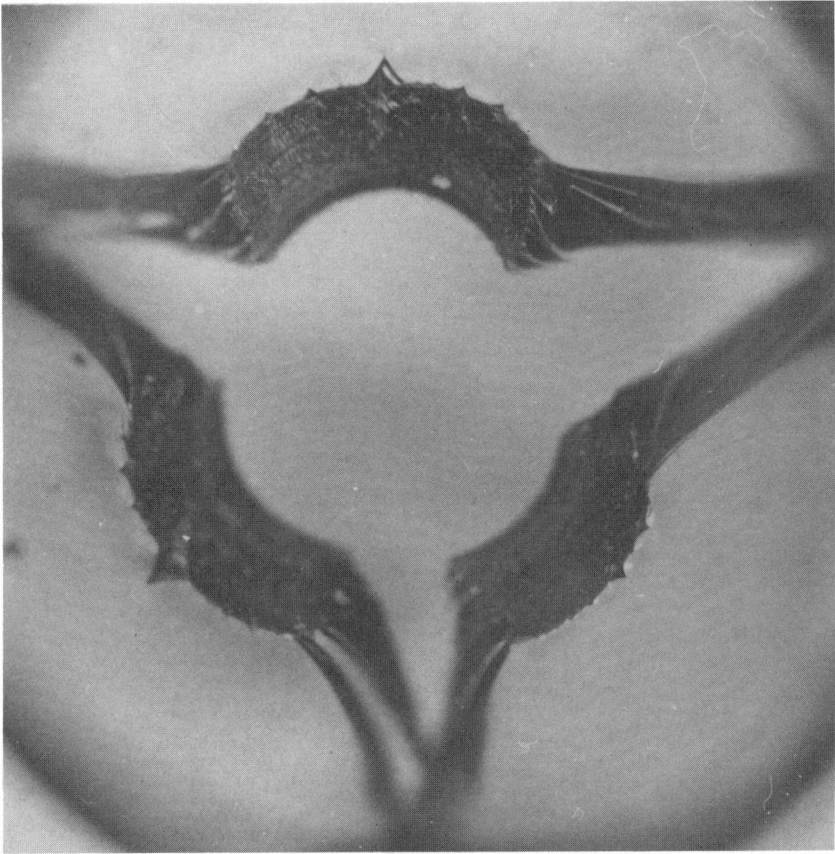

FIG. 3a—*Ductile failure—radial splits in the vicinity of the central hole, showing brittle characteristics on the surface of the split.*

through the specimen. By that stage most but not all of the cracks appear to have started at the edge of the hole, though many of them may not have done so. If a cracked specimen is subsequently given a second, or possibly a third, blow of the same energy as the priming blow, the specimen usually breaks. Further supporting evidence is that when the early part of the force-time curves for the ductile subset are expanded so as to be visually comparable with the force-time curves of the low-energy-brittle subset, they are found to be very similar in character; the brittle specimens and the ostensibly ductile ones apparently excite similar levels of vibration in the impactor. This was mentioned earlier in the different context of Fig. 1b.

In contrast to the ambiguously ductile specimens, the low-energy-brittle ones behaved very straightforwardly. Cracks initiated at, or close to, the hole

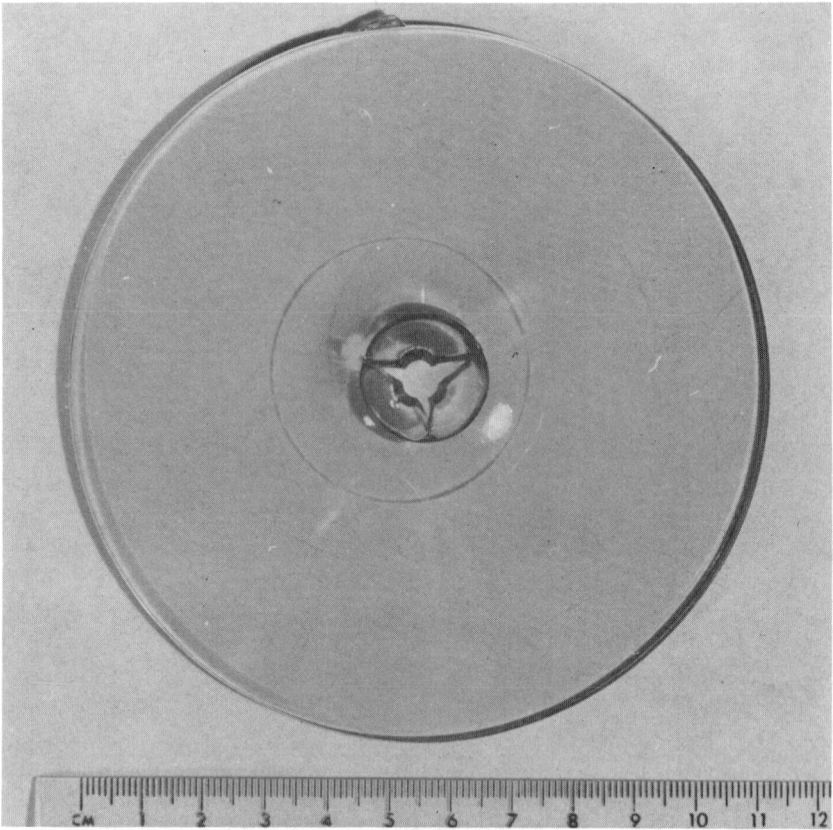

FIG. 3b—*Complete specimen detailed in Fig. 3a, showing typical ductile failure.*

and progressed radially with some branching but little of the fragmentation that is a feature of the high-energy-brittle specimens. A typical broken specimen is shown in Fig. 4. In such cases it is clearly correct to claim that the hole has induced embrittlement. The dominant fractures in the high-energy-brittle specimens did not start at or near the hole. The splits developed as described for the ductile failures, but before they could grow sufficiently for the impactor to penetrate the specimen, brittle failure initiated at some point near the support ring and grew along an approximately circular path as a highly splintered fracture (see Fig. 5).

Failures such as these and their associated force-time or force-deflection curves strikingly demonstrate a limitation of the flexed-plate method that tends to be obscured by the fact that the force-time or force-deflection curves are very similar in shape to the stress-strain curves of plastics materials; the

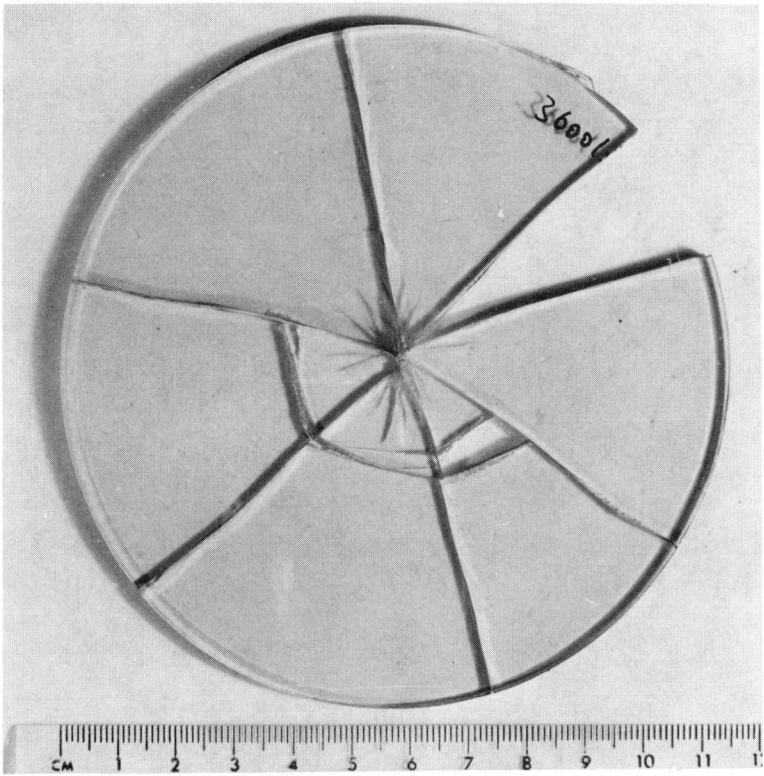

FIG. 4—*Brittle fracture initiated at the central hole.*

force recorded is that which arises where the impactor contacts the specimen, and the stress at the fracture site has to be inferred on the basis of various assumptions and deductive paths. Thus, the similarity between the loading portion of curves associated with ductile failure and those associated with high-energy-brittle failure is largely fortuitous, and one can conclude from the data merely that a force applied transversely at the center of the plate can cause either splitting and penetration or brittle failure near the support. It follows, of course, that brittle failure initiating nearer the center would entail a smaller force at the center. By the same argument, the designation of the brittle subsets as "high-energy-brittle" and "low-energy-brittle" must be recognized as an empirical convenience rather than a physically meaningful discrimination.

Since elastic analysis for plate deformation proved satisfactory for evaluation of the modulus during the initial stages of the loading, elastic analysis for the stress distribution in plates has been explored to see if this provides any

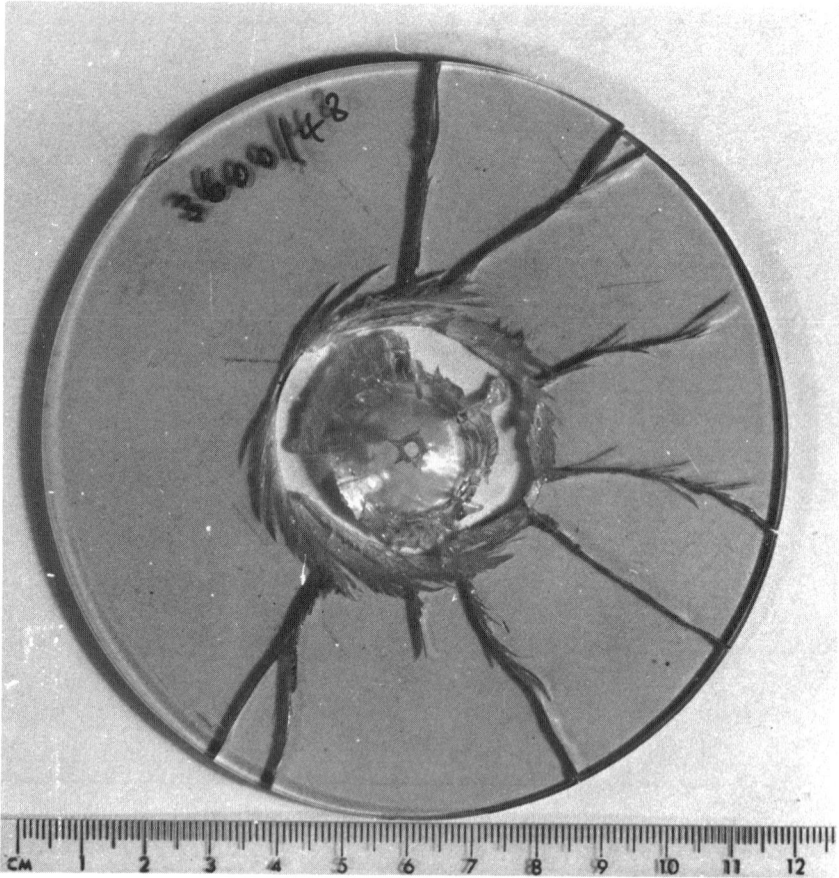

FIG. 5—*Brittle fracture with eventual failure initiated away from the hole.*

explanation of the failure mode phenomena. Figure 6a shows the stress distribution for unit force in a 3.34-mm thick, centrally loaded plate containing a 1-mm-diameter central hole. The analysis [7] assumes that the plate is unrestrained at its periphery. The circumferential stress, σ_θ, exceeds the radial stress, σ_r, at every position in the plate. Such a stress distribution clearly favors the propagation of radial cracks, driven by the larger circumferential stress, initiating at the central hole. The inclusion of edge restraint on the specimen due to either imposed clamping or excess material outside the support ring superimposes a radial bending moment. This serves to increase the radial stresses across the plate without modifying the circumferential stresses. Figure 6b shows the elastic analysis for the same case as in Fig. 6a, but with full clamping at the outer edge. Clamping serves to suppress the stresses in

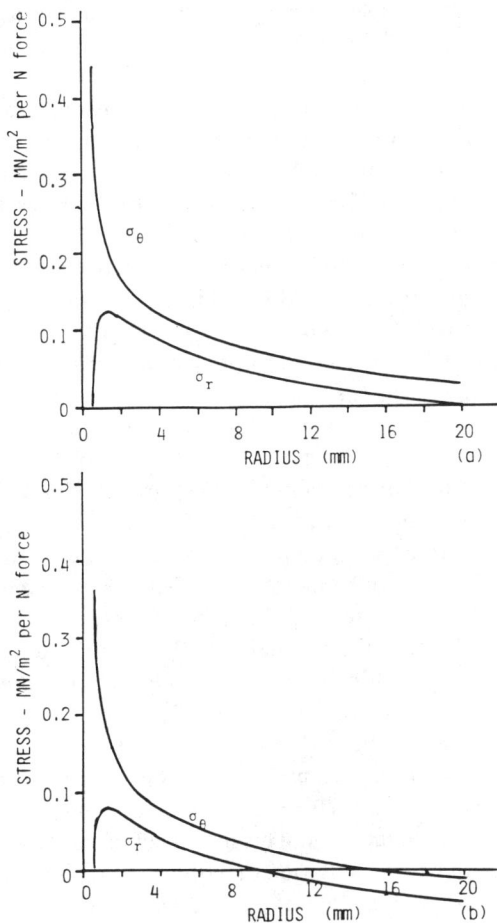

FIG. 6—*Theoretical radial and circumferential elastic stresses in a centrally loaded 3.34-mm-thick, 40-mm-diameter plate containing a 1-mm central hole: (a) outer edge unrestrained; (b) outer edge restrained by radial bending moment for full clamping.*

the plate for a given applied central load and alters the tensile surface to the struck side close to the clamping ring. The radial stress exceeds the circumferential stress at radii greater than 12 mm, although these values are small compared with those near the hole, where the circumferential stress still dominates.

The stress values given in Fig. 6 are for a force value of one at the center. During any impact test, the force increases with time to a peak value. The elastic analysis thus predicts that the stresses throughout the plate would increase pro-rata to the central force but maintain the same form of distribution. Typical peak force values for a plate of the dimensions assumed for Fig. 6 would be in the region of 1600 N for a low-energy brittle failure. Based on the elastic analysis, this would suggest a circumferential stress of 710 MN/m² at the surface of the specimen in the region of the central hole, or eight times the quoted tensile yield stress for the material. Considering the assumptions involved in the analysis and the disregard of inertia effects, a factor of eight is not particularly outlandish. The high circumferential stresses in the region of the hole could be the origin of the creases previously discussed, which form radially on the tensile face of the specimen. The creases could then be viewed as localized drawing at the skin of the specimen. Subsequent penetration of the striker into the specimen could result in plastic deformation in the region of the striker tip, which would render application of the elastic theory inappropriate in this zone at higher forces.

Figure 6*b* shows that the stresses in the outer regions of the plate are only about one tenth of the maximum stress at the hole according to elastic theory. Thus, if the force at the center is only just sufficient to cause yielding at the hole under σ_θ, the outer regions will still be deforming elastically. The effect of edge restraint serves to raise σ_r above σ_θ toward the periphery of the plate. Thus, if the central force can be increased sufficiently to bring the stresses in the outer regions of the plate toward the failure point, circumferential cracking could be favored rather than radial cracking in this zone. Higher forces occur with high-energy brittle behavior, in which circumferential cracking becomes a feature. The authors thus tentatively suggest that failure initiates in such cases at low loads in the region of the hole under the action of the high σ_θ value at the hole, but subsequent plastic deformation in the region of the striker prevents the failure zones from developing and allows the central resistive force to penetration to increase. This increased force causes the stresses in the peripheral regions to rise, and even though the actual edge constraint does not necessarily equate with a state of perfect clamping, the stresses may become such that secondary circumferential fracture is possible.

The "quality" of the hole did not appear to influence the fracture behavior. No special care was taken in preparation of the holes, but none of them was grossly imperfect. There was a slight birefringence associated with some of the holes, but there was no correlation between such strain and the incidence of brittleness. On the other hand, small nicks from a scalpel in the leading edge of the hole (that is, in the tensioned face) or light abrasion in that region invariably promoted brittle failure, as did surface notches (nicks) in specimens with no hole. In such cases the peak forces and the energies were significantly smaller than those quoted for the low-energy brittle subset in Table 1. These subsidiary results are in keeping with the PES reputation for notch sensitivity; they imply, however, that the hole, irrespective of its "quality," is not a particularly effective stress concentrator or stress intensifier under the particular test conditions adopted in these experiments. In contrast to that, an ethylene-propylene copolymer injection molded into disks of the same diameter and 3.18 mm thick was very effectively embrittled at an impact velocity of 3 m/s by a hole 2 mm in diameter.

It is obvious that the hole cannot have been directly influential in those cases in which the destructive fracture started at some point far from the impact site. Possibly the direct cause was some weakening inclusion, even though the moldings were apparently of very high quality and free from any overt defects. It follows that one should expect a proportion of brittle fractures in specimens without a hole, provided that they are impacted with sufficient energy. The safe working limit on the sensors currently in use on the equipment precluded a direct check on that, but a few moldings had material machined from one face down to various final thicknesses and were then impacted with the remaining molded surface in tension.

The consequences are set out in Table 2. It is quite clear, despite the high interspecimen variability evinced by the data in Table 1, that ordinary specimens of PES (that is, unnotched ones) may fail by brittle fracture rather than by ductile penetration when impacted at room temperature. This experimental fact is at variance with the perception of PES as tough but brittle when sharply notched, and it may be that the material should instead be regarded as strong but occasionally brittle. Many more data would be needed before such a suggestion could be rigorously supported because of the interspecimen variability, to which reference has already been made.

Insofar as Table 2 may be tentative evidence of a thickness effect *per se*, it is worth noting that the thinnest specimen failed rather differently from any other specimen. It was the only penetration failure out of dozens in which the mechanism approximates ductile tearing rather than splitting. The diameter of the flap was slightly less than 20 mm (the striker tip diameter), the tear progressed more than three fourths of the way around the circumference, and a shear failure along that periphery would have required a force of about

TABLE 2—*The mode of failure of thinned specimens, edge-gated disks thinned from 3.34 mm by removal of material from one face (impact velocity, 5 m/s).*

Thickness, mm	Mode of Fracture
3.36 (unthinned disk)	did not break
3.07	dome of biaxially drawn material formed under impact nose; penetration of impactor through diametral split across dome
2.55	formation of dome, major split and a minor one, brittle fracture initiated near support ring, and central section driven out intact
2.02	brittle fracture initiated at or near center
1.44	brittle fracture not initiated at center; central section driven out intact
1.37	ductile failure along a circumferential path; penetration of impactor nose through the circular flap so formed

3100 N, based on the quoted yield stress values. The peak force recorded was 4050 N, which is hardly close agreement, but, on the other hand, the assumptions for the simple calculation were crude and made no allowance for "cup drawing" distortion near the point of impact. This evidence of genuine ductile failure is tenuous, and the results in Table 2 will have to be extended before firm conclusions can be drawn, but, on balance, it seems possible that, at an impact speed of 5 m/s and about 20°C, PES has a tough-brittle transition in the thickness range of 1 to 2 mm.

Conclusions

Investigations of the impact resistance of a material, or of objects made from it, tend to require the testing of many specimens because the facts can be partially obscured by interspecimen variability. Experience shows that the coefficient of variation for a set of specimens breaking in a ductile manner is likely to lie between 5 and 10% and that, for a set breaking in a brittle manner, it will be twice as large; hence, inferences drawn from results on small samples can be erroneous. The relative paucity of the results published in this paper is not a deliberate flaunting of statistical principles; it arises, rather, because the program was conceived as an exploration of possibilities rather than as a mapping of certainties, and also because many of the pathfinding details have been omitted.

The conclusions to be drawn from the results fall into two groups relating, respectively, to the test method and to PES. They all relate to the one specific test geometry, an impact speed of 5 m/s, and an ambient temperature of between 15 and 20°C.

In relation to the test method, the results indicate the following:

1. A central hole induces failure in specimens that would not otherwise fail under the imposed conditions.

2. The mechanism of failure induction corresponds to that of a conventional notch (that is, by crack initiation at the stress concentrator) in only about one third of the specimens; in the others, the influence of the hole is indirect. Therefore, for this particular material the hole is not particularly effective; it corresponds in its effect to a blunt notch.

3. An assessment based solely on the force-time relationship (or on other data derived from it) could be misleading; cognizance must always be taken of the nature of the fracture and its location.

4. The peak force does not necessarily mark the point at which significant levels of damage develop; it merely marks the point at which the damage dominates the situation. The peak force may reflect only the onset of instability of the test piece as a whole rather than the initiation of failure in the material at some point.

In relation to the molding grade of PES, the results indicate the following:

1. PES plates are not overwhelmingly embrittled by a bored hole at the point of impact. They are embrittled by a surface scratch.

2. PES plates are not always tough when unnotched. They should perhaps be described as "strong and often tough."

3. There is some evidence from the current work that unblemished plates are tough if they are thin, say below 1.4 mm thick, and brittle if they are thicker than, say, 2 mm. However this conclusion requires further testing, and extrapolation of findings in the area of impact testing can be misleading.

Acknowledgments

The authors gratefully acknowledge the financial support by the Polymer Engineering Directorate and the provision of moldings by the Petrochemicals & Plastics Division of Imperial Chemicals Industries PLC.

References

[1] Sjoerdsma, S. D., Boyens, J. P. H., and Mooij, J. J., "Shear Deformation under Impact Conditions," *Polymer Engineering and Science*, Vol. 25, 1985, p. 250.

[2] Wnuk, A. J., Ward, T. C., and McGrath, J. E., "Design and Application of an Instrumented Failing Weight Impact Tester," *Polymer Engineering and Science*, Vol. 21, 1981, p. 313.

[3] Reed, P. E. and Turner, S., "Flexed Plate Impact Testing II—The Behaviour of Toughened Polystyrene," *Plastics and Rubber Processing and Application*, Vol. 5, 1985, p. 109.

[4] Mooij, J. J., "Instrumented Flat-Headed Falling Dart Test," *Polymer Testing*, Vol. 2, 1981, p. 69.

[5] Turner, S. and Reed, P. E., "Flexed Plate Impact Testing III—Results for a Polypropylene and Some Experiments with Notched Specimens," *Kunststoffe*, 23 Nov. 1985, in press.

[6] Roark, R. J. and Young, W. C., *Formulas for Stress and Strain*, McGraw Hill, Kogakusha, Tokyo, 1975.

[7] Timoshenko, S. P. and Wolnowsky-Kreiger, S., *Theory of Plates and Shells*, 2nd ed., McGraw Hill, New York, 1959.

Partial Impact Testing and Fatigue Response of Plastics

George C. Adams[1]

Impact Fatigue of Polymers Using an Instrumented Drop Tower Device

REFERENCE: Adams, G. C., "Impact Fatigue of Polymers Using an Instrumented Drop Tower Device," *Instrumented Impact Testing of Plastics and Composite Materials, ASTM STP 936,* S. L. Kessler, G. C. Adams, S. B. Driscoll, and D. R. Ireland, Eds., American Society for Testing and Materials, Philadelphia, 1987, pp. 281–301.

ABSTRACT: Using low and constant-impact energy, a relatively simple method of impact fatigue has been devised to provide toughness measurements on polymers. Crystalline polymers, particularly the nylons and polyacetal, appear to have better fatigue performance than noncrystalline polymers. Fatigue curves that are normalized with the fracture area can potentially detect the influence of crystalline morphology on fatigue. Using retained energy values and fracture area measurements from each impact, an estimate of the fracture energy is obtained. The higher fracture energy values obtained in fatigue, over those obtained in single-blow impacts, indicate that different energy absorption processes occur in single- and multiple-impact testing.

KEY WORDS: impact testing, fatigue, nylon, acrylonitrile-butadiene-styrene (ABS), high-impact polystyrene (HIPS), Charpy test, fracture energy

The toughness of polymers has been extensively studied by testing to bring about immediate failure, most often in a single-impact blow. However, there is increasing use today of polymers in such applications as hinges, gears, springs, and automated arms, which have made fatigue performance of increasing concern. Consequently, a fatigue test has been developed using repeated impacting with a known energy, called the impact energy. The method is an extension of the single-blow impact test on which is based a sizable understanding of material toughness.

Fatigue testing of polymers is not nearly as advanced as that for metals. Because the phenomenon of fatigue of polymers has received relatively little fundamental attention, it is one of the unsolved problems of polymer science. Recent reviews of polymeric fatigue include treatises by Beardmore and Ra-

[1]Senior research chemist, E. I. du Pont de Nemours & Co., Inc., Polymer Products Department, Experimental Station, Wilmington, DE 19898.

binowitz [1], Schultz [2], and Owen [3]. Hertzberg and Manson have reviewed the subject with emphasis on molecular structure and composition [4]. The latter workers made use of load and strain control fatigue data providing analysis by fracture mechanics principles to give the stress intensity factor, K. A large number of polymers have been so characterized including nylon 66 [5], polyethylene terephthalate (PET) [6], polyacetal [7], acrylonitrile-butadiene-styrene (ABS) [8], and high-impact polystyrene (HIPS). This conventional fatigue technique is in contrast to that for constant-impact energy fatigue, which is the subject of this paper.

Impact fatigue requires using a constant-impact energy device for repeated impacts. There has been very little reported work on the subject of impact fatigue. Ultrahigh-molecular-weight polyethylene was studied in the three-point bend geometry by Bhateja [9] using fewer than 20 impact cycles. Polycarbonate was studied by Ohishi in repeated impact giving rise to S-N curves [10]. Biaxially oriented polycarbonate film was studied by Takemori [11] to obtain fatigue crack propagation curves. These and other limited efforts [12], which result in destruction of the specimen in only a few impacts, indicate the need for a greater sophistication in an important area of mechanical behavior.

This work shows that a repeated-impact energy method merits consideration. Different material responses of a phenomenological nature emerge in fatigue loading that are entirely bypassed in single-blow impacts. Indeed, the author has found an advantageous effect of crystallinity in increasing fatigue life. Second, failure due to adventitious flaws is more evident under less severe loading conditions. These effects are related to the linear load-displacement response during each loading cycle.

These preliminary remarks show that the technique of impact fatigue is energy based and related to the single-blow (Izod and Charpy) toughness technique on which our nonfatigue knowledge is founded. An obvious gap, and the subject of this paper, is the study of fatigue by an energy-related technique.

Experimental Procedure

Equipment

The design of the instrumented impact assembly originates with Zoller [13] and has been further described by the author [14-16]. An overview is shown in Fig. 1. Because of the ease of adjusting the drop height, a drop tower is preferred over a pendulum device. It has been determined that identical break energies are obtained [14-16]. Since there must be no energy sinks in the apparatus, the design of Bluhm [17] is followed, and a 45-kg steel block serves to make the impact area rigid. Impact energies from 0.01 to 27 J are obtained using a 1-m-long drop tower and drop weights ranging from 300 to

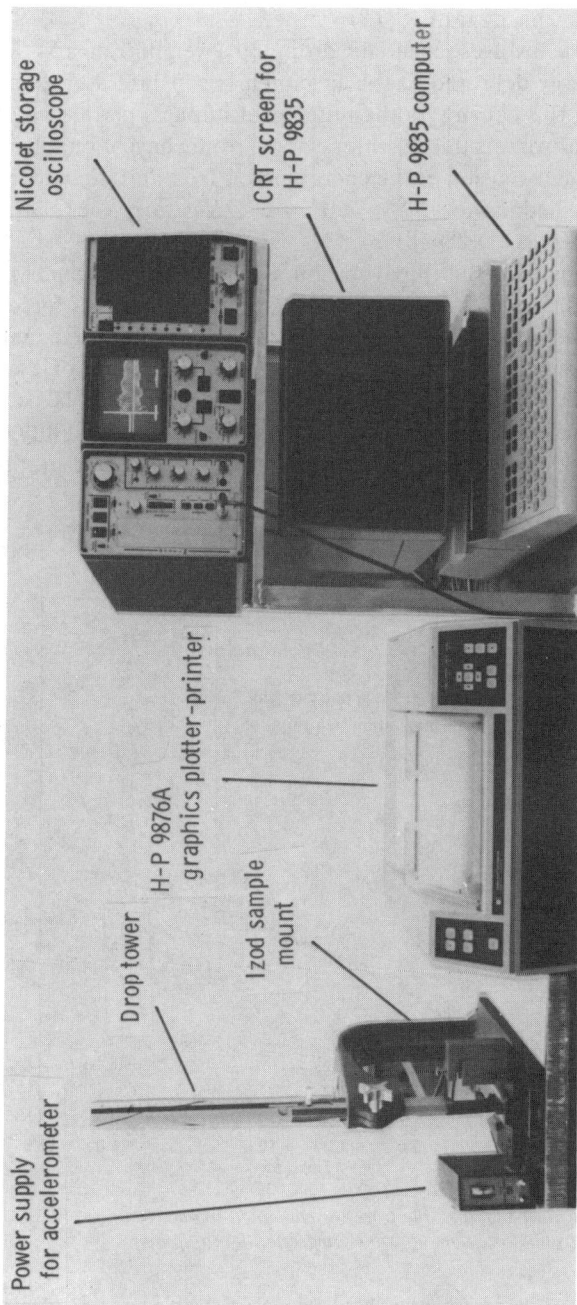

FIG. 1—*Overview of the instrumented impact apparatus.*

2800 g. In fatigue work the high values are seldom used. The details of equipment construction have been developed with the aid of reference to a similar effort in metals Charpy testing [*18*].

Impact fatigue involves repeatedly dropping a weight from a known height, so that the energy delivered to the specimen is constant and the maximum load generated will depend on the number of impacts performed. This constant energy feature sets this technique apart from conventional fatigue tests, but it is a logical extension of the energy-based Izod and Charpy techniques, which are described in the *ASTM Test for D-C Resistance or Conductance of Insulating Materials* (D 256–81).

An accelerometer is the signal-sensing device in this system. Mounted at the top of the drop weight, its use enables the calculation of force, velocity, displacement, and energy as a function of time. The millivolt signal from the accelerometer is converted to acceleration through the use of a calibration constant provided by the manufacturer (PCB Piezotronics). This makes all the quantities calculated from this signal absolute. The reliability of the displacement values has been checked by high-speed cinematography on impact, and Fig. 2 shows the comparison obtained.[2]

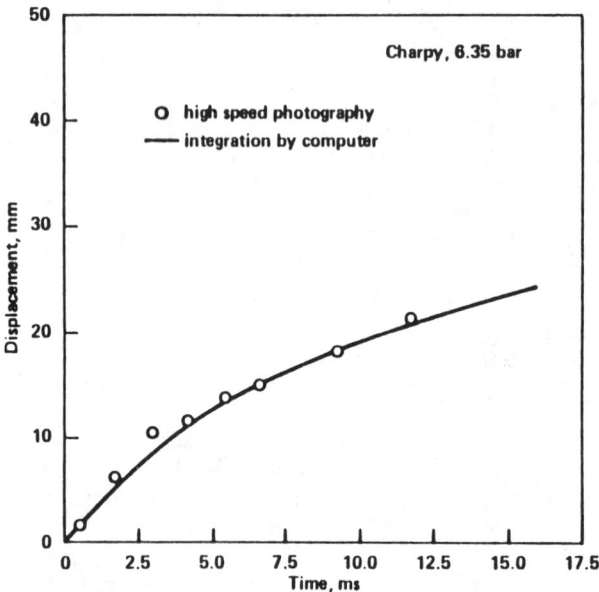

FIG. 2—*Displacement versus time from instrumented impact testing. The symbols indicate high-speed photography; the line indicates integration by computer.*

[2]Since this writing, the accelerometer has been replaced with a force transducer, which does not require any data smoothing.

The output from the accelerometer, which is proportional to acceleration, is first sent through a Krohn-Hite electronic variable filter, which removes signal above a preset frequency. Noise in the impact signal originates from specimen oscillation at a natural frequency based on the span length, as described by Williams [19]. After filtering, the signal is digitized and stored as up to 4096 discrete data points on a Nicolet Model 2090 oscilloscope. Impacts last from 0.5 to 50 ms and may be recorded through a selectable time interval setting between data points. The signal is sent on command to an HP 9836 microcomputer, which performs the necessary calculations. Multiplication of acceleration by the mass of the tup gives the force as a function of time of impact. Successive integration of the acceleration signal gives the velocity and displacement as a function of time of impact. The result is point-by-point tabulated results and a force versus displacement curve. Its accessibility to programming software during experimentation makes this design particularly attractive.

Sample Preparation

The fatigue work utilizes a Charpy or three-point bend geometry. The design follows that of ASTM Tests D 256-81, including a support span of 95.2 mm and specimen dimensions of 12.7 by 6.3 mm. In this initial effort, the interest is primarily on the energy required to propagate an existing crack through the cross section. Therefore, all specimens were notched in the 6.3-mm face to a depth of 1 mm by a sharp razor blade. For overhead impact, the notch is mounted on the underside. Commercial materials used in this study are identified in Table 1. Rubber-toughened Nylon A is an experimental toughened nylon. All the results of this work except for Figs. 6 and 8 are based on this material. Figures 6 and 8 are based on fatigue data for experi-

TABLE 1—*Sample designations.*

Name	Designation	Manufacturer
Rubber-toughened Nylon A	experimental toughened nylon resin	Du Pont
Rubber-toughened Nylon B	Zytel ST801 commercial nylon resin	Du Pont
Rubber-toughened acetal	Delrin 100ST acetal resin	Du Pont
ABS	Lustran 640	Monsanto
High-impact polystyrene	Lustrex 4300	Monsanto
Nylon 66	Zytel 101	Du Pont

mental (rubber-toughened Nylon A) and commercial (rubber-toughened Nylon B) nylons. As stated in Table 1, the latter is designated Zytel ST801 nylon resin.

Procedure

In this initial fatigue study, specimens mounted in the three-point bend configuration have been impacted with as low as 50% of the energy required to bring about failure in a single impact. The energy is sufficient to cause failure in less than 200 impacts. For ductile materials the deformation is clearly not linear, whereas for brittle materials linear loading is obtained.

Repeated impacting has been accomplished using manual techniques. That is, on release of the tup from a known height and after observing a bounce, the tup is caught manually to prevent impacts from lower and unknown drop heights. For a constant-impact energy the tup is repeatedly dropped from the specified drop height until the specimen fails. This is followed by selection of a different impact energy and repeated impacting to failure with a new specimen.

Results of Fatigue Studies

Single Impacts

Impacts that cause failure in one blow are necessary in order to determine conditions for repeated impacting. The analysis of single-blow impacts is described as follows. It is through the force-displacement curve that the impact event is portrayed. In single-blow impacts, this curve consists of a crack-initiation energy represented by the area under the curve and to the left of a vertical line drawn from the maximum force. The propagation energy is represented by the area under the curve and to the right of the maximum force value. The sum of these energies is energy absorbed in fracture, called energy-to-break. In single-blow tests, impact energies at least as great as this are required for failure. For brittle failure of nylon 66, a single blow causes unstable growth and failure, as shown in Fig. 3. There is a linear loading portion and low propagation energy. Failure occurs catastrophically once maximum load is attained. For this 12.7 by 6.3-mm cross section, 0.57 J is absorbed by the specimen in failure.

Ductile fracture, on the other hand, involves much more, as can be seen in Fig. 3. An impact energy of 7.6 J gives initiation and propagation energies of 1.9 and 2.8 J. The total energy absorbed in a fracture of 4.7 J involves 60% propagation energy, which comes from a transfer of energy from the drop weight during this phase of fracture. Toughening greatly alters both the qualitative and quantitative aspects of fracture.

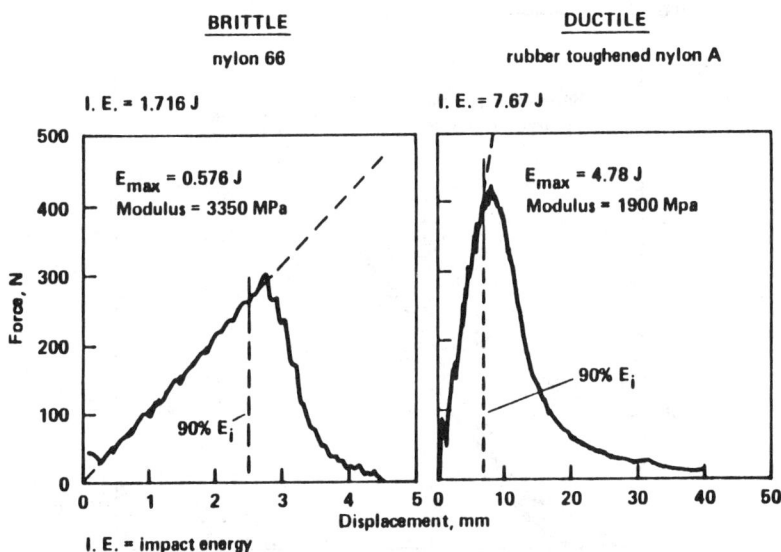

FIG. 3—*Force-displacement curves for* (left) *brittle fracture in nylon 66 and* (right) *ductile fracture in rubber-toughened Nylon A.*

Repeated Impacts

Repeated impacts are accomplished at selected energies less than that required to initiate a crack—hence, the need for single-blow impacts. Using an impact energy of 50 to 98% of the energy to maximum force, repeated impacts are performed until specimen failure. The results for nylon 66 are shown in Fig. 4. The bounce of the tup after each impact is shown by the decreasing force and displacement from some maximum value. For this material there is a retrace of the force–displacement curve for each impact until the failure impact. Figure 4 is an example of the retrace phenomenon, which involves 146 impacts to failure of nylon 66.

The greatly different fracture behavior for rubber-toughened Nylon A is evident in the force–displacement curves in Fig. 5. The force-displacement curve traces out a loop. The energy represented by the area inside the loop is the absorbed or retained energy for that impact. As expected, the retained energy increases with the greater number of impacts, and the curve deteriorates to a failure curve at low force and high displacement. The data that follow will show that the accumulated retained energy for all impacts is many times greater than the absorbed energy in a single impact to break.

In ductile fracture the absorbed energy per impact results in gradual crack movement through the specimen cross section. This is in contrast with brittle fracture in which a crack neither develops nor moves during impacting, as

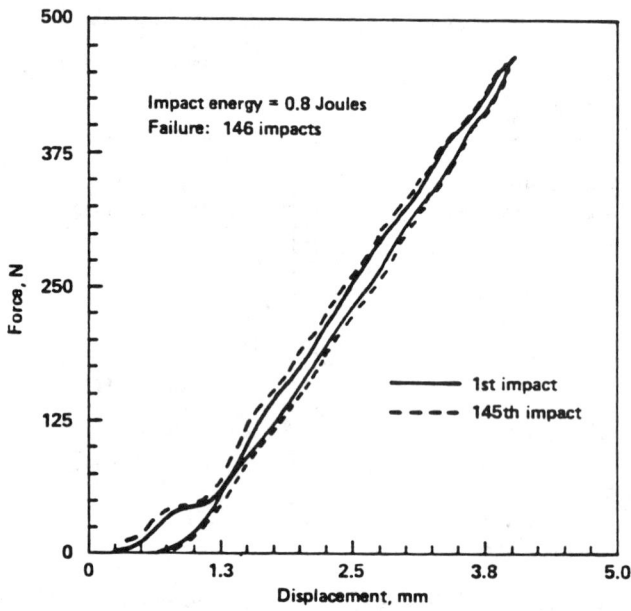

FIG. 4—*Force–displacement curve for nylon 66 obtained in impact fatigue.*

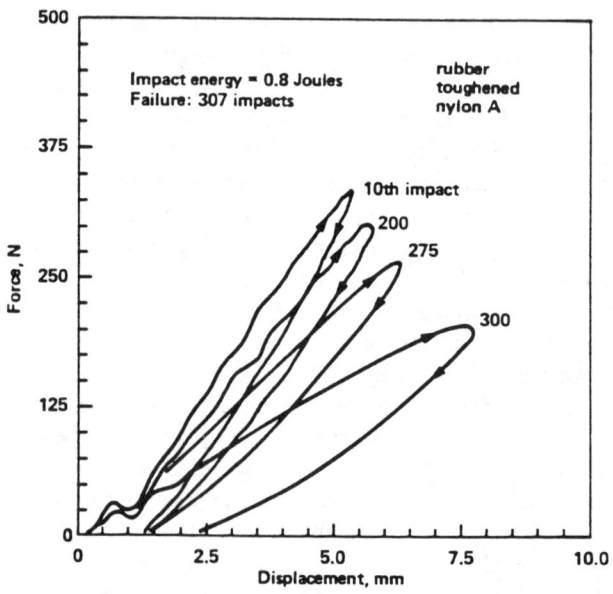

FIG. 5—*Force–displacement curve for rubber-toughened Nylon A obtained in impact fatigue.*

testified by an identical force maximum and absorbed energy for each impact. The repeated-impact test further emphasizes the contrasting natures of brittle and ductile fractures.

Fatigue Curves

Use of a single-impact energy is not sufficient to characterize a material in impact fatigue. Repeated impacts to failure are accomplished at several impact energies, each on a new specimen. The number of impacts to failure will increase as the impact energy decreases, and a plot of these quantities is the simplest means of displaying the fatigue behavior of a given material. For nylon 66 and rubber-toughened Nylons A and B, the fatigue curves are given in Fig. 6. Some of the utility of impact fatigue becomes evident from this figure.

Fatigue failure of nylon 66 is quite sensitive to the impact energy. A decrease of as little as 0.2 J in impact energy from the single-blow value of 0.8 J results in a large increase in the number of impacts to failure. This can be considered characteristic of brittle fracture. The toughened Nylons A and B,

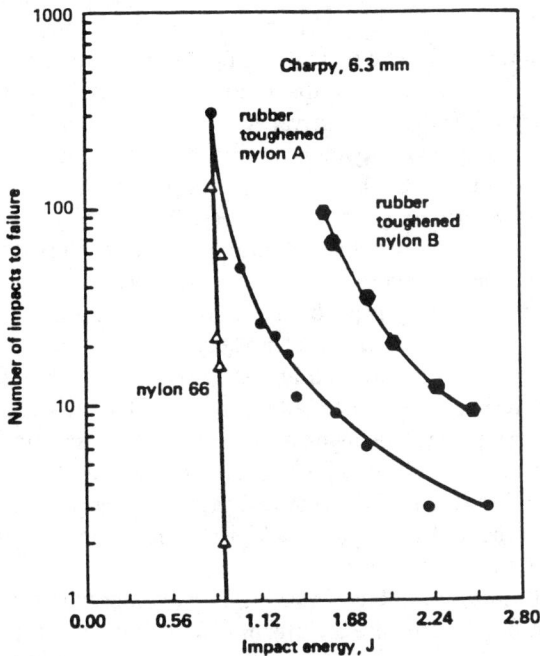

FIG. 6—*Fatigue curve for nylon 66 (△) and rubber-toughened Nylon A (○) and B (●).*

however, exhibit more than one impact to failure for impact energies between 0.8 and 4.7 J, indicating that the toughening process provides this additional fatigue resistance to nylon 66. The merging of the curves for nylon 66 and rubber-toughened Nylon A at about 200 impacts and 0.8 J indicates that fatigue in the toughened nylon is governed by the matrix. However, the process of failure is unstable and somewhat unpredictable in the brittle nylon but very predictable and by gradual crack advance in the toughened nylons.

The two toughened nylons, Nylons A and B, can be compared in fatigue by viewing their performance at a single-impact energy, such as an impact energy of 1.6 J. The 10 impacts required for failure of rubber-toughened Nylon A increase to seven times this number for rubber-toughened Nylon B. At lower impact energies the difference is even greater. This shows that two chemically similar materials can behave differently in fatigue.

While not evident in Fig. 6, the energy-to-break values for single-blow impacts for the two toughened nylons are remarkably similar—4.7 J for Nylon A and 5.4 J for Nylon B. This indicates the greater sensitivity of impact fatigue to differences in toughness.

Short and Long-Term Failure

The fatigue curve exhibits two regimes, one of long-term fatigue resistance characterized by more than several hundred impacts to failure for impact energies below about 0.8 J. The second regime exists above this impact energy level, requiring fewer than 200 impacts to cause failure. Figure 7 shows fatigue curves for high-impact polystyrene, ABS, and toughened polyacetal. In addition, rubber-toughened Nylon B and rubber-toughened acetal have similar and high fatigue performance. For all the toughened materials, two regimes, one of long-term and one of short-term failure, may be seen, although the impact energies are shifted and characteristic of each material. This figure suggests that the impact energy separation between short-term and long-term fatigue is partly improved by the material crystallinity or crystalline morphology, or both. Materials that are crystalline, whether toughened or not, generally have fatigue curves that reside to the right, that is, they require higher impact energies to produce failure in fewer than 200 impacts. Because they are highly crystalline, the nylon and acetal materials conform to this generalization. Using constant-load fatigue, Hertzberg and Manson arrive at the same conclusion, that materials most resistant to crack propagation are crystalline and that the lowest fatigue crack growth rates have been recorded for nylon 66, nylon 6, and polyacetal [4].

These results show the value of varying impact energy in fatigue tests. Impacts of low-impact energy are as informative as impacts of high-impact energy, but the foremost information coming from this work is the need to vary impact energy in a systematic manner in evaluating materials.

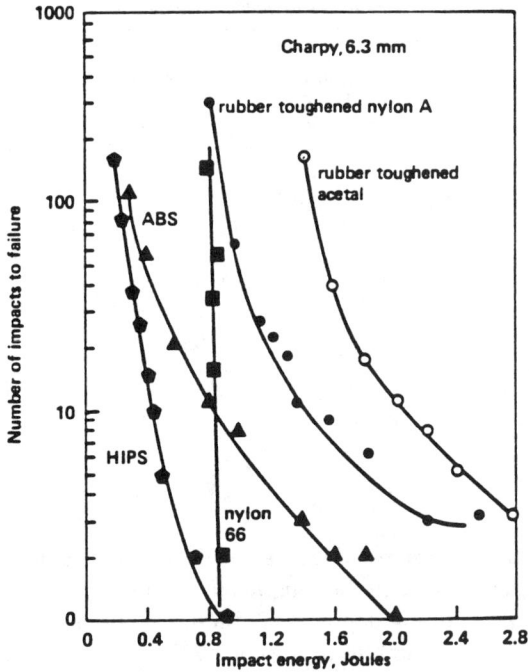

FIG. 7—*Fatigue curves for nylon 66 (■), high-impact polystyrene (●), ABS (▲), toughened Nylon A (●), and rubber-toughened acetal (○).*

Effect of Test Frequency

A common concern with this fatigue technique is whether the rate of impacting affects the results. To test this, ABS was repeatedly impacted with an impact energy of 0.4 J. Normally, 60 impacts would be required for failure at this impact energy. However, 40 impacts were completed, at both a slow and a fast rate on separate specimens, followed by one impact at 3.0 J, which was sufficient to produce failure. Table 2 summarizes the results. The crack advanced further in slow impacting, causing the maximum force in the failure impact to be lower. One explanation of these results is a temperature rise at the root of the notch in fast impacting and a consequent small-scale local viscous flow, which results in notch blunting. Many variations of this experiment are possible. The results are mentioned here to remind readers of a dependence on impacting rate.

Effect of Thickness

To this point the repeated impact test is found to be a useful comparison test in fatigue analysis. However, it is necessary that it be performed on speci-

TABLE 2—*Tabulation of results of frequency of impacting in fatigue for ABS.*[a]

Run No.	Number of Impacts	Impact Energy, J	Notch Length, mm	Time Interval Between Impacts, s	Force Maximum, N	Comments
1	1	3.0	1.0	15	500	break
2	40	0.4	2.2[b]	0.5	...	fatigue, no break
	1	3.0	...	0.5	390	break
3	40	0.4	3.5[b]	30	...	fatigue, no break
	1	3.0	...	30	320	break

[a]Each entry is the average of values for ten specimens. Impact energy selection is based on 60 impacts to failure at an impact energy of 0.4 J.
[b]After 40 impacts.

mens of identical dimensions. Because this is an obvious limitation, fatigue curves have been obtained on the toughened nylons at thicknesses of 6.3 and 3.2 mm, while maintaining depth at 12.7 mm. Figure 8 shows these results. However, it is more meaningful to normalize them in respect to the fracture areas, and this gives the results shown in Fig. 9. The curves show that for these materials normalized fatigue results are independent of the fracture area.

However, in some instances thicker moldings of a crystalline material may offer greater fatigue resistance. This may occur if the thicker moldings are sufficiently more crystalline or if crystalline morphology differs from the thinner counterparts. The normalized fatigue curves become a convenient means of detecting such improvements in fatigue performance, which may not show up in single-blow impacts. Conversely, materials exhibiting plane stress (greater) toughness in thin section and plane-strain (lower) toughness in thick section would not give identical normalized fatigue curves.

Results of Fracture Energy Studies

Returned and Retained Energy

It has been stated that the area inside the force–displacement curve is the energy retained by the specimen and the area under the curve is that returned to the tup. A representation of the partitioning of the retained and returned energies and the impact energy is given in Figs. 10 and 11. During the downward motion of the tup—that is, during the impact—the tup and specimen are in intimate contact, and the upper half of the force–displacement curve is

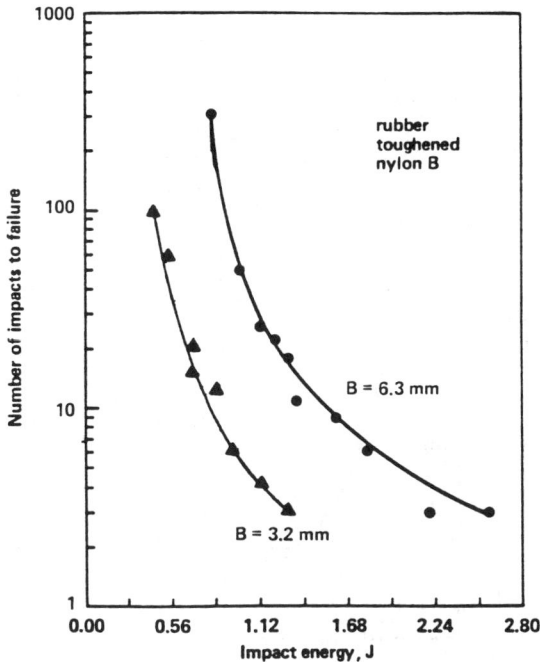

FIG. 8—*Fatigue curve of rubber-toughened Nylon B for thicknesses of 3.2 and 6.3 mm.*

established. Since a bounce occurs, contact is lost and the impact signal represents energy returned to the tup. The retained energy, represented in Figs. 10 and 11, for rubber-toughened Nylon A, is plotted versus the impact number in Fig. 12. An increase in retained energy signifies increasing damage to the specimen as fatigue proceeds.

It was stated previously that low impact energies are likely to invoke material responses that are not observed at high impact energies. This point is evident in Fig. 13. For a given impact energy the sum of the retained energy for all impacts is referred to as the accumulated retained energy. This energy is plotted versus impact energy in this figure. When low impact energy is used, the specimen can accumulatively absorb greater energy than that absorbed in a single impact to failure. For instance, rubber-toughened Nylon A will cumulatively absorb a value much greater than the 4.75 J absorbed in a single impact to break. Similarly, for high-impact polystyrene the accumulated retained energy of 8 J is greater than the 1 J obtained in a single blow to break. The accumulated retained energy curves have a shape similar to that of the number of impacts to failure or fatigue curves.

The fatigue curves of several common toughened polymers are given in Fig. 7, and the corresponding accumulated retained energy curves for two of these

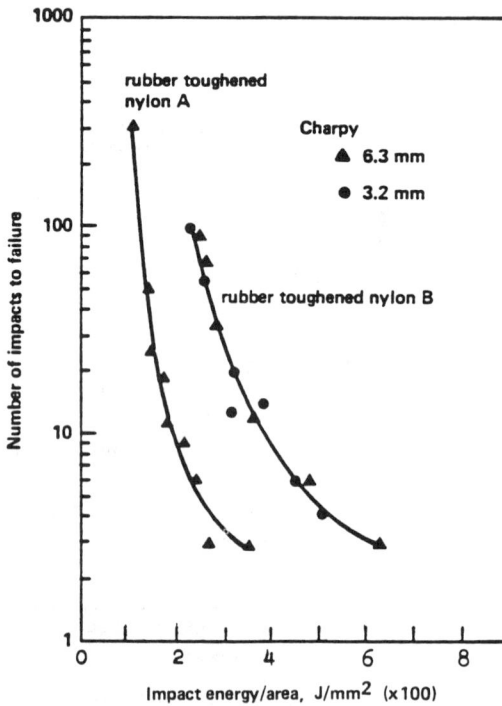

FIG. 9—*Fatigue curve of rubber-toughened Nylons A and B expressed as impact energy normalized with the fracture area (▲) 6.3 mm and (●) 3.2 mm.*

materials are given in Fig. 13. It is interesting that these energy curves are similar in shape to the familiar *S-N* or stress–number of cycles diagram for polymers tested in unnotched cyclic tension. In the latter the number of cycles is usually plotted horizontally. In the impact test, the test frequency is considerably lower than that of the tension tests, and the number of cycles is fewer than 1000. Because the impact fatigue procedure is relatively straightforward and is based on energy, curves from impact fatigue testing can be especially useful for comparative fatigue studies. In this fatigue work, manual drops were conducted, and the time between impacts is between 15 and 30 s. Therefore, a test frequency between 0.03 and 0.06 Hz is used, which is much lower than that of conventional fatigue testing. An automated impact fatigue apparatus could increase the test frequency [*14–16*].

Fracture Energy

The fatigue curves described thus far make no use of the extensive analysis in force and energy that results from the instrumented feature of the test.

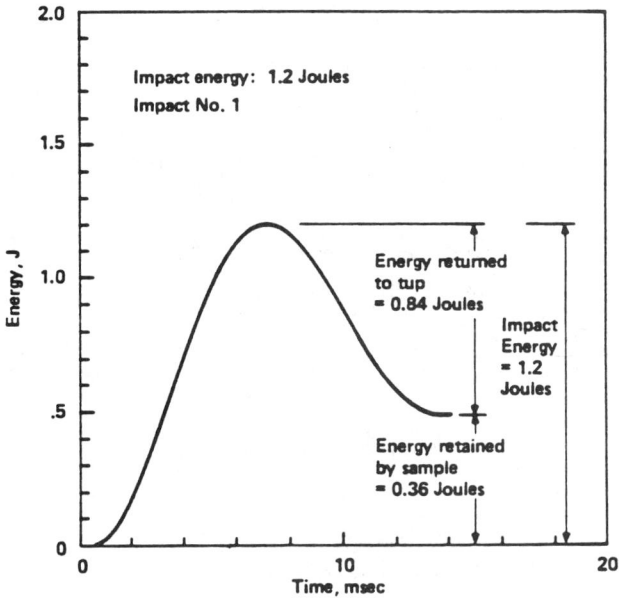

FIG. 10—*Energy transferred to the specimen versus the time of impact for rubber-toughened Nylon A.*

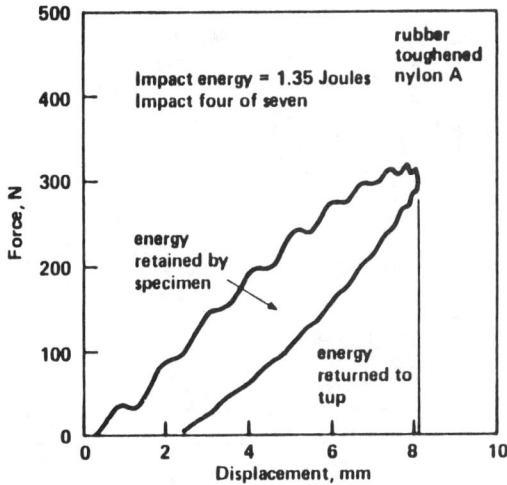

FIG. 11—*Force–displacement curve for rubber-toughened Nylon A obtained in impact fatigue testing.*

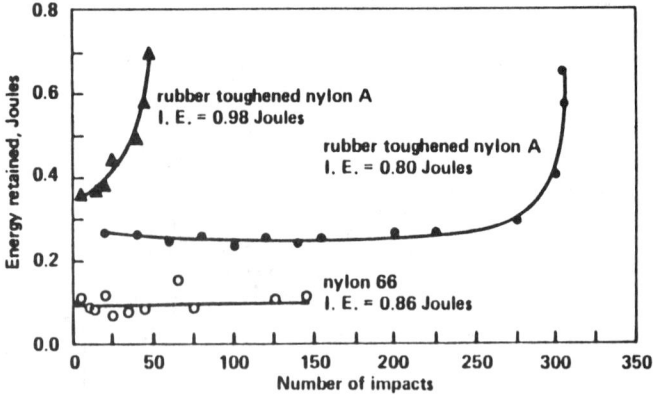

FIG. 12—*Energy retained by the specimen versus the number of impacts for rubber-toughened Nylon A (●) and nylon 66 (○).*

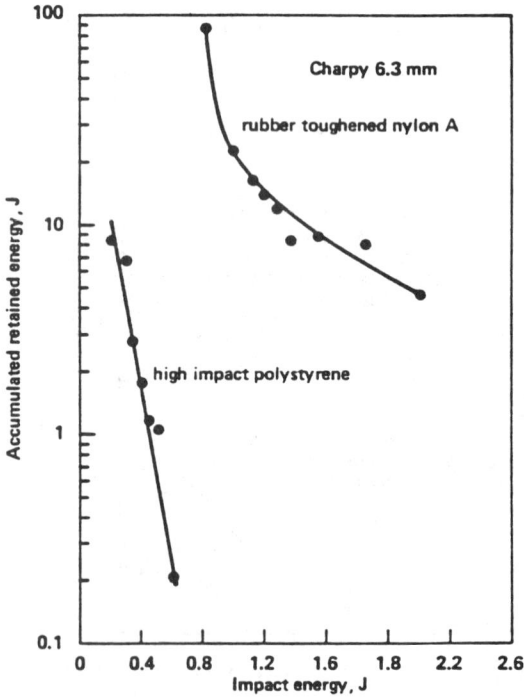

FIG. 13—*Accumulated retained energy versus the impact energy for rubber-toughened Nylon A (●) and high-impact polystyrene (●).*

Nevertheless, for each impact the retained energy provides the basis for a determination of fracture energy.

Conventional impact testing involves an energy to break the specimen divided by the ligament area. The fatigue curves may be analyzed in a similar manner using the retained energy associated with each impact and the incrementally increased cross-sectional areas resulting from the impact. The latter is calculated from scanning electron micrographs of the fracture surface. Figure 14 shows a portion of the fracture surface of tough nylon encompassing impact numbers 1 to 24 in which 27 impacts were required to complete fracture and an impact energy of 1.2 J was used.

The fracture area generated for the initial impacts is small but increases significantly as the number of impacts proceeds. Indeed, the last 2 to 3 out of 26 impacts can constitute nearly one third of the specimen cross-sectional area. Associated with each incremental area is a retained energy for that impact. It is calculated from the area inside the force–displacement loop. Plotting the retained energy per impact versus the corresponding fracture area per impact results in the relationship seen in Fig. 15 for HIPS ($J_c = 18 \text{ kJ/m}^2$) and the rubber-toughened nylon ($J_c = 100 \text{ kJ/m}^2$). The slope represents an approach to obtaining fracture energy from impact fatigue and provides another means of assessing toughness, which may be compared with the energy-area results in single-blow impact testing [19].

In single-blow impacting, specimens have been machine notched to several depths and fractured in one impact by this same instrumented device. The energy–ligament area plot, shown in Fig. 16 for rubber-toughened Nylon A, gives a fracture energy, J_c, of 64 kJ/m². The value, which is only 64% of that obtained by the fatigue test described earlier, results from the larger accumulated retained energy obtained in fatigue compared with the absorbed energy in single-blow fracture. Both these energies are absorbed in generating the same fracture area. The higher fracture energy in the fatigue situation reflects additional energy dissipation processes that occur during fatigue testing. All of these processes, including heat generation, are reflected in the retained energy, and no separation is possible without additional experimentation. Wu has studied these processes in single-blow impacts [20].

Discussion

The fatigue analysis herein described is related to conventional energy-based toughness testing such as Izod and other geometries. The study introduces a straightforward and understandable energy-based fatigue test through repeated impacts of constant impact energy. The comparison is through the number of impacts to failure versus the impact energy, that is, the fatigue curve. Use of the lower-impact energies and multiple impacts numbering fewer than 200 is the basis for a short-duration fatigue analysis. The initial results show a striking difference between brittle and ductile frac-

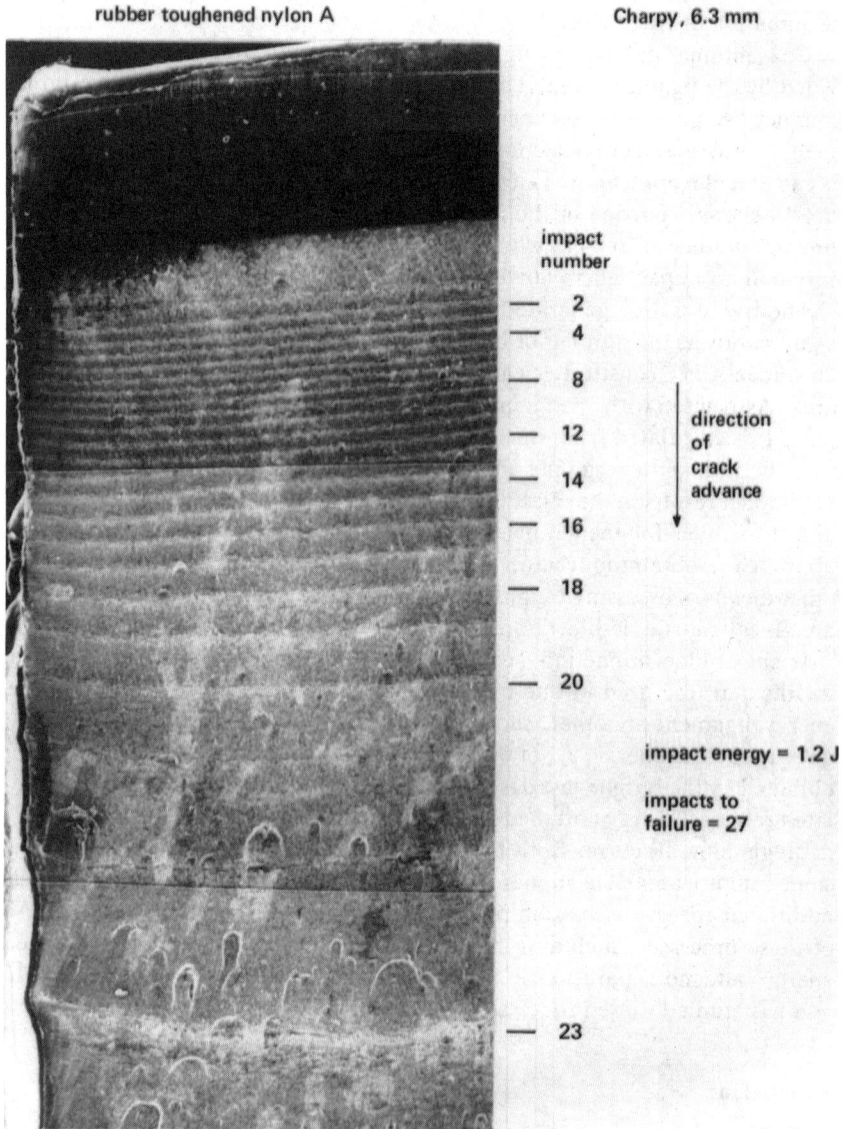

FIG. 14—*Scanning electron photomicrographs of the fracture surface of rubber-toughened Nylon A, indicating the location of the crack advance in impact fatigue.*

ture and enable qualitative assessment of energy absorption in various systems, including unmodified and modified or toughened polymers. This study shows that chemically similar polymers may be prepared with significantly different fatigue performances. Also, crystalline polymers appear to have better fatigue performance than noncrystalline polymers.

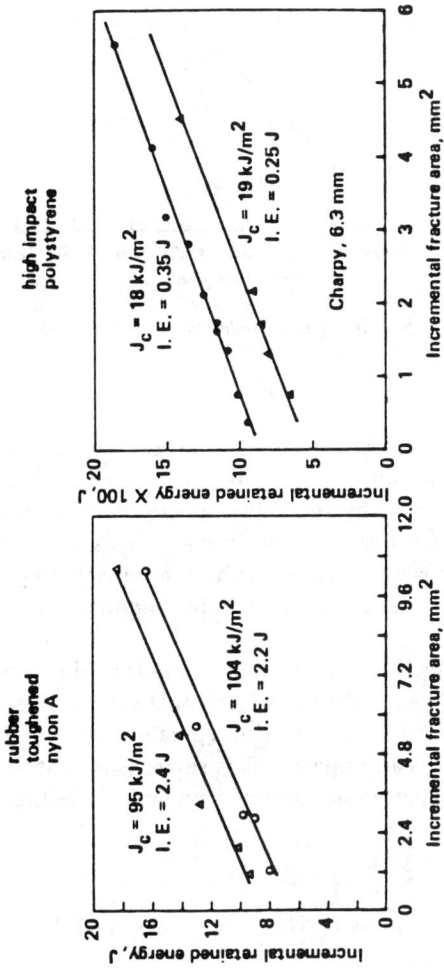

FIG. 15—*Energy-area curves showing determination of the fracture energy for rubber-toughened Nylon A (open symbols) and high-impact polystyrene (closed symbols).*

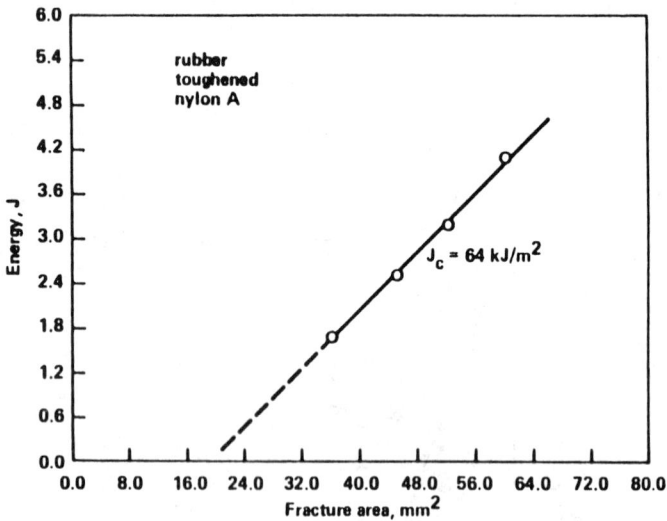

FIG. 16—*Energy versus the fracture area in single-blow-to-break impacts to obtain the fracture energy for rubber-toughened Nylon A.*

Impact energy may be normalized with respect to the specimen dimensions. Through such normalization, the possibility arises of detecting differences in crystalline morphology and skin-core effects in crystalline materials that are often overlooked in single-blow impacts. This simple normalization procedure can be improved, however, with an approach employing fracture mechanics to provide material parameters in toughness, albeit obtained in fatigue.

Since the test is energy based, an attempt is made to obtain fracture energy from specimen retained energy and fracture area values. In so doing a fracture energy is found that is nearly two times greater when obtained in fatigue than in single-blow-to-break impacts. This suggests that absorption mechanisms occur in fatigue that do not occur in single-blow testing.

Acknowledgment

Review of this manuscript by D. D. Huang is acknowledged with appreciation.

References

[1] Beardmore, P. and Rabinowitz, S., "Plastics Deformation of Materials," in *Treatise on Materials Science and Technology*, R. J. Arenault, Ed., Academic Press, New York, 1975.
[2] Schultz, J. M., "Properties of Solid Polymeric Materials," in *Treatise on Materials Science and Technology*, Part B, J. M. Schultz, Ed., Academic Press, New York, 1977.

[3] Owen, M. J., "Fracture and Fatigue," in *Composite Materials*, Vol. 5, Academic Press, New York, 1974, Chapters 7 and 8.

[4] Hertzberg, R. W. and Manson, J. A., *Fatigue of Engineering Polymers*, Academic Press, New York, 1980.

[5] Lang, R. W., Hahn, M. T., Hertzberg, R. W., and Manson, J. A., *Journal of Materials Science Letters*, Vol. 3, 1984, p. 224.

[6] Ramirez, A., Manson, J. A., and Hertzberg, R. W., *Polymer Engineering and Science*, Vol. 22, 1982, p. 975.

[7] Hertzberg, R. W., Skibo, M. D., and Manson, J. A., *Journal of Materials Science*, Vol. 13, 1978, p. 1038.

[8] Bucknall, C. B. and Stevens, W. W., *Journal of Materials Science*, Vol. 15, 1980, p. 2950.

[9] Bhateja, S. K., Rieke, J. K., and Andrews, E. H., *Journal of Materials Science*, Vol. 14, 1979, p. 2103.

[10] Ohishi, F., Nakamure, S., Koyama, B., Minabe, K., and Fujisawa, J., *Journal of Applied Polymer Science*, Vol. 20, 1976, p. 79.

[11] Takemori, M. T., *Journal of Materials Science*, Vol. 17, 1982, p. 164.

[12] Reike, J., Bhateja, S., and Andrews, E., *Industrial and Engineering Chemistry Process Research and Development*, Vol. 19, 1980, p. 601.

[13] Zoller, P., *Polymer Testing*, Vol. 3, 1983, p. 197.

[14] Adams, G. C. and Wu, T. K., *Society of Plastics Engineers, Annual Technical Conference '81*, Vol. 27, 1981, p. 185.

[15] Adams, G. C. and Wu, T. K., "Materials Characterization by Instrumented Impact Testing," *Failure of Plastics*, W. Brostow and R. D. Corneliussen, Eds., Hanser, Munich, 1986.

[16] Adams, G. C. and Wu, T. K., *Society of Plastics Engineers, Annual Technical Conference '82*, Vol. 28, 1982, p. 898.

[17] Bluhm, J. I. in *Symposium on Impact Testing, ASTM STP 176*, American Society for Testing and Materials, Philadelphia, 1955, p. 85.

[18] Abe, H., "Instrumented Charpy Impact Testing of Silicone Carbide," Ph.D. thesis, Pennsylvania State University, University Park, PA, 1976.

[19] Williams, J. G., *Fracture Mechanics of Polymers*, Ellis Harwood, London, 1984.

[20] Wu, S., *Journal of Polymer Science*, Vol. 21, 1983, p. 699.

W. M. Lee,[1] *William W. Predebon,*[2] *and Michael L. Jurosek*[3]

Impact Response of Polymeric Materials at Varying Depths of Penetration

REFERENCE: Lee, W. M., Predebon, W. M., and Jurosek, M. L., "**Impact Response of Polymeric Materials at Varying Depths of Penetration,**" *Instrumented Impact Testing of Plastics and Composite Materials, ASTM STP 936,* S. L. Kessler, G. C. Adams, S. B. Driscoll, and D. R. Ireland, Eds., American Society for Testing and Materials, Philadelphia, 1987, pp. 302–323.

ABSTRACT: Usually, an impact test is conducted to the point of fracture, so that the fracture energy and other fracture properties of a material under impact loading conditions can be assessed. To understand the fundamentals of and gain insight into impact phenomena, it would be valuable to monitor and quantify the deformation mechanics in a material leading to fracture and to relate the deformation profile history to the resulting impact traces. This task is accomplished in the present paper by carrying out impact tests at varying depths of penetration using a Rheometrics impact tester. The materials under investigation are a rubber-modified polystyrene [high-impact polystyrene (HIPS)] and a high-density polyethylene (HDPE).

The force-displacement traces of partial impacts of these two materials are presented and compared with those at fracture. The effects of impact speed and clamp ring size are also discussed. The two polymers, being ductile in nature, exhibit indentation deformation before complete puncture occurs. The deformation profile/impact force/ram displacement relationship is established.

The present study also includes a computational analysis of the impact response in polymeric materials. The impact model makes use of a finite-difference wave mechanics code. A wave mechanics approach is used in order to include the effects of inertia in the development of deformation and stress states under impact conditions. It incorporates a critical stress criterion for incipient damage and a damage assessment on the basis of stress-time accumulation. Comparisons are made between the calculations made at high impact speeds and those from the impact experiments conducted at low speeds. The computer simulation and experimental results are in qualitative agreement.

KEY WORDS: instrumented impact test, total penetration impact, partial impact, controlled depth of penetration impact, impact deformation profile, impact energy, impact

[1]Senior associate scientist, Dow Chemical Co., Midland, MI 48640.
[2]Professor, Department of Engineering Mechanics, Michigan Technological University, Houghton, MI 49931.
[3]Research engineer, Honeywell, Inc., Edina, MN 55424.

load, impact testing, polymeric materials, rubber-modified polystyrene, high-density polyethylene, impact fracture, computer simulation, wave mechanics approach

In recent years, polymeric materials have been widely used in molded and extruded articles. In such forms, they are expected to possess a number of common engineering properties such as modulus of elasticity, strength, fatigue resistance, creep resistance, and impact resistance. The impact resistance of a material is usually determined by the use of an impact tester and is expressed in terms of an impact energy. With the advent of automated, instrumented impact testers [1], the assessment of detailed impact characteristics in a polymeric material is possible. These impact characteristics include impact duration, the rate of impact force rise (which is related to part stiffness), peak impact force, part deflection or depth of penetration, and impact energy. These characteristics are obtainable from the impact trace generated by an instrumented impact tester.

Investigators have studied the impact response of polymeric materials using instrumented impact testers [1-11]. Most impact studies have examined the material response under total penetration testing conditions, in which fracturing or puncturing of the test specimen has occurred. Even though the load-displacement trace is available from impact testing, the progression of the contact surface, leading to irrecoverable deformation and eventually to fracture, is not readily observable because of the high-speed nature of an impact test. By the time the impact test is completed, the contact surface change or the deformation profile history during impact is lost to the investigator. Valuable information concerning the details of the deformation mechanisms operative in the material cannot be retrieved afterwards. Thus, if this deformation profile history can be captured and related to the recorded impact load development, one can establish a deformation profile/part deflection/ impact force relationship for the entire impact event. One way to accomplish such a goal is through the use of high-speed photography, which is a tedious and cumbersome technique. An alternative is to carry out impact tests at various depths of penetration or in short, partial impacts. In the present study, the impact responses of two polymeric materials at controlled penetration depths are probed by means of a Rheometrics high-speed impact tester. The materials under examination are a rubber-modified polystyrene [high-impact polystyrene (HIPS)] and a high-density polyethylene (HDPE). Both materials are ductile in nature, but the polyethylene is more ductile than the rubber-modified polystyrene.

Impact damage assessment from small indentation up to complete penetration involving sheet molding compound (SMC) or bulk molding compound (BMC) has been made by using instrumented falling dart impact testers [2,3]. In these studies, the impact speed was varied by adjusting the drop heights. Therefore, the conditions were quite different from the impact condition adopted in the present paper, in which the penetration depth is adjusted through the impact ram travel distance at a constant impact speed. The impact data on the rubber-modified polystyrene and the high-density polyethylene under investigation are presented and discussed in the first part of this paper.

The second part of this paper consists of a computer simulation of the impact response of polyethylene and polystyrene. Most attempts in the past to model the puncture testing and associated failure of polymers were quasi-static applications of elastic or viscoelastic theory. In these analyses, the wave propagation effects or acceleration effects were neglected. In this study a wave mechanics approach is used. Unlike a quasi-static analysis [12], a wave mechanics analysis allows a truly dynamic approach to the problem. It is felt that this approach may be important in this study because the velocities of interest are in the range of 1.27×10^{-2} to 0.508 m/s (30 to 1200 in/min), which is within the velocity range in which wave propagation or inertial effects may become significant [13].

Impact Responses of Rubber-Modified Polystyrene and High-Density Polyethylene

Experimental Procedure

Impact characteristics were obtained using a Rheometrics impact tester (RIT) in a total penetration mode as well as in a partial impact mode, in which the impact ram travel distance is preset. In general, a total penetration impact was made first (Fig. 1). Partial impacts were then conducted by presetting the ram travel distance short of the ultimate travel distance from the total penetration impact (Fig. 1). In the total penetration impact, the ultimate point is taken as the point in the force-displacement trace where the force drops off precipitously. The ram travel distance, impact force, and impact energy at this ultimate point are designated as the actual travel distance, the ultimate impact force, and the ultimate impact energy (the area under the impact load-deflection trace), respectively. In the partial impact case, the actual ram travel distance, which is determined from the ram velocity, is noted in addition to its corresponding impact force and impact energy. In general, the actual ram travel distance at the higher impact speed is larger than its set point, indicating the inability of ram stoppage at its present penetration depth because of the higher momentum involved. After an impact, the test specimen may or may not have fractured. In cases in which no fracture occurs, the specimen thickness at the center is measured and compared with that before the impact. This change in specimen thickness is employed to assess the extent of irrecoverable deformation after impact. After each impact, including the partial impacts, the moving ram returns to its original position, not maintaining any physical contact with the test specimen.

Impact tests were run in the speed range of 0.0127 and 0.508 m/s. Beyond

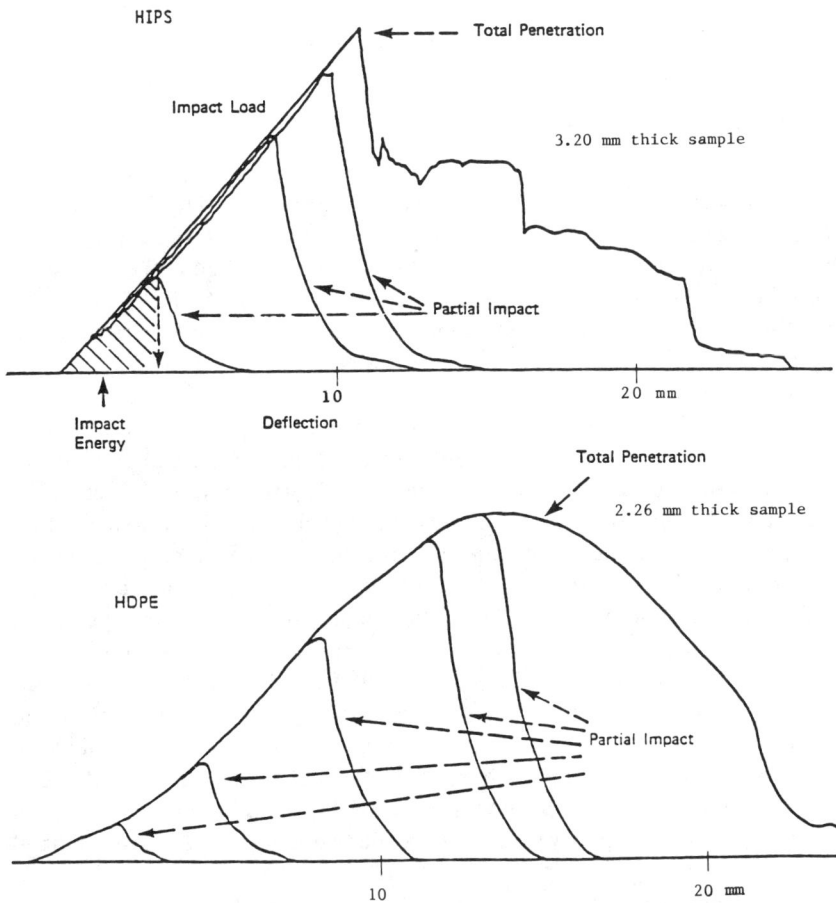

FIG. 1—*Typical impact traces for HIPS (top) and HDPE (bottom).*

a 0.508-m/s impact speed, the ram would significantly overshoot its pre-scribed travel distance; therefore, impact speeds greater than this were not used. The test apparatus is capable of higher impact rates, however; in the total penetration test mode, one can achieve impact speeds of up to 8.5 m/s.

Two specimen sizes were employed in the impact tests: 0.0508 by 0.0508 m and 0.102 by 0.102 m. For the smaller specimens, a 0.0381-m-diameter clamp ring was used, and a 0.0889-m-diameter clamp ring was used for the larger specimens. The impact ram head is a hemisphere with a 0.0127-m di-ameter. The specimen thickness ranged from 2.26 to 6.35 mm (0.089 to 0.250 in.). All of the impact tests were carried out in a laboratory with 50% relative humidity and 296 K temperature.

Two thermoplastics were investigated in this study: a rubber-modified polystyrene and a high-density polyethylene (at 0.954 g/cc or 954 kg/m³). The impact test specimens were compression molded.

Results and Discussion

Both the rubber-modified polystyrene and the high-density polyethylene exhibited ductile deformation upon impact. The two materials, however, displayed different modes of failure when complete penetration occurred. For the former, the impacted area of the test specimen cracked and broke away from the rest of the test specimen. On the other hand, the latter showed a punch-like fractured surface. At times, the punched-out cap, even though highly drawn out by the impact ram head, was still attached to the test specimen on one side.

The distinctively different failure modes of the two materials were also reflected in the total penetration impact traces. Typical total penetration impact traces for these two kinds of materials are shown in Fig. 1. The cracking fracture in the rubber-modified polystyrene was manifested by a sharp drop in impact force. On the other hand, the drawing of material and its eventual punch-out in the high-density polyethylene were accompanied by a more gradual decline in impact load. When the ram head punctured the specimen, the impact load dropped off sharply. Thus, information on the mode of failure is usually contained in the impact trace. In fact, the impact trace contains additional information concerning the load buildup with depth of penetration leading to failure. This information (which is not and cannot be obtained by a single total penetration impact, because of the high-speed nature of the test) is the progression of the specimen deformation profile with penetration depth. The deformation profile can be determined by successive partial impacts.

Figure 1 depicts typical partial impact traces, as well as total penetration impact traces, for the two materials investigated. In both instances, the partial impact traces follow the complete puncture impact traces, as expected. The specimen deformation/ram displacement/ultimate impact force/ultimate impact energy data under the various impact conditions for HIPS and HDPE are summarized in Tables 1 and 2, respectively. In these tables, the specimen thicknesses before and after impact are listed. Furthermore, the specimen thickness after impact is expressed in terms of the percentage of the original specimen thickness in Column 5 of both tables. Both the set ram travel distance (from the contact point) and the actual ram travel distance are included.

In general, the specimen thickness after impact decreased with increasing penetration depths. At the same time, the ultimate impact force and ultimate impact energy for partial impacts increased. Specimen thickness change from impact can be expressed in terms of a percentage of the thickness reduction, and then plotted against the actual ram travel distance.

TABLE 1—Rheometrics impact data on HIPS at room temperature.

Impact Speed, m/s	Test Specimen Size, m	Specimen Thickness Before Impact, mm	Specimen Thickness After Impact, mm[a]	Thickness After Impact, %	Set Ram Travel Distance, mm	Actual Ram Travel Distance, mm	Ultimate Impact Force, N	Ultimate Impact Energy, J
1.27×10^{-2}	5.08×10^{-2} by 5.08×10^{-2}	3.20	3.12	98	2.54	2.54	312	0.1
		3.20	3.12	98	5.08	5.08	743	1.2
		3.20	3.05	95	8.89	8.89	1509	5.8
		3.25	broke[b]	...	...	12.70	1891	9.3
	1.02×10^{-1} by 1.02×10^{-1}	3.30	3.20	97	3.81	4.01	102	0.1
		3.25	3.18	98	8.89	8.89	797	2.2
		3.33	broke[b]	...	12.46	11.66	1233	5.1
		3.35	broke[b]	...	...	16.08	1856	12.0
1.27×10^{-1}	5.08×10^{-2} by 5.08×10^{-2}	3.30	3.25	98	2.90	3.51	503	0.6
		3.28	3.23	98	5.51	6.17	1268	3.3
		3.25	3.15	97	6.63	7.47	1544	4.9
		3.33	broke[b]	...	...	8.08	1891	6.3
	1.02×10^{-1} by 1.02×10^{-1}	3.20	3.20	100	5.08	5.69	294	0.6
		3.23	3.20	99	7.62	8.15	743	1.9
		3.23	3.15	98	10.16	10.82	1299	4.8
		3.25	broke[b]	...	...	13.89	1962	10.3

[a]Specimen thickness at the center.
[b]Total penetration impact and brittle failure.

TABLE 2—Rheometrics impact data on HDPE at room temperature.

Impact Speed, m/s	Test Specimen Size, m	Specimen Thickness Before Impact, mm	Specimen Thickness After Impact, mm[a]	Thickness After Impact, %	Set Ram Travel Distance, mm	Actual Ram Travel Distance, mm	Ultimate Impact Force, N	Ultimate Impact Energy, J
1.27×10^{-2} (Section A)	5.08×10^{-2} by 5.08×10^{-2}	2.24	2.24	100	1.91	1.91	120	0.0
		2.24	2.24	100	3.81	3.81	396	0.6
		2.21	2.13	96	6.35	6.35	935	2.2
		2.21	2.01	91	8.89	8.89	1442	5.3
		2.24	1.96	88	10.16	10.16	1598	7.4
		2.21	broke[b]	...	...	17.75	1580	7.9
1.27×10^{-1} (Section B)	5.08×10^{-2} by 5.08×10^{-2}	2.26	2.26	100	1.91	2.46	189	0.1
		2.26	2.21	98	3.81	4.32	538	0.8
		2.26	2.13	94	6.35	6.81	1161	2.9
		2.26	1.96	87	8.89	9.25	1700	6.6
		2.26	1.80	80	10.16	10.29	1825	8.5
		2.26	broke[b]	...	...	22.20	1856	8.9
4.23×10^{-1} (Section C)	5.08×10^{-2} by 5.08×10^{-2}	6.63	6.35	96	3.81	7.11	5474	22.6
		6.58	6.02	91	7.62	11.23	6675	49.7
		5.84	0.58	10	10.16	15.04	4940	59.9
		6.05	broke[b]	...	12.19	17.45	4050	72.3
		6.45	broke[b]	...	...	16.79	4584	82.5
5.08×10^{-1} (Section D)	5.08×10^{-2} by 5.08×10^{-2}	5.51	5.51	100	3.81	7.85	4495	17.0
		5.28	5.11	97	3.81	7.80	4050	14.7
		5.23	0.58	11	11.43	17.96	3382	65.5
		5.18	broke[b]	...	11.68	16.87	3961	61.0
		5.71	broke[b]	...	...	18.34	3649	75.7

[a]Specimen thickness at the center.
[b]Total penetration impact and puncture failure.

Figure 2 shows such a plot for a 0.127-m/s impact speed and a specimen size of 0.0508 by 0.0508 m. It can be seen that the percentage of thickness reduction for the rubber-modified polystyrene is small compared with that of the high-density polyethylene, clearly illustrating the effect of material ductility. In the former, the material fractured after only a 3% reduction in specimen thickness, whereas specimen fracture occurred after a 20% thickness reduction in the latter. In the case of rubber-modified polystyrene, the three preset ram travel distances were 0.0029, 0.0055, and 0.0066 m. The corresponding actual travel distances were 0.0035, 0.0062, and 0.0075 m, indicating a slight ram travel overshoot at the 0.127-m/s (300-in./min) impact speed.

The ultimate impact energy increased progressively from 0.6 to 3.3 J, and then to 4.9 J. The actual ram travel distance for a total penetration impact run was 0.0081 m, slightly greater than the 0.0075-m travel distance from the last partial impact. The material fractured at an impact energy of 6.3 J. In the case of the HDPE material, five partial impacts were run. They gave rise to the following actual ram travel distances and ultimate impact energy values: 0.0025, 0.0043, 0.0068, 0.0093, and 0.0103 m, and 0.1, 0.8, 2.9, 6.6, and 8.5 J, respectively. When the material was impacted in the total penetration

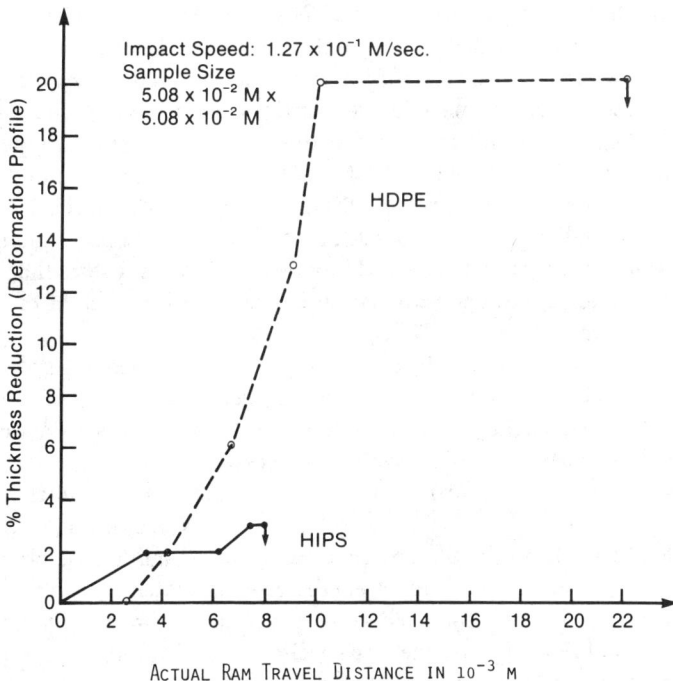

FIG. 2—*Effect of material ductility on the deformation profile from partial impacts.*

mode, the actual travel distance was found to be 0.0222 m, and the ultimate impact energy was 8.9 J. Thus, the partial impacts permit one to ascertain the extent of material deformation as a function of the ram displacement, the impact load buildup, and the impact energy associated with such a deformation. It should be noted that the specimen deformation (in terms of its reduction in thickness) is small in comparison with the actual ram travel distance. For instance, a ram travel distance of 0.0103 m at a speed of 0.127 m/s gave rise to a 20% reduction in thickness in a 2.26-mm-thick HDPE specimen. The actual material deformation at the center of the specimen is only 4.52×10^{-4} m. Hence, most of the ram displacement came from the bending of the clamped specimen as a circular plate; only a very small portion was from the irrecoverable deformation of the material. At the low impact speed of 0.0127 m/s (30 in./min), the actual ram travel distance coincided with the preset ram travel distance. At higher impact speeds, the former exceeded the latter, indicating that the higher momentum of the moving ram caused it to overshoot the preset point.

Partial impacts of polystyrene as well as polyethylene revealed the possibility of the specimen fracturing before total penetration occurred, resulting in a smaller impact energy. For example, in the polystyrene case, it occurred at 0.0127-m/s impact speed for the 0.102 by 0.102-m-size specimen. When a partial impact of 0.012 46 m ram travel was carried out, the specimen fractured at an ultimate impact force of 1233 N and at an ultimate impact energy of 5.1 J. In comparison, the corresponding impact force and impact energy from complete penetration were 1856 N and 12.0 J. Similar situations arose in the HDPE tests, when the material was partially impacted at speeds of 0.423 m/s and 0.508 m/s. The ultimate impact energy values at these two speeds, at preset ram travel distances of 0.1219 m and 0.1168 m, were 72.3 J and 61.0 J, respectively. However, the corresponding total penetration impact energies were 82.5 J (at 0.423 m/s) and 75.7 J (at 0.508 m/s). The observation that a material could fracture under partial impact conditions suggests that the total penetration impact test conditions could have overestimated the energy for material fracture.

Examination of the partially impacted polystyrene specimens displayed progressively enlarged stress whitening areas accompanied by a reduction in specimen thickness as the penetration depth increased. At the same time, the indentation surface areas increased. At successively increasing penetration depths, small cracks developed over the back side of the contact surface. Eventually, these cracks caused the specimen to break, exhibiting a deformation mode change from ductile deformation to brittle failure. Before the development of visible cracks, impact energy was dissipated through bending, crazing (stress whitening), and indentation of the specimen. Depending on the impact condition (that is, the test specimen's size and thickness and the impact speed), the impact energy before specimen fracturing could amount to a large fraction of the ultimate impact energy for total penetration. For instance, for the rubber-modified polystyrene, at an impact speed of 0.127 m/s and a specimen size of 0.0508 by 0.0508 m, the partial impact energy just before total penetration was 4.9 J, in comparison with 6.3 J from total penetration. The actual ram displacement of 0.007 47 m was fairly close to the 0.008 08 m for total penetration. Thus, this partial impact is probably on the verge of fracturing the material as a total penetration impact would have done. For this case, the partial impact energy accounts for 78% of the total penetration impact energy. Therefore, it could be said that the energy required to enlarge the small cracks to the catastrophic failure point accounts for only approximately 22% of the total energy. Consequently, it may not constitute a major energy absorption mechanism for this rubber-modified polystyrene.

A study of the deformation profile from partial impacts of HDPE also showed the indentation surface areas to grow with increasing depths of penetration. Usually, a protrusion developed in the reverse side of the impact surface when the contact circle enlarged. The thinning of the indented surface eventually led to total specimen penetration by the ram head. During partial impacts, no stress whitening was induced. HDPE, being a ductile material, gave rise to a larger indentation than the less-ductile rubber-modified polystyrene did at a similar penetration depth. The former also imparted a larger part deflection than the latter under identical impact conditions.

One of the limitations in using a hydraulically driven impact tester, such as the Rheometrics impact tester, to conduct partial impacts is the tester's inability to stop at a preset penetration depth at high impact speeds. The actual stopping distance will vary with the stiffness of the material being tested. Thus, it may be difficult to compare two materials at the same preset penetration depth under these circumstances. As a result, the use of a hydraulic impact tester is best suited to low-speed partial impacts. Nevertheless, if a comparison of two materials at a given penetration depth is essential, trial runs can be made to match their actual ram travel distances.

For partial impacts, increasing impact speed produced an increase in impact force and impact energy at a given preset ram travel distance. For example, in the case of polystyrene at a preset ram displacement of 0.0051 m (see Table 1), the impact force and impact energy developed at 0.0127 m/s were 743 N and 1.2 J, respectively. When the material (at the specimen size of 0.0508 by 0.0508 m) was impacted at 0.127 m/s, this tenfold increase in impact speed induced an increase in both impact force (1268 N) and impact energy (3.3 J). Polyethylene also behaves in a similar manner. However, in the case of total penetration impacts, the impact speed effects on impact force and impact energy varied with the actual ram displacements. If the actual ram travel distance decreased with increased impact speed, as in the polystyrene case, the ultimate impact energy decreased as a result. When the actual ram travel distance increased with increasing impact speed, as in the polyethylene case, the ultimate impact energy increased accordingly. This impact speed effect is illustrated in Fig. 3 for the 0.0508 by 0.0508-m-size specimen.

The test specimen size also affected the impact characteristics. In the

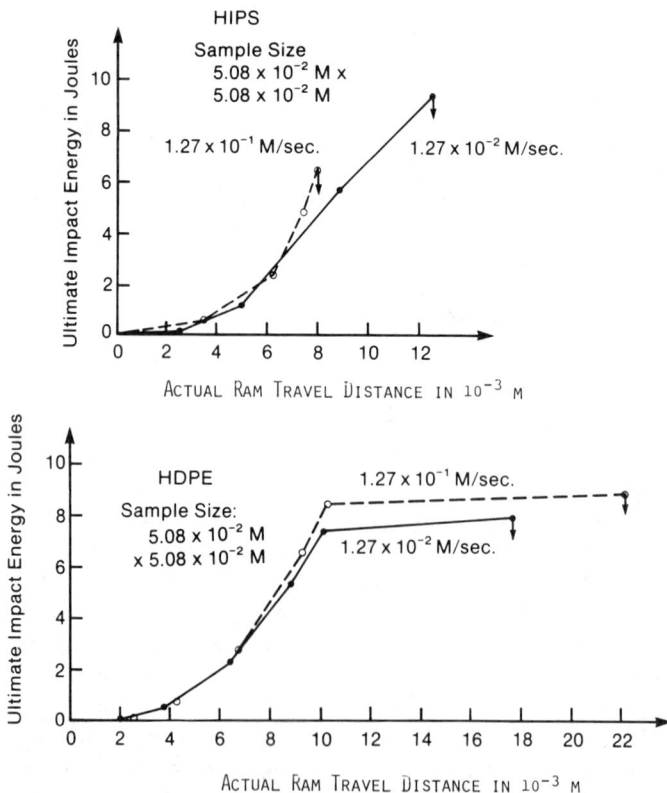

FIG. 3—*The effect of impact speed on the ultimate impact energy.*

present study, the large test specimen (that is, the 0.102 by 0.102-m specimen) was held with a large clamp ring (that is, the 0.0889-m-diameter ring). The small test specimen (that is, the 0.0508 by 0.0508-m specimen) used a smaller, 0.0381-m-diameter clamp ring. The clamp ring size ratio for these two size specimens is 2.33. Thus, the larger specimen was less rigidly held than the smaller one. Under the total penetration impact condition, the large specimen produced a greater part deflection and a higher ultimate impact energy than the small one. Figure 4 depicts the effect of test specimen size, or the effect of clamp ring size, for polystyrene at 0.127 m/s. At a given ram displacement, the partial impact of a more rigidly held specimen produced a larger impact force, because of the higher flexural stiffness, and hence a larger impact energy. However, because of this high plate stiffness, the part deflection from total penetration was reduced. As a consequence, the corresponding ultimate impact energy was low. Therefore, the higher impact en-

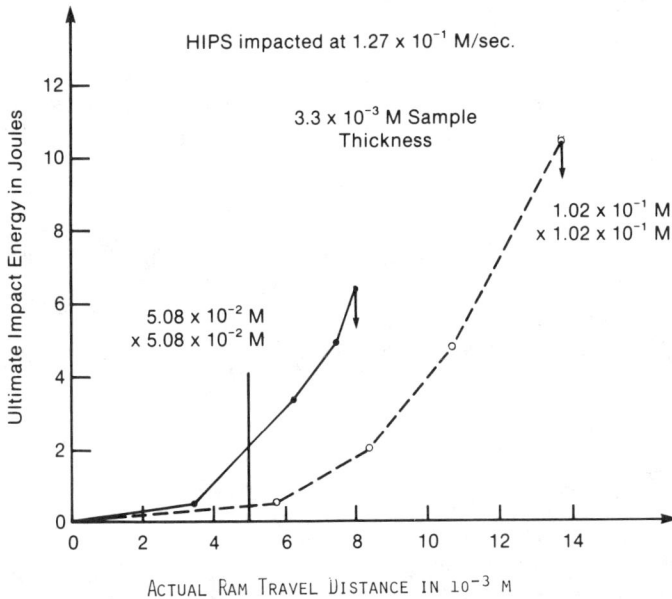

FIG. 4—*The effect of the test specimen size on the ultimate impact energy.*

ergy obtained from the larger clamp ring size (or test specimen size) resulted mainly from its greater plate flexibility or lower flexural stiffness.

Computer Simulation of the Impact Response of Rubber-Modified Polystyrene and High-Density Polyethylene

Computational Model

The computer code used in the simulation is the HEMP (hydrodynamic-elastic-magneto-plastic) code [14], originally developed at the Lawrence Livermore Laboratories in California and modified by Jurosek [15] for this investigation. HEMP is a two-dimensional, multimaterial, Lagrangian, finite-difference wave propagation code. The HEMP code solves approximately the partial differential equations of motion in continuum mechanics, together with the appropriate constitutive equation and initial and boundary conditions, using a finite-difference scheme.

To model the dynamic response and wave propagation effects of polymeric materials in compression, Hugoniot (equation of state) data [15,16] for the HDPE (density = 954 kg/m^3, yield stress = 2.50 × 10^7 Pa) and for the rubber-modified polystyrene (density = 1050 kg/m^3; yield stress = 1.76 × 10^7 Pa) were incorporated into the HEMP code. The behavior in tension was modeled as elastic-plastic, using a von Mises yield condition without rate effects. The rate effects could not be included in the constitutive model because of insufficient experimental data for the materials considered. The failure model used in this investigation is a two-dimensional extension of the accumulative damage model of Tuler and Butcher [17], given by

$$D = \int_{t_0}^{t} (\sigma - \sigma_c)^\alpha dt$$

where

D = damage,
σ = one-dimensional stress,
σ_c = critical threshold stress for damage,
α = damage accumulation parameter, and
t = time.

Their model was originally developed for one-dimensional applications, but in this study, the state of stress is two-dimensional. The model was extended to two dimensions by replacing the one-dimensional stress with the principal stress at a point. This model describes the time-dependent nature of the dynamic fracture of impact-loaded materials. It calculates the damage parameter, D, which is indicative of a decrease in material strength. The material is considered to be damaged (with unspecified voids or cracks) when it experiences a tensile stress larger than the critical threshold stress for damage. The amount of damage produced by a tensile stress depends on the magnitude of the tensile stress along with the time duration of stress. Fracture occurs when the accumulated damage reaches a critical value.

At present, insufficient data exist to determine the damage parameters for polyethylene and polystyrene, so values had to be assumed. For polystyrene, crazes develop at stresses of approximately 50 to 60% of the yield strength around intrinsic flaws in the material. These flaws act as stress concentrators [18] and can eventually cause failure in their vicinity. A value of 60% of the yield stress was assumed to be the critical threshold stress for polystyrene. Polyethylene, in contrast with polystyrene, is a very ductile material and can sustain large deformations without fracture. Under biaxial loading conditions of polyethylene, it has been observed that stress whitening occurs during plastic deformation. Consequently, the critical threshold stress for polyethylene was assumed to be the yield stress in tension [19]. Since the damage parameters are approximate, the damage model will be used only to predict incipient damage. Consequently, while the quantitative value of the damage may not be accurate, it can be used qualitatively to compare the amount of damage occurring in different zones and predict the location of areas of significant damage. Because only incipient damage will be compared, the value of the damage accumulation parameter is arbitrary and was assumed to be two (which, as can be shown, represents an energy-type criterion).

Results and Discussion

The computational model just described is now used to study the effects of target thickness, impact velocity, and clamp diameter on the impact responses of polystyrene and polyethylene. When appropriate, the computer simulation results will be compared with the experimental results reported in the first section of this paper.

Before the parameter study was conducted, a zoning study was performed to determine the effect of zone size on the simulation results and, in this way, to determine an acceptable zone size for the calculations. The zone size was reduced and compared until the difference between the impact simulation results for two different zone sizes was acceptable. The zone size used for all of the simulations presented here was 0.000 203 cm on each side (initially, square zones).

The effect of the impact velocity on the deformation behavior of polyethylene (HDPE) and polystyrene for a 3.2×10^{-3}-m (125-mil)-thick target and 7.6×10^{-2}-m (3.0-in.)-diameter clamp is shown in Fig. 5, Parts a and b, respectively. The diameter of the punch is 1.27×10^{-2} m (0.5 in.) and is the same for all the simulations presented in the paper. Also, in all the simulations shown, only the results about the axis of symmetry are given because the problem is axisymmetric. In Fig. 5a and b, it can be seen that, at the slower impact speed, the response is more one of plate bending, whereas at the higher speed, more penetration occurs. These effects, although present, are less pronounced in the simulations using the smaller, 3.8×10^{-2}-m (1.5-in.)-diameter clamp. A second observation from the simulation results is that, as the impact velocity increases, more penetration occurs in both polyethylene and polystyrene. This is what has been observed experimentally (see Fig. 3) at the slower speeds, as discussed in the first part of this paper and elsewhere. It is apparent that the speeds chosen for the calculations are considerably higher than those used in the experiments described in this paper. The choice of a velocity of 17 m/s (40 000 in./min = 40 mph) was arrived at as a compromise between achieving approximate steady-state results and minimizing computer time and cost. The 17-m/s-velocity case took approximately 6 h to run and was at the limit of the available memory of our computer facility (UNIVAC 1110). However, the large computer time and memory requirements can be overcome with modern wave propagation codes and making use of supercomputers (such as a CRAY).

The effect of material thickness on the impact response of polyethylene is

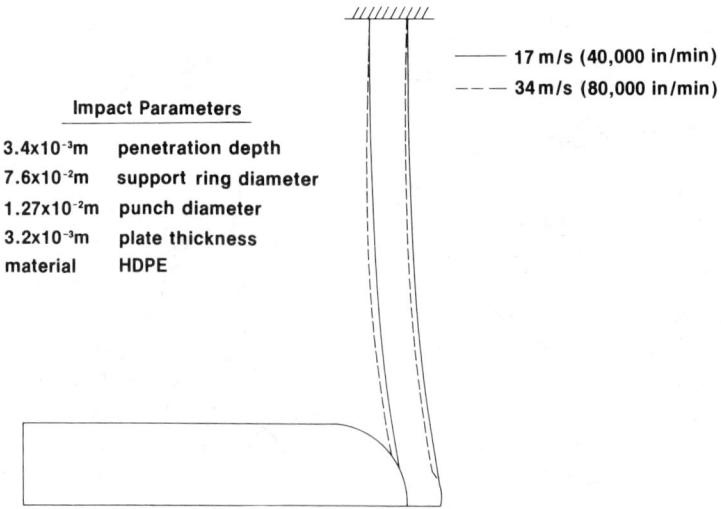

Impact Parameters

3.4×10^{-3}m	penetration depth
7.6×10^{-2}m	support ring diameter
1.27×10^{-2}m	punch diameter
3.2×10^{-3}m	plate thickness
material	HDPE

——— 17 m/s (40,000 in/min)
– – – 34 m/s (80,000 in/min)

(a)

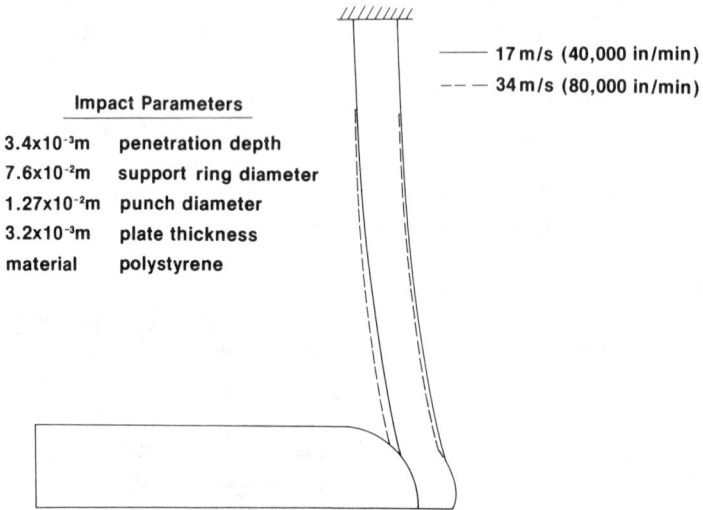

Impact Parameters

3.4×10^{-3}m	penetration depth
7.6×10^{-2}m	support ring diameter
1.27×10^{-2}m	punch diameter
3.2×10^{-3}m	plate thickness
material	polystyrene

——— 17 m/s (40,000 in/min)
– – – 34 m/s (80,000 in/min)

(b)

FIG. 5—*Deformation profiles at 17-m/s and 34-m/s impact velocities for* (a) *high-density polyethylene and* (b) *polystyrene.*

shown in Fig. 6. In these simulations, the impact velocity and clamp diameter were kept constant at 17 m/s and 3.8×10^{-2} m, respectively. For the 5.4×10^{-3}-m (214-mil) case, the impact process is characterized mainly by penetration of the specimen by the punch. This penetration results in the formation of a lip at the edge of the crater, as shown in Fig. 6. These results were also observed for polystyrene but are not shown here. This phenomenon is common in high-velocity impact situations but was not observed in the experimental results at the lower impact speeds discussed in this paper.

The effect of the clamp ring diameter on the impact response is shown in Figs. 7 and 8. As would be expected, the smaller-diameter clamp gives the specimen more rigidity and, therefore, puncture occurs rather than plate bending. This is shown in Fig. 7. These results were also observed experimentally at a lower impact speed (see Fig. 4). In Fig. 8a and b are shown the damage plots for 3.20×10^{-3}-m thick polystyrene specimens at an impact velocity of 17 m/s and using clamp diameters of 3.8×10^{-2} m and 7.6×10^{-2} m, respectively. The damage plots are shown at a very early computational time to illustrate the progress of damage for the two different clamp diameter cases. As shown, the damage opposite the impact point has progressed farther toward the clamped end in the smaller clamp diameter case than in the larger one at the same computational time. Also, damage has begun at one corner of the clamped end for the smaller clamp diameter and has not yet begun there in the case of the larger-diameter clamp. This same trend continues at later computational times, with damage eventually occurring at the clamped end for the larger-diameter clamp also. The authors conclude that the increased support of the specimen provided by the smaller-diameter clamp causes damage to occur sooner than it does with the larger-diameter clamp.

It can also be noted from the stress plots, which are now shown here, that in all cases, a tensile state of stress exists at the clamped end on the punch side, and a compressive state exists on the opposite side. This trend is reversed at the area around the axis of symmetry, where there is a state of compression just under the punch and a tensile state in the specimen directly opposite the punch. This is what one would expect for a simple plate-bending problem. Finally, the damage model predicts the occurrence of damage at locations that correspond with what is observed experimentally at the lower-impact velocities discussed earlier in this paper, that is, at the support clamp and in the polymer opposite the point of impact.

Conclusions

For both the rubber-modified polystyrene and the high-density polyethylene investigated in the present study, the impact traces from partial impacts followed the complete penetration impact traces. The deformation profile history generated from these partial impacts made possible the establishment

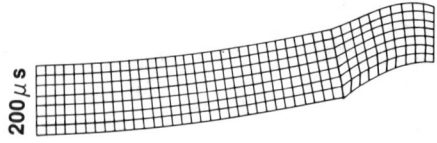

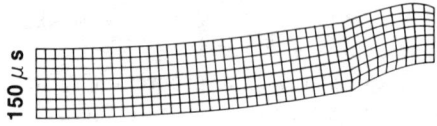

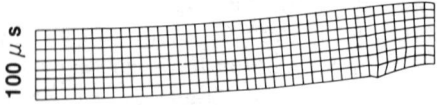

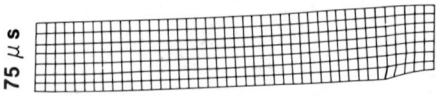

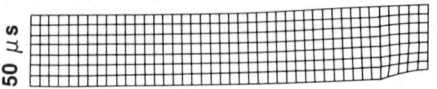

(a)

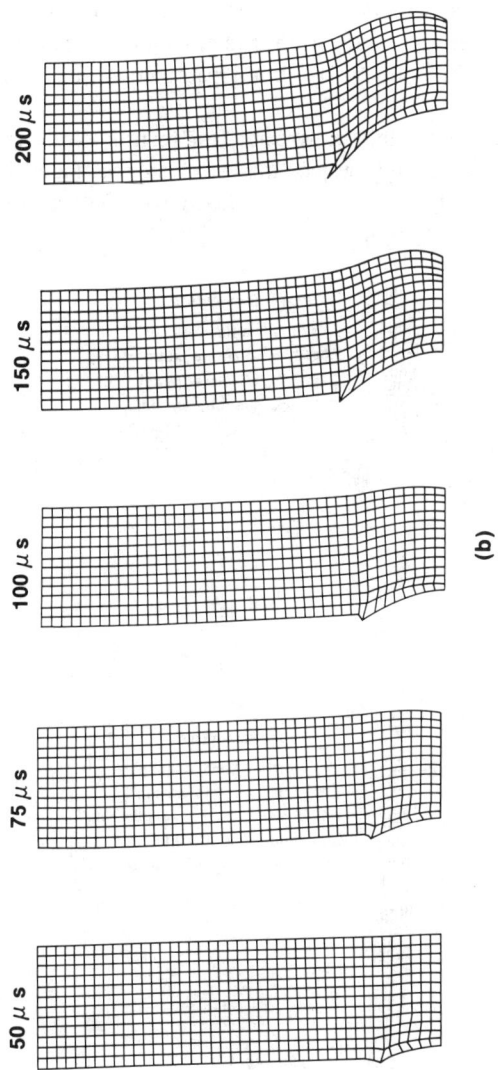

(b)

FIG. 6—Effect of thickness on the impact response of polyethylene for two different thicknesses (a) 3.2×10^{-3} m and (b) 5.4×10^{-3} m. The impact conditions in both cases were impact velocity of 17 m/s, a 3.8×10^{-2}-m-diameter support ring, and a 1.27×10^{-2}-m-diameter punch tip.

of a deformation profile/ram displacement/impact force/impact energy interrelationship. This interrelationship provides valuable information concerning the impact deformation mechanics, the development of impact load from the deformation profile and part deflection, the governing fracture mechanisms, and the development of the resulting impact energy. It also furnishes the necessary experimental evidence for any theoretical analysis of impact response in polymeric materials. In addition, the partial impact of these two materials discloses the possibility of the specimen fracturing prior to total penetration impacts.

For the rubber-modified polystyrene, the impact test results exhibited a constant fracturing impact load, strongly suggesting the existence of a critical fracture stress for this material. On the other hand, the high-density polyeth-

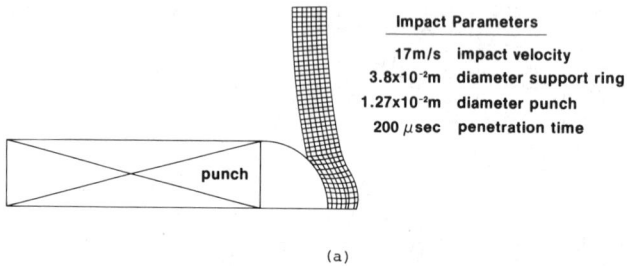

Impact Parameters

17m/s	impact velocity
3.8×10^{-2}m	diameter support ring
1.27×10^{-2}m	diameter punch
200 μsec	penetration time

punch

(a)

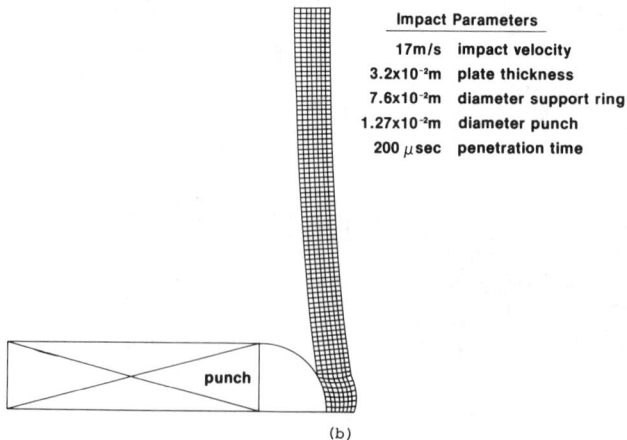

Impact Parameters

17m/s	impact velocity
3.2×10^{-2}m	plate thickness
7.6×10^{-2}m	diameter support ring
1.27×10^{-2}m	diameter punch
200 μsec	penetration time

punch

(b)

FIG. 7—*Effect of clamp diameter on polyethylene for two different clamp support ring diameters:* (a) 3.8×10^{-2} m; (b) 7.6×10^{-2} m.

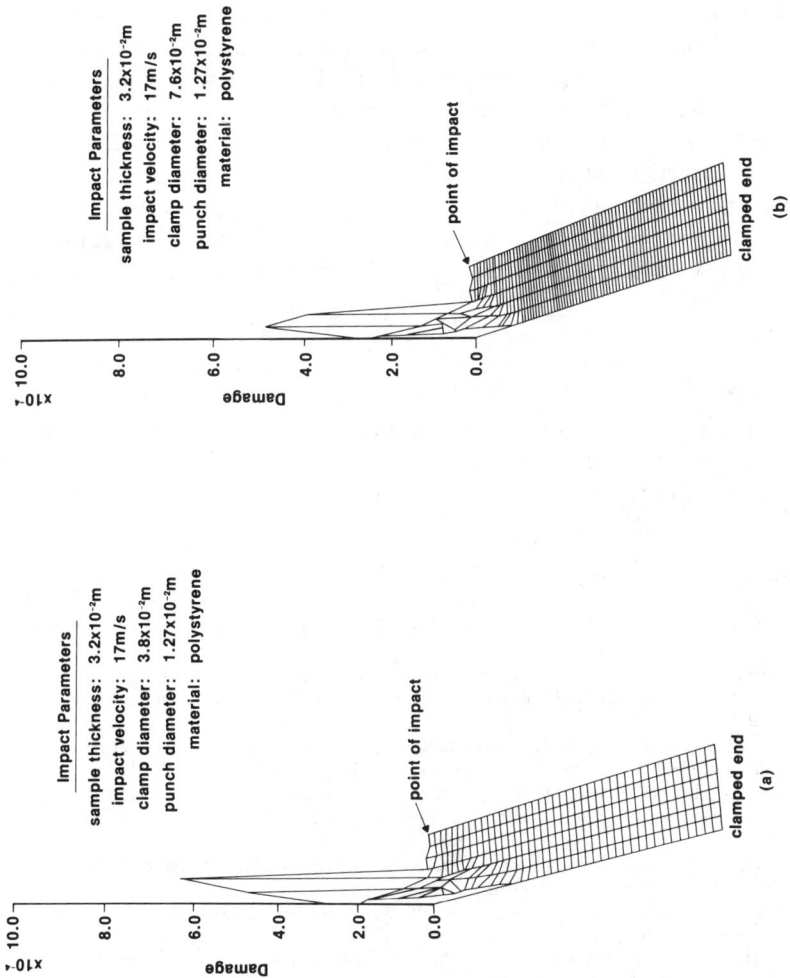

FIG. 8—*Damage contours for polystyrene at two different support ring diameters:* (a) *3.8 × 10⁻² m;* (b) *7.6 × 10⁻² m.*

ylene material did not show such a characteristic. The effect of increasing impact speed was, in general, a reduction of part deflection, and thus a decrease in impact energy under total penetration impact conditions. Increasing the test specimen size facilitated the part deflection of a specimen because it reduced flexural stiffness and produced an impact energy increase.

One of the main purposes of the computational study was to determine if the inclusion of wave propagation effects was needed to model deformation and failure at the velocity range of the experiments. Unfortunately, because of the limitations previously discussed, the computer simulations could not be conducted at the lower impact velocities. Instead, they were conducted at a somewhat higher velocity than were the experiments. However, the computational results at the higher velocities indicate that the inclusion of the wave propagation effects is important to the prediction of damage. These results are in qualitative agreement with the experimental results at the lower velocity range used in the first part of this paper, and it is, therefore, the authors' belief that the wave propagation effects may be important at this lower velocity range as well.

Acknowledgment

The authors wish to express their gratitude to L. G. McKay and T. W. Toyzan for collecting the impact data. The permission of the Dow Chemical Co. to publish this study is also appreciated.

References

[1] DeTorres, P., *Society of Plastics Engineers, Annual Technical Conference,* 1980, pp. 218–221.

[2] Johnson, A. E. and Jackson, J. R., *Society of the Plastics Industry, Reinforced Plastics/Composites Institute,* Annual Conference, Washington, DC, 11–15 Jan. 1982, Session 5 B, pp. 1–9.

[3] Myers, R. A., *Society of the Plastics Industry, Reinforced Plastics/Composites Institute,* Annual Conference, Washington, DC, Session 1C, 11–15 Jan. 1982, pp. 1–12.

[4] Dawson, J. R. and Shortall, J. B., *Society of the Plastics Industry, Reinforced Plastics/Composites Institute,* Annual Conference, Washington, DC, Session 5C, 11–15 Jan. 1982, pp. 1–4.

[5] Sing, P., "Effect of Velocity on Impact Strength Evaluation of Plastic Materials," General Motors Institute, Flint, MI, 30 April 1982.

[6] Boden, H. E. and Menges, G., *Society of Plastics Engineers, Annual Technical Conference,* 1982, pp. 7–8.

[7] Rice, D. M. and Dominguez, R. J. G., *Society of Plastics Engineers, Annual Technical Conference,* 20–22 Oct. 1982, pp. 185–191.

[8] Adams, G. C. and Wu, T. K., *Society of Plastics Engineers, Annual Technical Conference,* 1983, pp. 541–543.

[9] Gilwee, W. J. and Nir, Z., "Impact Properties of Rubber-Modified Epoxy/Graphite Fiber Composites," *Preprint,* 186th National Meeting of the American Chemical Society, Washington, DC, 28 Aug.–2 Sept. 1983, pp. 228–237.

[10] Englehardt, J. T. and Nichols, C. S., *Society of Plastics Engineers, National Technical Conference,* Detroit, MI, 20–22 Sept. 1983, pp. 72–75.

[11] Fernando, P. L., *Society of Plastics Engineers, National Technical Conference,* Detroit, MI, 20-22 Sept. 1983, pp. 161-168.

[12] Nimmer, R. P., "Analysis of the Puncture of a Biophenol—A Polycarbonate Disc," *Polymer Engineering and Science,* Vol. 23, 1983, pp. 155-164.

[13] Lindholm, U. S., "Dynamic Deformation of Metals" in *Behavior of Materials Under Dynamic Loading,* N. J. Huffington, Jr., Ed., American Society of Mechanical Engineers, New York, 1965, pp. 42-61.

[14] Wilkins, M. L., "Calculation of Elastic-Plastic Flow," Report UCRL-7322, Rev. 1, Lawrence Radiation Laboratory, Livermore, CA, 1969.

[15] Jurosek, M. L., "An Investigation of the Impact Response of Polyethylene and Polystyrene Using a Wave Mechanics Computational Approach," M.S. thesis, Michigan Technological University, Houghton, MI, 1984.

[16] Kohn, B. J., "Compilation of Hugoniot Equations of State," Air Force Weapons Laboratory Report No. AFWL-TR-69-38, Kirtland Air Force Base, NM, 1969.

[17] Tuler, F. R. and Butcher, B. M., "A Criterion for the Time Dependence of Dynamic Fracture," *International Journal of Fracture Mechanics,* Vol. 4, No. 4, 1986, pp. 431-437.

[18] Andrews, E. H., "Cracking and Crazing in Polymeric Glasses" in *The Physics of Glassy Polymers,* R. N. Haward, Ed., Applied Science Publishers, London, 1973.

[19] Cherry, B. W. and Hin, T. S., "Stress Whitening in Polyethylene," *Polymer,* Vol. 22, No. 12, 1981, pp. 1610-1612.

W. M. Lee[1] and J. K. Rieke[1]

Assessment of Impact Characteristics for Incipient Crack Formation in Polymeric Materials

REFERENCE: Lee, W. M. and Rieke, J. K., **"Assessment of Impact Characteristics for Incipient Crack Formation in Polymeric Materials,"** *Instrumented Impact Testing of Plastics and Composite Materials, ASTM STP 936,* S. L. Kessler, G. C. Adams, S. B. Driscoll, and D. R. Ireland, Eds., American Society for Testing and Materials, Philadelphia, 1987, pp. 324–334.

ABSTRACT: For many material applications, the evidence of a visible flaw (for example, a crack) can be regarded as the start of failure. It is important to determine the impact conditions for which incipient crack formation occurs. From the material design standpoint, it is essential to identify the key material properties governing this crack formation process. Therefore, its assessment and characterization become indispensable to polymeric material manufacturers as well as design engineers in material selection.

Typically, a Gardner impact tester has been employed to determine the first crack formed in a material. However, Gardner methods lack the capability of evaluating the crack initiation process as a function of impact speed. The Rheometrics impact tester can be used to determine the crack formation point by adjusting the ram travel distance into a test specimen short of complete penetration. Nevertheless, the large momentum of the impact ram at increasingly high speeds makes its stoppage at a preset travel distance extremely difficult. Hence, this mode of testing is limited to low impact speeds. In order to facilitate crack formation study at high speeds, the authors of this paper have developed a testing technique using a Dynatup impact tester. This technique involves the use of variable-thickness shims to raise the test specimen height from its lowest impact point, where contact between the tup and test specimen is just made. This allows us to control the distance the tup travels into the specimen.

The impact characteristics (that is, impact energy, impact force, and ram displacement) at crack formation are analyzed in connection with material selection. One of these three quantities is suggested for impact toughness evaluation, and its underlying rationale and significance are discussed. Also, the impact characteristics at crack initiation and at complete puncture are compared in relation to material design and selection.

KEY WORDS: impact, instrumented impact, acrylonitrile-butadiene-styrene (ABS) copolymers, crack formation, fracture, polymer, impact testing

[1]Senior associate scientist and research associate, respectively, Central Research, Dow Chemical Co., Midland, MI 48674.

The impact properties of polymeric materials are often determined in a total penetration mode by the use of instrumented impact test apparatus [1-9]. The impact energy at ultimate failure is usually employed to assess the impact toughness of a polymeric material or composite. However, in end-use applications, a plastic article, on impact, may first develop a visible crack without undergoing catastrophic failure. Subsequently, this crack may grow under environmental, thermal, or mechanical stress and lead to gross failure of the article.

In addition, there are circumstances under which the plastic object is subjected to impact fatigue treatment. Upon repetitive impact blows, a visible crack may develop, which can grow into multiple unstable cracks that can lead to gross article failures. Under these circumstances, it is important to understand the material properties that govern the resistance to crack formation and assess which are the more important.

The objective of the present study is to develop a technique for studying incipient crack formation in polymeric materials using instrumented impact test apparatus. Typically, Gardner impact test equipment has been employed to determine the conditions of first crack formation in polymeric materials. Gardner impact methods not only lack the capability of evaluating the crack formation process at prescribed impact speeds, but are also time-consuming and provide no information on the load-deformation behavior of the specimen being tested. Under most conditions, the weight is raised and lowered until an observable failure occurs in a series of specimens of similar kind.

Instrumented impact testers have been used to study crack initiation in composites [10-12] and to examine impact fatigue resistance [13-16]. In most instances in which instrumented falling dart impact test equipment has been utilized, the methods are similar to those utilized in standard Gardner impact studies. In this study the authors propose a simple method for studying incipient crack formation using a falling weight impact test machine (the Dynatup apparatus) with a simple modification so that the depth of penetration can be adjusted while a constant velocity at impact is maintained. In principle, the technique can be used on most falling weight impact testers.

Servohydraulic machines (such as the Rheometrics machine) also have the capability of assessing incipient crack formation by adjusting the ram travel distance. However, because of the relatively high speed, valve opening and closing times, and ram momentum it is very difficult to preset the travel distance accurately. Used at lower speeds, the Rheometrics impact tester has been satisfactory.

Experimental Procedure

A Dynatup 8000A drop tower impact tester was employed for the incipient crack formation study reported here. The measuring circuit has been modified so that the load-time trace is recorded on a Nicolet 3092 digital oscilloscope. The stored impact data are passed batchwise to a Digital Equipment PDP 11/44 computer for subsequent analysis and display. When released, the dropping weight is arrested by metal blocks at a predetermined point. In order to control the penetration of the tup (fastened to the dropping weight) into the specimen, the anvil height (specimen holder) is made adjustable through the use of a variety of metal shims. The shim thicknesses used are multiples of 1.27×10^{-5} m. With this method, both the drop height and the degree of penetration of the specimen are controlled accurately and independently.

Figure 1 illustrates typical load-time traces for a specimen that was subjected to a total penetrating impact blow and also for a specimen that was subjected to partial impact. Since the small perturbations shown on the traces occur regularly and reproducibly, we believe they are real effects. At this time, however, because of the general accuracy of these impact measurements, we do not believe a more detailed interpretation is warranted.

In this study a standard dropping mass (including the tup) of 62.82 kg was

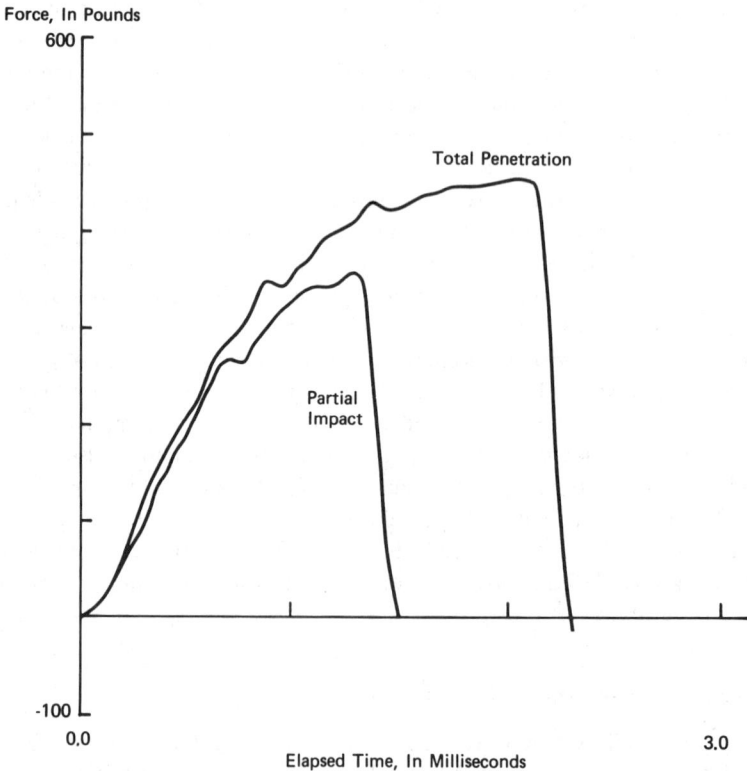

FIG. 1—*Typical force-time trace for ABS at* $-20°C$ *and an impact speed of 2.45 m/s.*

employed. Three different drop heights were used to provide a measure of speed variation. These were 3.05×10^{-1} m, 1.52×10^{-1} m, and 0.762×10^{-1} m. The corresponding velocities at impact obtained from individual calibrations were 2.45, 1.72, and 1.09 m/s, respectively. All the tests were run using a tup with a hemispherical end whose diameter is 1.59×10^{-2} m. The specimens were clamped, and the anvil had a circular opening of 3.18×10^{-2} m. The test specimens were injection-molded acrylonitrile-butadiene-styrene (ABS) disks of 5.08×10^{-2} m in diameter and were approximately 3.23×10^{-3} m thick.

Impact tests were carried out at several different temperatures, that is, -40, -20, 0, 23, and 40°C. Subambient measurements were made by placing specimens in a closed glass jar which was, in turn, placed in a dry ice bath. An identical specimen, except with an embedded thermocouple, was also placed in the bath. The test specimens were allowed to equilibrate at the temperature of the bath ($-73°$C) for approximately 1 h. The specimens were removed from the ice bath, placed on the anvil, clamped, and allowed to warm until a predetermined surface temperature was achieved. The drop weight and tup were then released and allowed to impact on the specimen in the holder. Previously, a calibration curve for warming was obtained by monitoring the temperature of the specimen containing the thermocouple when it was placed on the anvil, clamped, and allowed to warm. Similar measurements have been made when the thermocouples were mounted in the center of the specimen and near the top surface. In general, the temperature at the center of the specimen lagged behind the surface temperature at subambient temperatures. The specimens run above room temperature were preheated in an oven, mounted, and allowed to cool to a predetermined test temperature. Specimen surface temperature was again selected as the reported impact test temperature. While these methods are not as accurate as would be obtained if the anvil and specimen were allowed to equilibrate in a thermostatically controlled arrangement, they do provide for speed and generally provide no undue complications in data interpretation.

Results and Discussion

At a given drop height and test temperature, impact measurements were made while the number (hence the thickness) of shims under the anvil was increased. This, in turn, permitted the tup to penetrate deeper into the specimen being tested. As the shim thickness increased, both the measured impact force and the impact energy increased (see Table 1) until, at some penetration, failure occurred. In the test temperature range studied, the particular ABS used exhibited mainly two types of failure behavior. As expected, the change in failure mechanism was more strongly dependent on temperature than on test speed. (Only a very limited speed range was available with this apparatus.)

TABLE 1—*Dynatup drop tower impact test data: effect of penetration on failure at constant temperature.* [a]

Drop Height, m	Impact Speed, m/s	Specimen Temperature, °C		Penetration Depth, m × 10⁻⁵	Impact	
		Surface	Center		Force, N	Energy, J
0.305	2.45	−20	−31	28	1630	7.6
				38	1850	8.6
				51	1850	9.5
				81	1980	10.4
				97	2200	11.3
				103	2270	11.7
				104	cracked	cracked

[a]Test conditions:
  Material: ABS.
  Specimen dimensions: 5.08 × 10⁻²-m-diameter disk, 3.23 × 10⁻³ m thick.
  Tup: 1.59 × 10⁻²-m-diameter hemispherical end.
  Anvil: 3.18 × 10⁻²-m opening, equipped with retaining ring.
  Drop mass: 62.82 kg.
  Drop height: 0.305 m.
  Impact speed: 2.45 m/s.
  Specimen temperatures: −20°C, surface; −31°C, center.

At subambient temperatures, ABS initially displayed stress whitening at small depths of penetration. The size of the stress-whitened area increased as the level of penetration increased, until the specimen "fractured" by a cylindrical blowout or by crack generation. The change from stress whitening to gross failure occurred quite suddenly at a given level of complex strain. Blowout refers to the breaking away of a portion of the specimen, generally cylindrical in shape, directly under the tup; crack generation is either a single large crack—often splitting the specimen into two halves—or a star-shaped series of cracks emanating from the impact point. A change of as little as 2.54×10^{-5} m penetration was sufficient to change the failure mode profoundly.

If one follows the observed values of impact energy as the depth of penetration increases, these values can be seen to increase until gross failure or fracture occurs. At this point, the values for impact energy decrease rapidly. For example, at $-20°C$, with a velocity of 2.45 m/s and a shim height of 103×10^{-5} m, the impact force was 2270 N and the impact energy was 11.7 J. This consumed energy is small compared with the total energy available for impact (that is, mass × acceleration × height = 62.8 kg × 9.8 ms⁻² × 0.305 m = 187.71 J). When the shim height was elevated by 1×10^{-5} m, the material failed by blowout, and the impact energy was considerably less than 11.7 J, as indicated in Table 1. (It should be noted that the drop height is the distance from the tip of the tup to the top surface of the specimen prior to deformation.

When shims are added to the base of the anvil, the drop height is decreased slightly but, for all intents and purposes, changes the drop height an insignificant amount. We have ignored this change in any discussions.)

At ambient temperatures and above, a procedure similar to that used at subambient temperatures was utilized. Observations made on the impacted specimens show more of a localized drawing around the tup—a kind of cap was formed around the tup—but the switch from ductile yielding to gross failure occurred just as quickly as in the subambient situation. However, the observed impact energy continued to increase as the depth of penetration was increased into the gross failure region. Observations made on failed specimens show that the cracking occurred after the drawn cap was formed, hence the total energy dissipated in failure would be the sum of the energy to draw the cap plus that to produce the crack.

The depth of penetration for incipient crack formation for ABS has been determined under a variety of impact test conditions. These results are summarized in Table 2. It is readily apparent that the amount of penetration for crack initiation increased as the test temperature was increased. For example, at a velocity of 2.45 m/s, the crack penetration depths were 33, 104, 204, 1034, and 1133×10^{-5} m at specimen surface temperatures of -40, -20, 0, 23, and 40°C, respectively. At low-impact temperatures, the material behaves in a brittle manner when subjected to the complex stresses and strains imposed by a tup impact. Hence, it would not take a large penetration to induce crack formation. As the specimen temperature is raised, the material becomes less brittle and the impacting tup travels further into the specimen before cracks occur. A large jump in the crack penetration depth took place as the test temperature rose from 0°C to room temperature but increased only slightly at temperatures above room temperature. The ductile deformation characteristics of these ABS materials at 23 and 40°C should be specifically noted and contrasted with the brittle behavior observed at 0°C.

It is reasonable to state that, in this geometry and under the specified test conditions, a brittle-ductile transition occurs between 0 and 23°C. As the impact velocity was varied from 2.45, 1.72, to 1.09 m/s, similar effects were observed. Over the very limited velocity range investigated, the ductile-brittle temperature is approximately independent of velocity. The cracking penetration depth and its relationship to the test temperature are plotted in Fig. 2.

When the effects of penetration on incipient crack formation (in contrast with those of the ductile-brittle transition discussed) are examined as a function of test velocity, the effect of test velocity becomes more pronounced. For instance, at $-40°C$ the incipient cracks are formed at penetration depths of 33, 104, and 152×10^{-5} m as the tup velocity varied from 2.45, to 1.72, to 1.09 m/s, respectively—a factor of 5—while the velocity was varied by a little more than 2. When the same effects are examined at room temperature, the effect is a factor of about 1.1. However, the difference in penetration depth is essentially the same whether the test is carried out at $-40°C$ or at room tem-

TABLE 2—*Dynatup drop tower impact test data: effect of impact speed.* [a]

Drop Height, m	Impact Speed, m/s	Specimen Temperature, °C		Penetration Depth, m × 10⁻⁵	Impact Energy, J
		Surface	Center	m × 10^{-5}	J
0.305	2.45	−40	−53	31	9.4
				33	cracked
		−20	−31	103	11.7
				104	cracked
		0	−1	239	19.1
				246	cracked
		23	23	958	53.0
				1034	56.6 [b]
		40	42	1090	52.1
				1133	54.2 [b]
0.152	1.72	−40	−54	102	7.8
				104	cracked
		−20	−27	150	9.5
				155	cracked
		0	−1	412	20.9
				417	cracked
		23	23	1085	55.1
				1092	56.6 [b]
0.076	1.09	−40	−53	140	5.7
				152	cracked
		−20	−28	216	9.2
				229	cracked
		0	−1	508	24.5
				516	cracked
		23	23	1171	58.1 [b]
		40	. . .	1232	51.3 [b]

[a] Test conditions:
  Material: ABS.
  Specimen dimensions: 5.08 × 10^{-2}-m-diameter disk, 3.23 × 10^{-3} m thick.
  Tup: 1.59 × 10^{-2}-m-diameter hemispherical end.
  Anvil: 3.18 × 10^{-2}-m opening, equipped with retaining ring.
  Drop mass: 62.82 kg.
[b] Test specimen cracked on the drawn cap.

perature (that is, 119 versus 147 × 10^{-5} m). This effect is easily lost if the material undergoes a brittle-ductile temperature transition under the test conditions used. These facts are not too surprising if one remembers that no large dynamic transitions occur in the temperature regime −40 to 23°C, and even if they did, the energy dissipation capabilities are small compared with the possibility of grossly distorting the rubber-reinforcing phases.

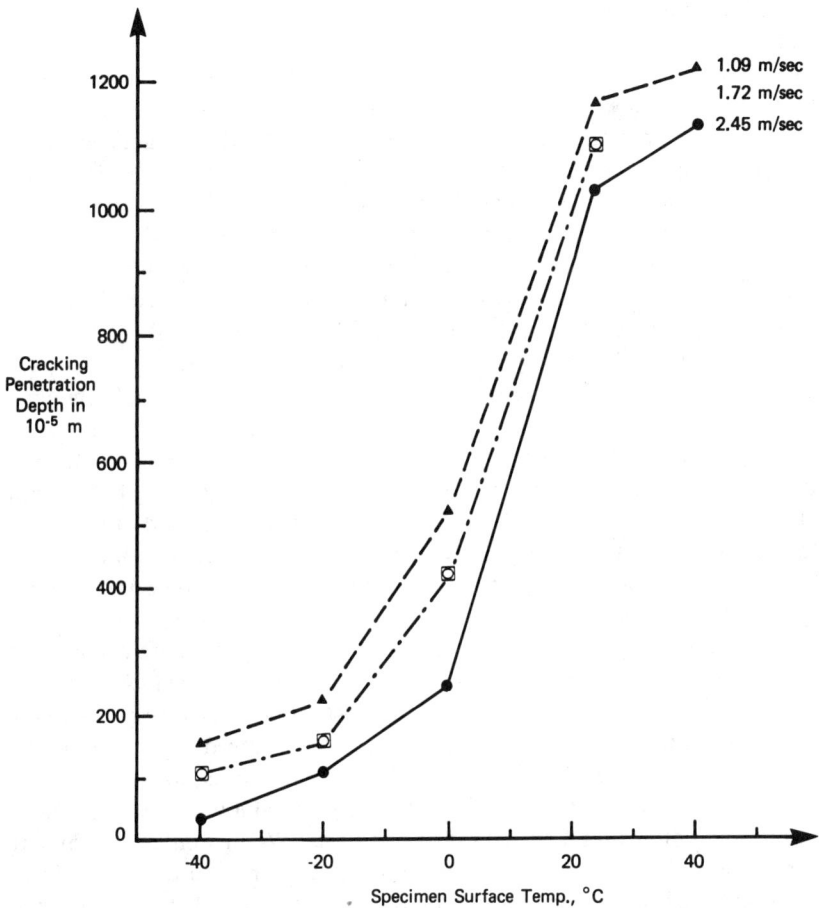

FIG. 2—*Penetration depth for cracking versus impact test temperature. Test conditions: ABS clamped specimen, 3.23 × 10⁻³ m thick; hemispherical tup, 1.59 × 10⁻² m in diameter; anvil, 3.18 × 10⁻² m.*

The measured impact energy values for penetration depths just prior to the onset of cracking are included in Table 2 for the various impact conditions studied. At a given impact velocity, the test data show that the impact energy increases with increasing test temperature. Close to and above room temperature, the impact energy became relatively insensitive to the test temperature—again, probably because of the large energy dissipation that occurred during the ductile deformation of the ABS specimens. As an example, the impact energy values at 2.45 m/s for temperatures of −40, −20, 0, 23, and 40°C were 9.4, 11.7, 19.1, 53.0, and 52.1 J at 31, 103, 239, 958, and 1090 × 10⁻⁵-m penetration depths, respectively. The corresponding depths at which incipient cracks were observed were 33, 104, 246, 1034, and 1133 $\times$ 10^{-5} m, respectively.

The effect of impact velocity on the impact energy varies with the test temperature. At very low test temperatures, the impact energy decreases with decreasing impact tup velocity. For instance, at $-40°C$, the impact energy values—just prior to crack formation—were 9.4, 7.8, and 5.7 J at velocities of 2.45, 1.72, and 1.09 m/s, respectively. The inverse relationship (greater energy dissipation at higher speeds) was attributed to the ability of the specimen to exhibit greater overall impact resistance throughout the impact region (that is, it loads up better). As the test temperature was increased, the impact energy became virtually insensitive to velocity. For example, the 23°C crack formation impact energies at 2.45, 1.72, and 1.09 m/s were 56.6, 56.6, and 58.1 J, respectively. These large values are probably attributable to the large amount of energy dissipated while the ABS undergoes a highly ductile deformation prior to cracking. The impact characteristics developed with this ABS material, and obtained under various test conditions, suggest that the incipient crack formation determination provides a valuable method of determining the impact resistance of polymer objects. While different from the total penetration methods, it should be a useful technique to determine failure when a single crack constitutes failure.

For comparison purposes, the total penetration impacts were run on ABS using the Dynatup drop tower without shims under the anvil. At 2.45-m/s velocity and 23°C, complete penetration of specimens gave rise to a highly drawn cap, which eventually split into two parts. Observed impact energy was 62 J. This value is slightly greater than the value measured to first crack formation (57 J). This difference (5 J) is largely the amount of work done in tearing the highly stretched ABS into two parts. When similar experiments were done at low temperature ($-40°C$), the specimen broke into several pieces and the impact energy was somewhat greater than the value measured at first crack formation. Again, this difference can be interpreted as the energy consumed as small cracks become large cracks which lead to shattering.

Conclusions

A useful technique, using a modified Dynatup drop tower impact test machine, has been developed for assessing the incipient crack formation in polymeric materials when impact loaded. Using this method, partial impacts—which we believe are important to initial crack formation studies—were achieved by using metal shims of various thicknesses under the specimen anvil. Hence, the depth of specimen penetration could be controlled independently of the velocity of impact. The technique was utilized to test injection-molded ABS under a variety of conditions. Generally speaking, the effects of temperature and impact velocity were as expected.

This method of testing offers design engineers and material users a basis for material selection when the occurrence of a visible crack is critical to the specific application. Under these circumstances, the impact energy at first crack, or even the degree of penetration of the specimen at first crack, may be used as a basis for material selection. Similar measurements can be made in a fatigue mode by applying the technique of partial impact repetitively until failure occurs.

The Dynatup drop tower has been used successfully to study crack formation when impact velocities were 1.27×10^{-1} m/s or greater. At shorter drop distances (lower velocities), the velocity variations are considerable. On the other hand, the Rheometrics tester has been used successfully at lower speeds at which the ram can be quickly stopped. At higher speeds, the system is too soft for accurate determinations of first cracks.

Acknowledgments

The authors wish to express their appreciation to E. D. Davis, who injection molded the specimens used in this study, and to D. Bank for carrying out the impact determinations. The permission of the Dow Chemical Co. to publish this paper is greatly appreciated.

References

[1] Silverwood, H. A., "Performance of Flexible RIM Urethane Fascia Composites," Report from Davidson Rubber Division, Division of Ex-Cell-O Corp., Dover, NH.

[2] de Torres, P. D., *Proceedings,* Annual Technical Conference, Society of Plastic Engineers, New York, 1980, pp. 218-221.

[3] Shortall, J. B. and Dawson, J. R., *Proceedings,* 37th Annual Conference, Reinforced Plastics/Composites Institute, Session 5-C, Society of the Plastics Industry, Washington, DC, 11-15 Jan. 1982, pp. 1-4.

[4] Sing, P. J., "Effects of Velocity on Impact Strength Evaluation of Plastic Materials," Report from General Motors Institute, Flint, MI, 30 April 1982.

[5] Menges, G. and Boden, H. E., *Proceedings,* Annual Technical Conference, Society of Plastic Engineers, Boston, MA, 1982, pp. 7-8.

[6] Rice, D. M. and Dominquez, J. G., *Proceedings,* Annual Technical Conference, Society of Plastic Engineers, Boston, MA, 1982, pp. 185-191.

[7] Gilwee, W. J. and Nir, Z., *Proceedings,* 186th National Meeting of American Chemical Society, 1983, Washington, DC, pp. 228-237.

[8] Englehardt, J. T. and Nichols, C. S., *Proceedings,* National Technical Conference, Society of Plastic Engineers, Detroit, MI, September 1983, pp. 72-75.

[9] Fernando, P. L., *Proceedings,* National Technical Conference, Society of Plastic Engineers, Detroit, MI, September 1983, pp. 161-168.

[10] Myers, F. A., *Proceedings,* 37th Annual Conference Reinforced Plastics/Composite Institute, Session 1-C, Society of the Plastics Industry, Washington, DC, 11-15 Jan. 1982, pp. 1-12.

[11] Johnson, A. E. and Jackson, J. R., *Proceedings,* 37th Annual Conference, Reinforced Plastics/Composite Institute, Session 5-B, Society of the Plastics Industry, Washington, DC, 11-15 Jan. 1982, pp. 1-9.

[12] Miller, A. G., Hertzberg, P. E., Rantala, V. W. *SAMPE Quarterly,* Vol. 12, January 1981, p. 36.

[13] Adams, G. C. and Wu, T. K., *Proceedings,* Annual Technical Conference, Society of Plastic Engineers, San Francisco, CA, 1983, pp. 541-543.

[14] Bhateja, S. K., Rieke, J. K., and Andrews, E. H., *Journal of Materials Science,* Vol. 14, 1979, p. 2103.
[15] Rieke, J. K., Bhateja, S. K., and Andrews, E. H., *Industrial and Engineering Chemistry Product Research Development,* Vol. 19, 1980, p. 601.
[16] Bhateja, S. K., Rieke, J. K., and Andrews, E. H., *Industrial and Engineering Chemistry Product Research Development,* Vol. 19, 1980, p. 607.

Fracture Toughness

John M. Hodgkinson[1] and J. Gordon Williams[2]

Analysis of Force and Energy Measurements in Impact Testing

REFERENCE: Hodgkinson, J. M. and Williams, J. G., "**Analysis of Force and Energy Measurements in Impact Testing,**" *Instrumented Impact Testing of Plastics and Composite Materials, ASTM STP 936,* S. L. Kessler, G. C. Adams, S. B. Driscoll, and D. R. Ireland, Eds., American Society for Testing and Materials, Philadelphia, 1987, pp. 337–350.

ABSTRACT: Complete velocity-time (v-t) relationships for a pendulum striker during impact testing of sharp-notched high-density polyethylene have been recorded by means of laser-Doppler velocimetry. The velocity data have been processed using a microcomputer to give force-deflection curves and hence absorbed energy. Toughness is expressed in terms of the critical strain energy release rate, G_c, and the critical stress intensity, K_c. The plane stress G_c from pendulum energy was 11.3 kJ/m², and that from areas under force-deflection curves was 9.9 kJ/m². The authors observed that an initiation G_c of the order 2.3 kJ/m² might be expected for this material below its glass transition. By defining crack initiation in these room temperature tests as the point of maximum recorded force, an initiation G_c of 2.0 kJ/m² is obtained. The critical stress intensity factor, K_c, is also defined from the maximum force, and a value of 2.1 MN/m³/² is calculated. The elastic modulus is computed from compliance measurements to be 1.84 GN/m², whereas from $E = K_c^2/G_c$ the plane-stress modulus is 1.93 GN/m², and the introduction of Poisson's ratio gives a plane-strain modulus of 1.69 GN/m².

KEY WORDS: instrumented impact test, fracture mechanics, fracture toughness, force measurement, laser-Doppler, velocity measurement, impact testing

Most methods currently available for determining applied forces during impact involve the physical attachment of a transducer to the striker, with consequent problems arising from stress waves within the striker/transducer assembly, resonances, and attached cables. Under these circumstances it is common practice to filter the output signal electronically, in order to obtain traces that appear reasonable. Since the level of filtering affects the shape of

---

[1]Senior research officer, Centre for Composite Materials, Imperial College of Science and Technology, London SW7 2BY, England.

[2]Professor, Department of Mechanical Engineering, Imperial College of Science and Technology, London SW7 2BX, England.

the signal, particularly the peak, the question arises as to how much filtering is reasonable. A further problem is associated with transducers that require static calibration, leading to uncertainty at impact rates.

To try to circumvent these problems the authors have developed a noncontacting system, which makes use of laser-Doppler, to measure striker velocities during impact. This technique has more usually been applied in the areas of fluid flow or gas velocity measurement. Various laser-Doppler techniques have been described in detail elsewhere [1]; the instrumentation, experimental approach, and some analysis for this particular application have also been the subject of previous publications [2-4]. The analysis performed previously has been done by hand calculation and measurement, a tedious and time-consuming occupation requiring the determination of the periodic nature of a signal varying in frequency up to 200 kHz, over a 4-ms time span. Since the Doppler shift recorded during a test is stored digitally in a transient recorder, the next obvious step, after demonstration that the approach used was viable, was to prepare suitable software to analyze the signal using a microcomputer. This signal analysis from the digitized information in the transient recorder to the derived fracture mechanics data is the subject of the present work.

Previous publications [2-4] have investigated the fracture of Charpy-type specimens with a falling weight striker. Here use has been made of a pendulum impact machine, so that an independent record of absorbed energy during fracture was obtained for direct comparison with data derived from velocity measurements.

Experimental Arrangement and Velocity Measurements

A series of bar-shaped high-density polyethylene (HDPE)[3] specimens 50 mm long with 6 by 6-mm cross sections were machined from flat sheet. Each specimen was subsequently notched centrally using a single point cutter with 60° vee and tip radius approximately 15 μm. The notch depth was varied from specimen to specimen in the range of 0.2 to 3 mm. All the machining operations and impact tests were carried out at room temperature.

Figure 1 shows the impact pendulum arrangement and specimen configuration, and Fig. 2 shows the laser optical system. The laser beams referred to in Fig. 2 were focused on the pendulum in its vertical position, so that as the pendulum struck a specimen, a signal was generated of a frequency proportional to the pendulum velocity. As the pendulum velocity changed during impact, the frequency of the signal developed in the photomultiplier altered accordingly, allowing the complete velocity-time curve to be derived by measurement of consecutive periods.

A typical velocity-time curve for the pendulum during the fracturing of a specimen is shown in Fig. 3. Each point represents one period of the Doppler

[3]Compression-molded sheets of HDPE Grade 006-60 were kindly supplied by BP Chemicals Ltd., Grangemouth, Scotland.

FIG. 1—*The pendulum impact machine and specimen configuration.*

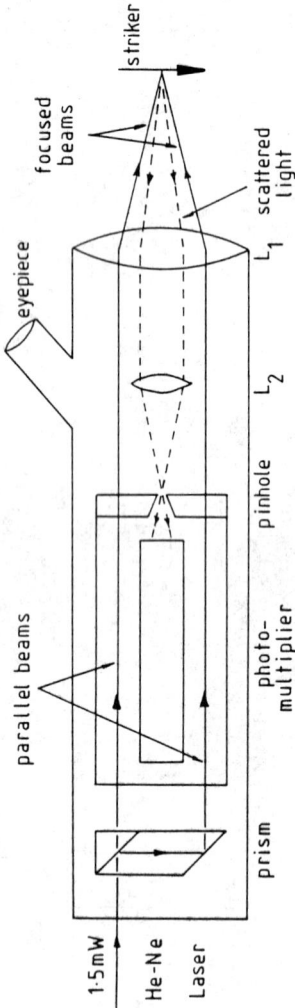

FIG. 2—The laser-Doppler optical system.

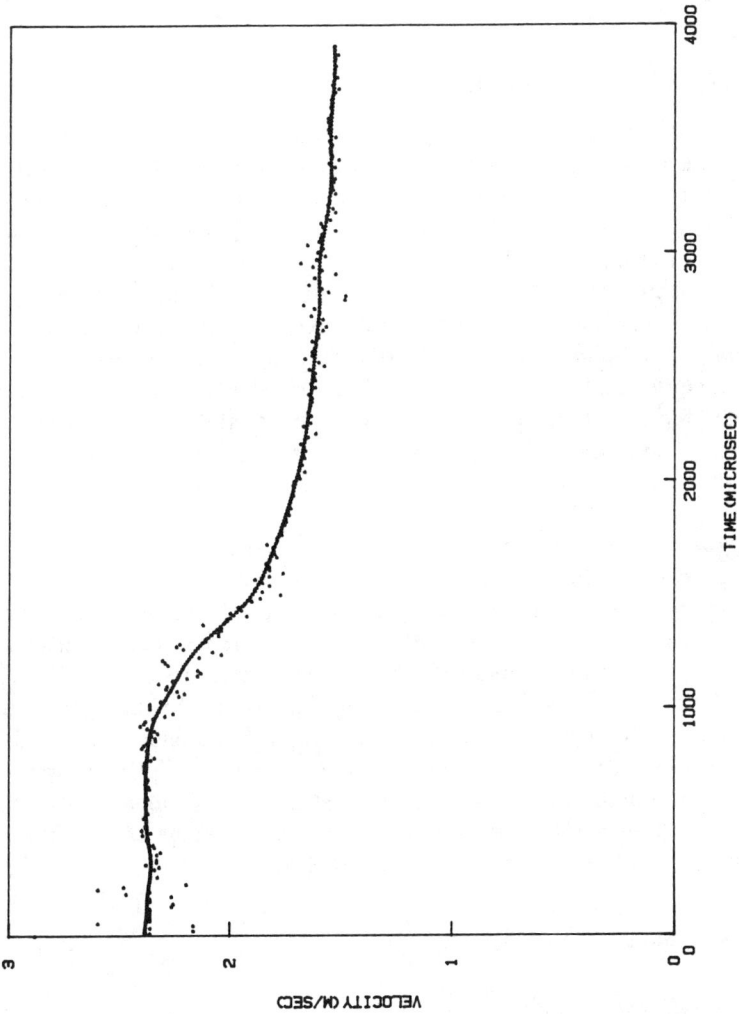

FIG. 3—*Pendulum velocity change with time during the impact period.*

signal, and clearly a fair degree of scatter exists within the data. Scatter of this order, some $\pm 2\%$, is quite common in laser-Doppler anemometry, but it is not normally critical since the velocities of interest are usually constant. Here, however, the necessity of differentiation of the velocity-time curves to obtain acceleration, and hence force, preferably point-to-point, requires a curve with virtually no scatter since positive slopes are meaningless in this context.

Processing the Velocity-Time Data

Various attempts were made to reduce the scatter on the velocity-time curves by means of digital analysis, and the most successful to date has been a repeated averaging system. Under this system the number of points to be averaged is chosen, and velocity averages are calculated along the curve, indexing one point at a time, with the average value in each case being plotted at the midpoint time. It is then possible to calculate the difference between the current value for a particular point and its previous value, allowing the realization of a smoothing factor. In order to reduce the scatter to the point where a smooth curve results, the rolling averaging procedure is repeated until the smoothing factor reduces to a predetermined small value, which can be common for each specimen. The smooth line in Fig. 3 is that resulting from this approach, and it clearly represents a good average for the scattered data points.

A degree of care is required in the choice of how many points to average. If too few are chosen, the smoothing is so fine that the process becomes interminable. On the other hand, if too many points are used, smoothing may be so coarse that the peak force derived from the velocity curve is reduced in a manner similar to that encountered when transducer signals are electrically overfiltered. Passing the data through the smoothing program too many times has a similar effect. By experimentation it was found that a choice of nine data points and five passes through the smoothing routine gave optimum smoothing. This combination of points and passes through the smoothing routine gave a smoothing factor, expressed in terms of the average change in force from one pass to the next, of approximately 5 N.

Fracture Mechanics Analysis

Strain Energy Release Rate

Once the smooth velocity-time (v-t) curve has been generated, the curve may be subjected to differentiation to obtain dv/dt, and hence, by assumption of the equation of motion with the striker of mass, m, and specimen in contact, the applied force, $m(dv/dt)$, as a function of time. This relationship for one particular specimen is shown in Fig. 4. Subsequent integration of the

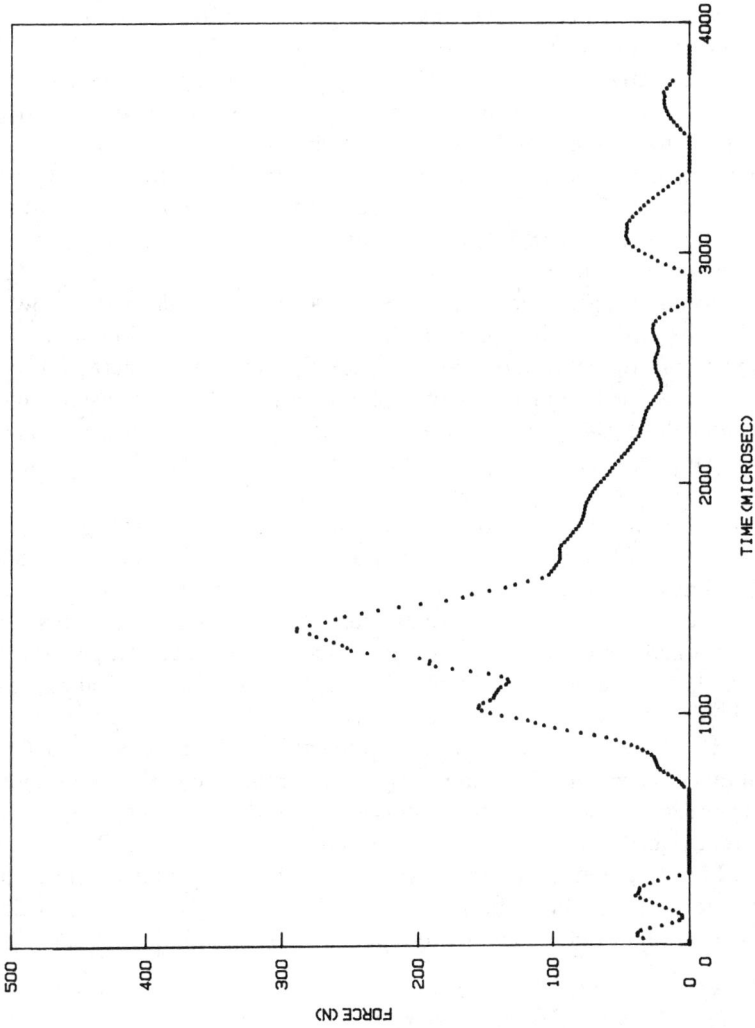

FIG. 4—*Applied force as a function of time.*

velocity-time curve results in data relating deflection and time, so that force-deflection curves may be generated; an example is shown in Fig. 5. The area under the force-deflection curve represents the energy absorbed during the fracture event, and these energy data may be plotted as a function of $BD\phi$ where B and D are the specimen cross-section dimensions, shown in Fig. 1, and ϕ is a compliance calibration function [5]. There are clearly a number of possibilities here; it is reasonable to assume that crack initiation occurs at the maximum force,[4] and lower peaks result from specimen vibration and loss of contact with the striker [6].

An initiation energy may then be determined, but a gross energy, inclusive of initiation, propagation, and any ductile effects, will be represented by the total energy under the curve, and this should be similar to that energy loss recorded by the pendulum. This total energy under the loading curve is plotted as a function of $BD\phi$ in Fig. 6 for comparison purposes with the pendulum energy loss record.

Quite reasonable agreement is achieved between the two sets of data. Both curves show extensive curvature as the initial crack is reduced in length, an indication of plastic deformation at the crack tip. The strain energy release rate, G_c, is obtained from the slope of the energy/$BD\phi$ relationship, and when corrected for crack-tip plastic zone effects [7], the indicated G_c values for these two sets of data are quite similar at 11.3 kJ/m² for the pendulum record and 9.9 kJ/m² for the areas under the loading curves.

These toughness values do, however, describe the propagation toughness of this relatively ductile material. Previous work [8] has shown that the plane-strain toughness would be significantly lower, approaching the initiation G_c that would result if the material were brittle. The plane-strain toughness has been determined for this material using the technique described previously [8] and suggests a value of 3.6 kJ/m². Although tests have not been carried out at low temperature here, to measure brittle initiation G_c, previous work [9] suggests a value of 2 to 3 kJ/m² to be typical for HDPE below −60°C.

Given this background, it is interesting to determine the initiation G_c indicated by the force-deflection curves of this work. To this end, the energy up to the maximum force was measured and plotted as a function of $BD\phi$. This is shown in Fig. 7, and although there is scatter, a mean line may be drawn to quantify the initiation G_c. The linear regression slope gives a value of 2.0 kJ/m², and no data fall outside the limits of 1.2 to 4.75 kJ/m².

Critical Stress Intensity Factor for Impact

The force record with time allows the identification of a peak force, P, which must be associated with crack initiation, so that stresses applied at the

---

[4]Kinloch, A. J., Department of Mechanical Engineering, Imperial College, personal communication referring to results of crack initiation gage applied to polymer impact specimen, February 1985.

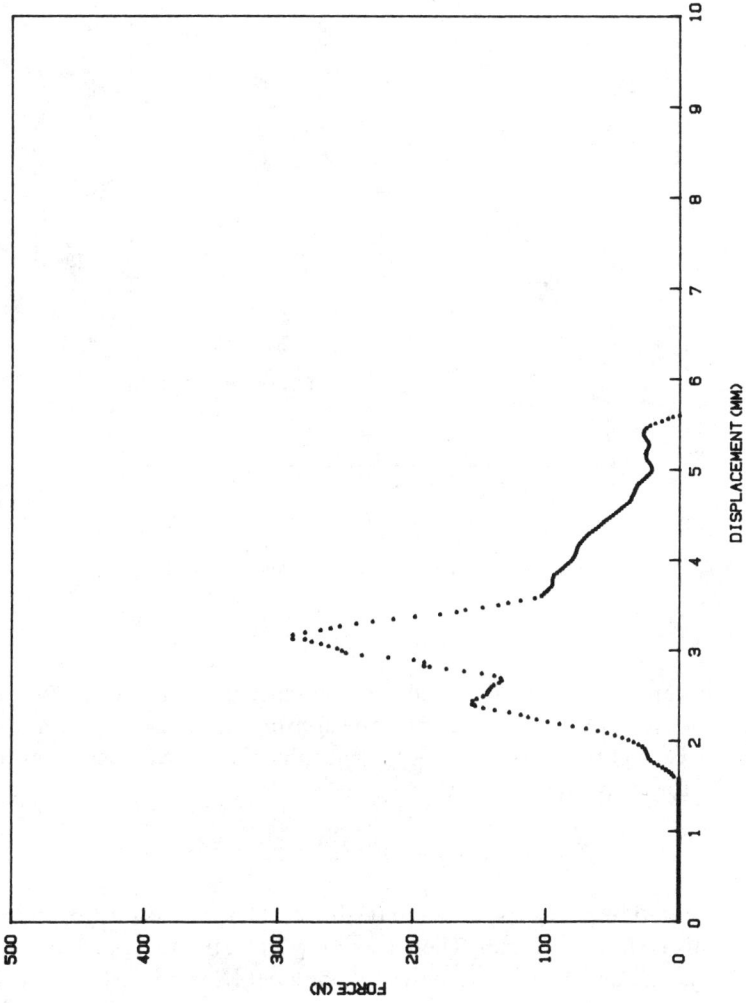

FIG. 5—*Applied force as a function of deflection.*

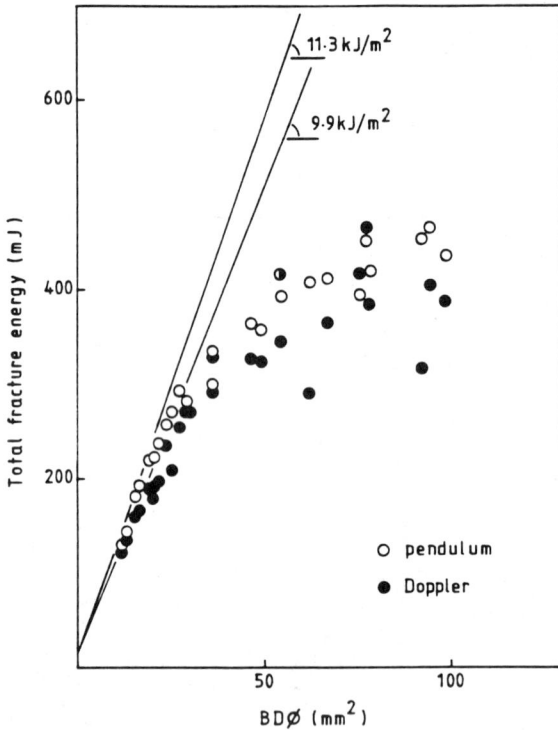

FIG. 6—*Total impact fracture energy as a function of* BDϕ.

point of initiation may be calculated for each specimen. These stresses, σ, may then be used in Eq 1, together with the appropriate crack length, a, and finite width correction factor, Y [10], to determine the critical stress intensity factor, K_c, for individual specimens.

$$K_c^2 = \sigma^2 Y^2 a \qquad (1)$$

For presentation purposes, it is interesting to plot $\sigma^2 Y^2$ as a function of a^{-1}, and this is shown in Fig. 8. There is scatter, but a trend exists to define a mean slope, K_c^2, and hence a value for K_c of 2.1 MN/m$^{3/2}$, determined by linear regression, with no values outside the range of 1.62 to 2.75 MN/m$^{3/2}$.

Impact Modulus

The compliance, C, of an impacted specimen is defined as the slope of the force-deflection curve so that

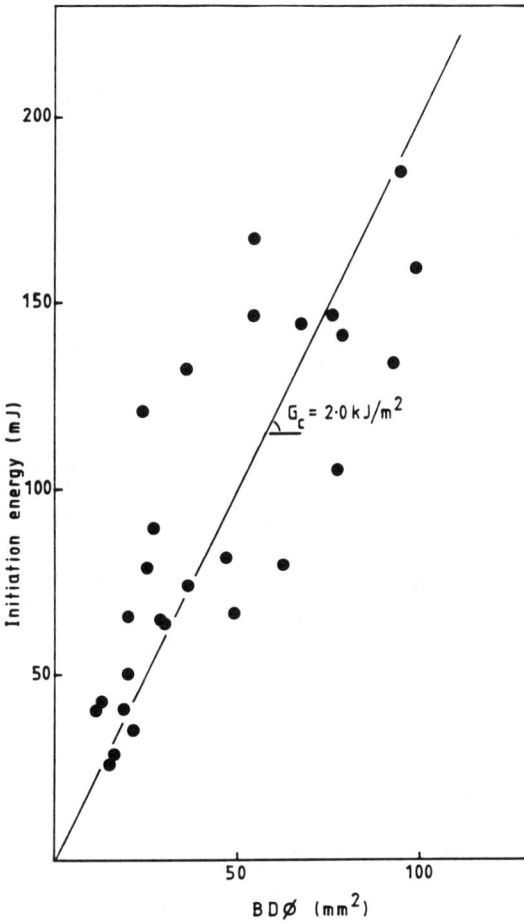

FIG. 7—*Initiation impact fracture energy as a function of* BDϕ.

$$C = \frac{\Delta}{P}$$

It can also be shown that the elastic modulus, E, is related to the compliance so that

$$E = \frac{9}{2} \cdot \frac{L^2}{BD^2} \cdot \frac{[\phi Y^2 x]}{C} \qquad (2)$$

where $x = a/D$, the crack length to specimen depth ratio. The modulus has

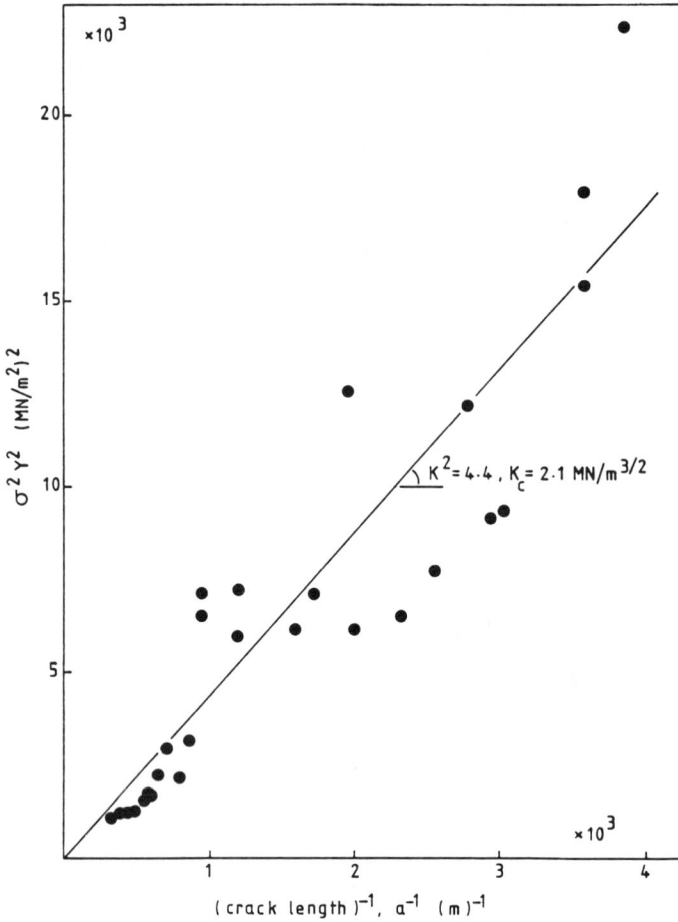

FIG. 8—$\sigma^2 Y^2$ as a function of a^{-1}.

been calculated in this way for each specimen, and the average value is found to be 1.84 GN/m^2, with scatter in the range of 0.69 to 4.05 GN/m^2.

A further means of quantifying E is that offered by the relationship [6] between K_c, G_c, and E shown in Eq 3.

$$K_c^2 = EG_c \tag{3}$$

K_c and G_c have already been determined independently for each specimen. When these data are used in Eq 3 to calculate the plane stress E, the average value is 1.93 GN/m^2, with a distribution for individual specimens of 0.83 to 3.62 GN/m^2. While the variation in modulus values determined in this way is fairly wide, the average value is very close to that obtained using Eq 2 (1.84 GN/m^2). A plane-strain modulus may also be determined by the introduction of Poisson's ratio into Eq 3, resulting in a value of 1.69 GN/m^2. Plane-stress and plane-strain moduli using the values for K_c and G_c determined by linear regression are 2.21 and 1.94 GN/m^2, respectively. Data from slow rate tests [11] and the use of the ASTM criterion for minimum thickness [ASTM Test for Plane-Strain Fracture Toughness of Metallic Materials (E 399-83)], suggest that plane-strain conditions prevail. Substitution of the impact rate K_c and yield stress[5] confirms this condition.

Conclusions

Characterization of the impact behavior of materials by velocity measurement using laser-Doppler is certainly technically feasible. While period-to-period variations in the Doppler signal currently necessitate a smoothing routine, to facilitate differentiation of the velocity-time curve, improvements in the system should allow analysis without smoothing. Such devices as preferential amplitude enhancement and frequency to voltage conversion, coupled with modifications to the light-scattering surface, are currently under consideration and evaluation. With or without improvement, the system has considerable potential where other methods cannot cope, such as at high striker velocities.

Differentiation and integration of the velocity-time curves allows determination of applied force and deflection throughout the impacting period. Subsequent integration of the force-deflection curves gives the energy absorbed. Comparisons between the total energy absorptions measured by the pendulum and those recorded from force-deflection traces show a good correlation and produce similar strain energy release rates, G_c, of the order 10 kJ/m^2. The G_c inferred here does, however, include not only crack initiation energy but also the energy necessary to propagate the crack in this relatively ductile material, plus any energy required to deform the specimen plastically. A critical value of G_c can be determined by considering the energy absorbed up to the point of crack initiation, which is taken as the peak force on the force-deflection curve. While there is a fair degree of scatter on the energy-$BD\phi$ graph, linear regression gives a value for G_c of 2.0 kJ/m^2.

The critical stress intensity factor, K_c, is computed from the maximum applied force, and linear regression through data plotted on the basis of $\sigma^2 Y^2$ versus a^{-1} gives a value of 2.1 $MN/m^{3/2}$.

A compliance approach to determine the elastic modulus produced an average value of 1.84 GN/m^2 for the specimens tested here, whereas the more conventional relationship between K_c, G_c, and E gave an average plane-stress modulus of 1.93 GN/m^2; inclusion of Poisson's ratio effects resulted in a plane-strain modulus of 1.69 GN/m^2.

[5]E. Q. Clutton, BP Chemicals, personal communication, Feb. 1985.

While one of the features of the results and analysis presented here is the degree of scatter from specimen to specimen, the average values obtained do correlate quite well using the fracture mechanics relationships developed to deal with brittle material behavior. Further, it has been shown quite clearly that for materials that are not brittle and using devices that record only gross energy absorbed, the energy release rate measured is well in excess of the initiation value. These high G_c values are largely due to plane stress and propagation effects, which can be allowed for up to a point. However, with instrumentation to identify the force involved, these effects are dealt with more readily and confidently.

Acknowledgments

The authors wish to acknowledge financial support from the Polymer Engineering Directorate. Also, John Hodgkinson wishes to thank Dr. P. Prentice for his interest and practical assistance.

References

[1] Durst, F., Melling, A., and Whitelaw, J. H., *Principles and Practice of Laser-Doppler Anemometry*, Academic Press, London, 1976.

[2] Hodgkinson, J. M., Vlachos, N. S., Whitelaw, J. H., and Williams, J. G., *Proceedings of the Royal Society*, London, Vol. A379, 1982, pp. 133–144.

[3] Hodgkinson, J. M. and Williams, J. G., *Physics in Technology*, Vol. 13, No. 4, 1982, pp. 152–158.

[4] Hodgkinson, J. M. and Williams, J. G., *Proceedings of the International Union of Theoretical and Applied Mechanics Symposium on Optical Methods in the Dynamics of Fluids and Solids*, Springer-Verlag, Berlin, 1985.

[5] Plati, E. and Williams, J. G., *Polymer Engineering and Science*, Vol. 15, No. 6, 1975, pp. 470–477.

[6] Williams, J. G., *Fracture Mechanics of Polymers*, Ellis Horwood, Chichester, England, 1984.

[7] Birch, M. W. and Williams, J. G., *International Journal of Fracture*, Vol. 14, 1978, pp. 69–84.

[8] Hodgkinson, J. M. and Williams, J. G., *Institute of Physics Conference Series*, No. 47, 1979, pp. 233–241.

[9] Williams, J. G. and Hodgkinson, J. M., *Proceedings of the Royal Society*, London, Vol. A375, 1981, pp. 231–248.

[10] Brown, W. F. and Srawley, J. E., *Current Status of Plane Strain Crack Toughness Testing of High Strength Metallic Materials, ASTM STP 410*, American Society for Testing and Materials, Philadelphia, 1966.

[11] Chan, M. K. V. and Williams, J. G., *Polymer Engineering and Science*, Vol. 21, No. 15, 1981, pp. 1019–1026.

Roger D. Goolsby[1] and Chunliang Lin[1]

Effect of Loading Rate on the Impact Fracture Toughness of Acetal and Polymethyl Methacrylate

REFERENCE: Goolsby, R. D. and Lin, C., **"Effect of Loading Rate on the Impact Fracture Toughness of Acetal and Polymethyl Methacrylate,"** *Instrumented Impact Testing of Plastics and Composite Materials, ASTM STP 936,* S. L. Kessler, G. C. Adams, S. B. Driscoll, and D. R. Ireland, Eds., American Society for Testing and Materials, Philadelphia, 1987, pp. 351–367.

ABSTRACT: This investigation is concerned with evaluating the suitability of fracture mechanics testing and analysis methods for characterizing the impact fracture resistance of polymers. The polymers characterized in this study were acetal and polymethyl methacrylate. A precracked three-point bend specimen configuration was selected for fracture characterizations, with tests being conducted at rates from 2.12×10^{-5} to 1.06 m/s. Tests at each loading rate were conducted with specimens having three different crack-length-to-specimen-width ratios, a/W. Specimens were instrumented with a clip gage at the notch opening, and a potential drop technique was also employed during testing to delineate critical crack instability. All the tests were performed on an Instron servohydraulic instrumented impact test system specifically designed and instrumented for impact testing of polymers. Plane-strain fractures were observed for all the specimens tested, and linear elastic fracture mechanics methods were employed for data analyses. Plane-strain fracture toughness was observed to be relatively constant with increasing a/W ratio for both polymers, except for those specimens tested at the highest loading rate. It was also found that fracture toughness was significantly influenced by loading rate, with transitions in toughness occurring for both polymers at particular rates of loading. The authors postulate that these transitions are associated with molecular relaxation mechanisms and a change of thermal state at the crack tip.

KEY WORDS: impact testing, dynamic tests, instrumented impact tests, fracture toughness, polymers

With the ever-growing use of polymeric materials in structures subjected to dynamic impact loading, it has become increasingly important to characterize and understand the fracture resistance of these materials properly under

[1]Associate professor and graduate student, respectively, Mechanical Engineering Department, University of Texas at Arlington, Arlington, TX 76019.

these severe conditions. One of the most attractive and promising approaches for dealing with this problem involves a fracture mechanics evaluation. Historically, a linear elastic fracture mechanics (LEFM) methodology has proven very successful in describing and predicting the failure of metals. However, its applicability to polymeric materials, which are viscoelastic and often exhibit considerable plasticity, has been limited. Although use of an LEFM approach has been less successful in general for polymers, its use for treating fracture of polymers under dynamic loading appears more likely to be appropriate, since large-scale plasticity is not usually observed at these higher rates.

In recent years several attempts have been made to extend LEFM into the polymer regime for impact fracture toughness measurement. Many of these efforts have been confined to somewhat lower loading rate analyses, since inertial effects occurring at very high rates complicate the evaluation of the applicability of LEFM. In general, load/time or load/clip gage displacement records are often used to ascertain the applicability of LEFM. However, the observed failure mode (plane strain versus plane stress) is also essential in this evaluation. Additionally, if LEFM is applicable, the fracture toughness determined experimentally should be independent of specimen size and crack size. Thus, determining the plane-strain fracture toughness, K_{Ic}, versus crack-length-to-specimen-width ratio, a/W, for a given specimen geometry is another means of evaluating the appropriateness of an LEFM approach. Although several works have been conducted to evaluate the fracture toughness over a range of crack-length-to-width ratios [1-7] or loading rates [8-10] for polymeric materials, little has been done to study the fracture mechanics approach systematically under variations of both test parameters [11]. In the present investigation these two parameters were evaluated concurrently.

Two polymeric materials were used in this study, acetal (polyoxymethylene) and polymethyl methacrylate (PMMA). These materials were chosen because both are used in structural applications in which dynamic loading occurs. Additionally, acetal and PMMA have significantly different molecular structures, with acetal being semicrystalline and PMMA being amorphous. The instrumented impact fracture toughness of these polymers was evaluated over a wide range of loading rates (2.12×10^{-5} to 1.06 m/s) and crack-length-to-width ratios (from about 0.15 to 0.50). The primary objective of this study was to assess the validity of fracture mechanics methodology for each material and loading condition. The methods used to accomplish this objective included (1) interpretation of load/time and load/clip gage displacement records, (2) evaluation of the effects of loading rate and crack-length-to-width ratio on fracture toughness, and (3) determination of the fracture modes under various loading conditions.

The variations in fracture toughness of acetal and PMMA are presented as a function of loading rate and crack-length-to-width ratio. It is proposed that the variations of fracture toughness with loading rate are associated with molecular relaxation mechanisms, a change in the crack-tip thermal condition, and morphological changes on the specimen fracture surfaces.

Experimental Procedure

The materials characterized in this investigation were Delrin 500 (acetal), manufactured by Du Pont Engineering Plastics Corp., and Acrylite FF (PMMA), supplied by Cyro Industries. Both materials were obtained in 6.3-mm-thick extruded sheets. Fracture characterizations were made using edge-notched, three-point bend fracture toughness specimens. The specimen size was 12.7 by 2.54 by 0.63 cm. Specimens were cut from the extruded sheets so that the lengthwise dimension of the specimen was parallel to the extrusion direction. Each specimen had a 0.16-cm-wide edge notch machined into it at midlength. These notches were machined to lengths of 3.8, 7.6, and 10.2 mm. Three specimen replicates were prepared for each notch length.

Prior to fracture testing, the notch in each specimen was sharpened by the following technique. Each specimen was cooled in liquid nitrogen, then removed from the coolant, and a sharp crack was introduced by lightly tapping a razor blade into the notch tip. This resulted in a sharp crack approximately 1.3 mm long at the end of each machined notch. This procedure resulted in specimens with nominal a/W ratios of 0.2, 0.35, and 0.45, where crack length, a, includes notch length plus the sharped notch tip crack.

Subsequent to the cracking of the specimens, two 2.0-mm-thick aluminum tabs were glued on the specimen edge on either side of the notch opening. These tabs served as attachment sites for a specially designed clip gage similar to that described in the ASTM Test for Plane-Strain Fracture Toughness of Metallic Materials (E 399-83). Additionally, conductive silver paint was applied to the specimen just ahead of the crack tip. This paint strip was used as a part of a direct-current (D-C) electric circuit to determine when the crack began to propagate. A schematic diagram illustrating this potential drop circuit technique and clip gage arrangement is given in Fig. 1.

Fracture toughness testing of the precracked edge-notched specimens was conducted at room temperature using three-point bend loading. A support span of 10.2 cm was used, resulting in a support-span-to-specimen width ratio of 4. Tests were run on an Instron servohydraulic test system specifically configured for instrumented impact testing. This included a modified actuator, oversized servovalue, quartz load transducer, and transient recorder (Nicolet digital oscilloscope). The tests were conducted under stroke control at five test rates (0.127, 12.7, 127, 1270, and 6350 cm/min). For all tests except those at the 6350-cm/min rate, the load and clip gage displacement were recorded on the transient recorder. For the highest test rate, the load and voltage across the conductive paint strip were recorded.

Fracture toughness values were calculated from the load records using the following relation from ASTM Method E 399-83:

$$K_{\text{Ic}} = \left[\frac{P_m S}{B W^{3/2}} \right] f\left(\frac{a}{W} \right)$$

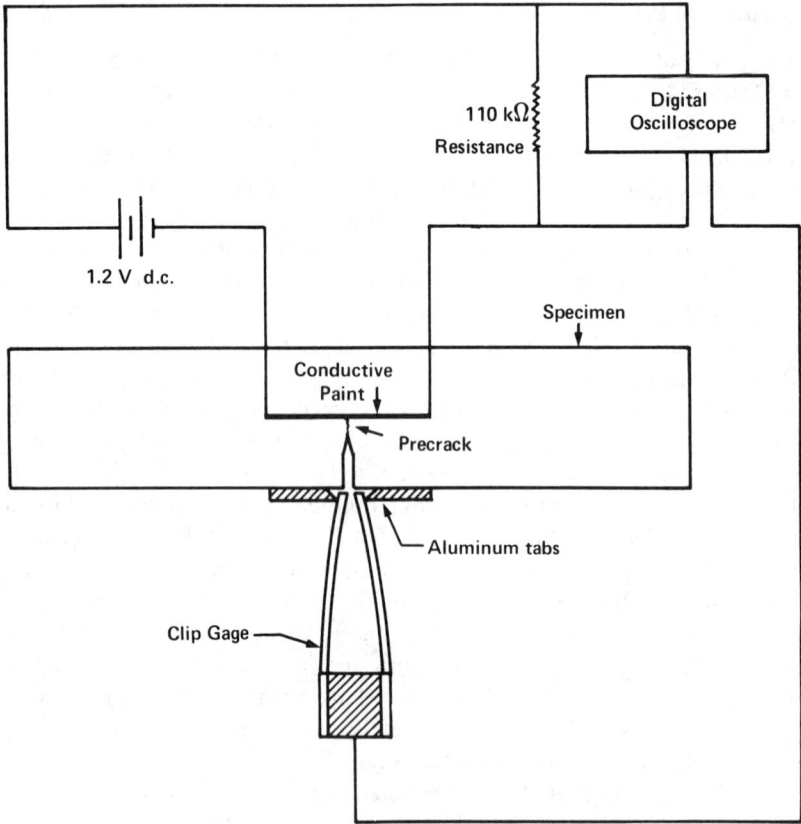

FIG. 1—*Three-point bend fracture specimen, with conductive paint D-C circuit and clip gage used for detecting crack instability.*

where

$$f\left(\frac{a}{W}\right) = \left[\frac{3\left(\dfrac{a}{W}\right)^{1/2}}{2\left(1 + \dfrac{2a}{W}\right)\left(1 - \dfrac{a}{W}\right)^{3/2}}\right.$$

$$\left. \cdot \left[1.99 - \left(\frac{a}{W}\right)\left(1 - \frac{a}{W}\right)\left(2.15 - 3.93\frac{a}{W} + 2.7\frac{a^2}{W^2}\right)\right]\right.$$

and

K_{Ic} = plane-strain fracture toughness,

S = specimen support span,
B = specimen thickness,
P_m = maximum load,
W = specimen width, and
a = crack length.

The crack lengths were measured on the broken specimens with a traveling stage optical microscope at the specimen surfaces and at midthickness. An average of the three values was used in calculating fracture toughness.

The fracture surfaces of failed specimens were studied by using a JOEL JSM-35 scanning electron microscope (SEM) and low-power optical microscopes. The SEM analysis was performed on gold-plated surfaces at 12 kV to avoid damage to the polymer from electron beam scanning.

Results and Discussion

Loading Rate Effects

Tests performed at rates of 2.12×10^{-5} to 0.212 m/s were instrumented with a clip gage, and the test records consisted of the load and clip gage displacement versus time. For all but the 0.212-m/s tests, these records exhibited the linear load versus clip gage displacement to failure for both material types. Thus, the maximum load was used in all calculations for fracture toughness. However, test records for the 0.212-m/s rate exhibited inertial loading effects, as evidenced by oscillations in the recorded signals, as shown in Fig. 2 for acetal. It was observed that several successive oscillations of load signal occurred before fracture, but the amplitude of the oscillations decreased with time and damped out before fracture occurred. As for the lower-rate tests, the maximum load was used to calculate fracture toughness.

The oscillations that become apparent at the 0.212-m/s test rate are characteristic of high-rate tests, and these oscillations become extremely pronounced at the 1.06-m/s test rate, the highest rate used in this program. It was also found that the clip gage performed unsatisfactorily at this rate, so the conductive paint potential drop measurement was used for these tests to provide a definitive indication of crack instability necessary for interpretation of the load/time record. A typical load record for the 1.06-m/s test rate is shown in Fig. 3. The load peak corresponding to the sudden change in potential across the conductive paint strip was used in fracture toughness calculations for these high-rate tests. This peak load was used directly in calculations without any modification for inertial effects.

One of the main goals of this research was to document the variation of fracture toughness with loading rate for acetal and PMMA. Figures 4 and 5 (for acetal) and 6 (for PMMA) are representative plots of how fracture toughness varied with loading rate for approximately constant a/W specimens.

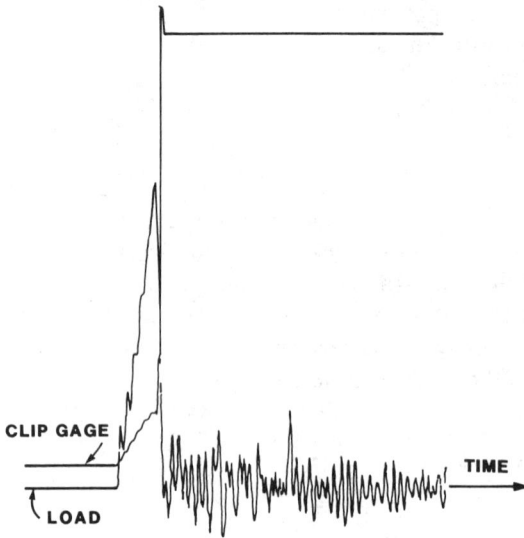

FIG. 2—*Typical load/time and clip gage displacement/time records for acetal specimen tested at 0.212 m/s.*

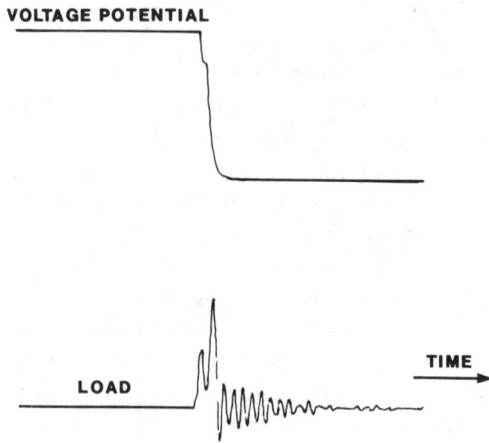

FIG. 3—*Typical load/time and conductive paint voltage potential/time records for acetal specimen tested at 1.06 m/s.*

The toughness values were found to be slightly sensitive to specimen a/W, particularly at the highest testing rate, an observation that will be discussed subsequently. The significant features of these curves are the two transitions in toughness that occur for each polymer with increasing loading rate. For acetal, one transition occurs between impacts rates of 2.12×10^{-5} and 2.12

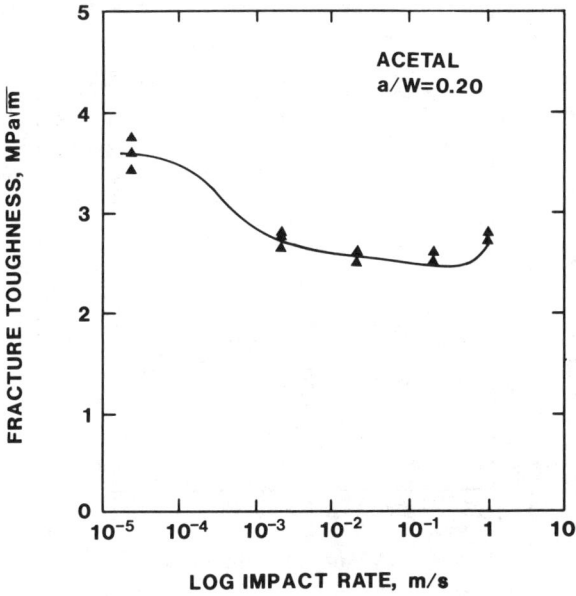

FIG. 4—*Fracture toughness versus logarithm impact rate for acetal specimens having a nominal crack-length-to-width ratio of 0.20.*

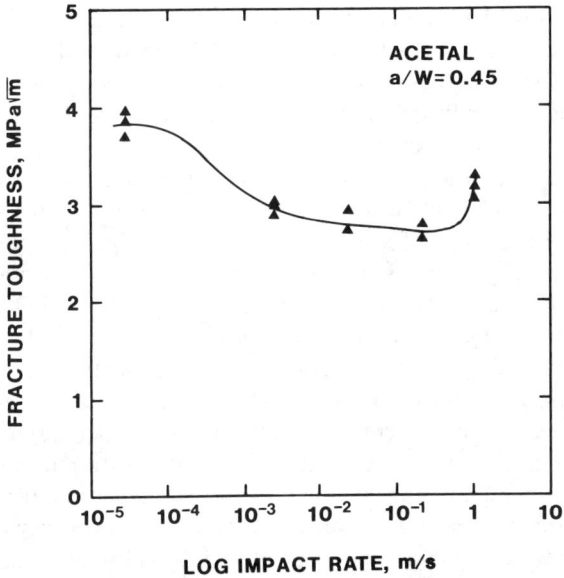

FIG. 5—*Fracture toughness versus logarithm impact rate for acetal specimens having a nominal crack-length-to-width ratio of 0.45.*

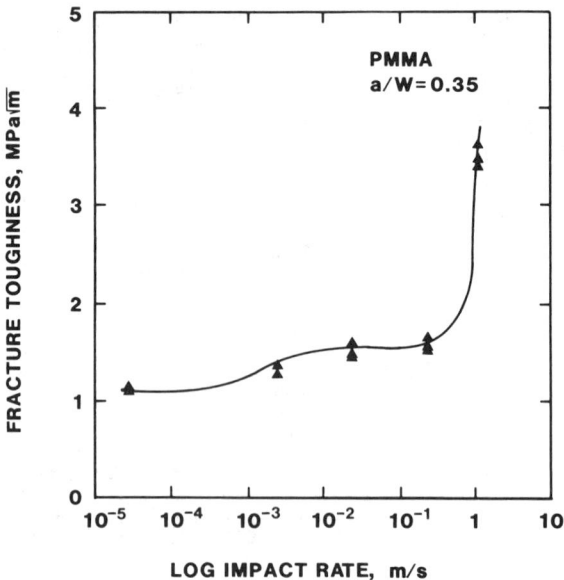

FIG. 6—*Fracture toughness versus logarithm impact rate for polymethyl methacrylate specimens having a nominal crack-length-to-width ratio of 0.35.*

$\times$ 10^{-3} m/s, while a second transition occurs at the highest impact rate of 1.06 m/s. For PMMA the first transition occurs between rates of 2.12×10^{-5} and 2.12×10^{-2} m/s, while the second is exhibited at the 1.06 m/s rate. The authors postulate that these transitions are associated with molecular relaxation mechanisms (for PMMA) and with a change in the thermal state at the crack tip (for both polymers).

Molecular relaxation phenomena in polymers have long been associated with mechanical behavior, especially impact properties. PMMA, in particular, has been the subject of numerous studies [12–17] of the interactions between mechanical properties, temperature, loading rate, dynamic mechanical tests, dielectric tests, and the molecular relaxations that occur in the polymer. These previous studies have identified two principal transitions peculiar to PMMA. One is the β transition, which is usually associated with the rotation of main chain segments (two to four repeats units) about the chain's longitudinal axis or movement of the ester groups. The second transition is the γ relaxation, and it is believed to be associated with movement of the α-methyl groups pendant from the main chain. By using time-temperature superposition and comparing the frequencies of toughness transitions of this study with previous dynamic mechanical and dielectric loss test results [16], good correlations between these results were obtained. It was found that the frequency of the first toughness transition for PMMA related directly to dynamic mechanical loss maxima at 1 Hz or dielectric loss maxima at 10 Hz at 25°C, which had been attributed to the β process. Similarly, the frequency of the second toughness transition was found to agree with dielectric loss data at 10^6 Hz and 25°C, which had been attributed to the γ relaxation process.

It should be noted that the fracture toughness transition occurring at the highest loading rate of 1.06 m/s can also be correlated to phenomena other than molecular relaxation transitions. Specifically, another significant factor contributing to the sudden increase in fracture toughness at this rate (for both PMMA and acetal) is a change in the thermal condition at the crack tip. It is believed that at the lower test rates crack movement is operating under essentially isothermal conditions, with no significant temperature rise occurring at the crack tip. However, at the very high impact rate, crack velocity is extremely rapid and the crack-tip region is operating under an adiabatic condition, with a significant temperature rise likely. The higher temperature results in a softening of the polymer locally with a resultant increase in toughness. This effect has been documented previously on PMMA [10,18,19] as well as other polymers [20]. In particular, Williams and Hodgkinson [10] have attributed increases in toughness with increasing loading rate to a temperature rise at the crack tip resulting in crack-tip blunting.

While considerable effort has been given to the study of mechanical behavior and molecular relaxation mechanisms for PMMA, few studies have been made on acetal. It has been shown [7] that secondary molecular transitions in acetal can be related to transitions in impact resistance as a function of temperature. That study of shear modulus and damping attributed an increase in impact resistance at −70°C to the γ transition, which was related to movements within the main chain. Comparing this with the present investigation, it is doubtful that either toughness transition in acetal can be related to this molecular mechanism, since the frequencies and temperatures involved do not relate to the observations made in Ref 17. As noted earlier, the increase in toughness that occurs at the 1.06-m/s rate is probably due to a local temperature rise at the crack tip. On the other hand, the decrease in toughness that occurs for acetal at the lower test rate is probably traceable to loss mechanisms unique to semicrystalline polymers, perhaps entailing a complex mechanism involving the crystallites.

Specimen Crack Length Effects

Another important aspect of this investigation concerned the determination of specimen geometry effects on fracture toughness. Specifically, toughness was measured at all loading rates employing a range of crack sizes. If conventional fracture mechanics procedures are applicable to these materials, K_{Ic} should be constant with changes in a/W. Figure 7 shows K_{Ic} versus a/W for the 0.212-m/s tests of acetal, which is typical of the acetal tests. As can be seen from this figure, there is a slight increase in K_{Ic} with increasing

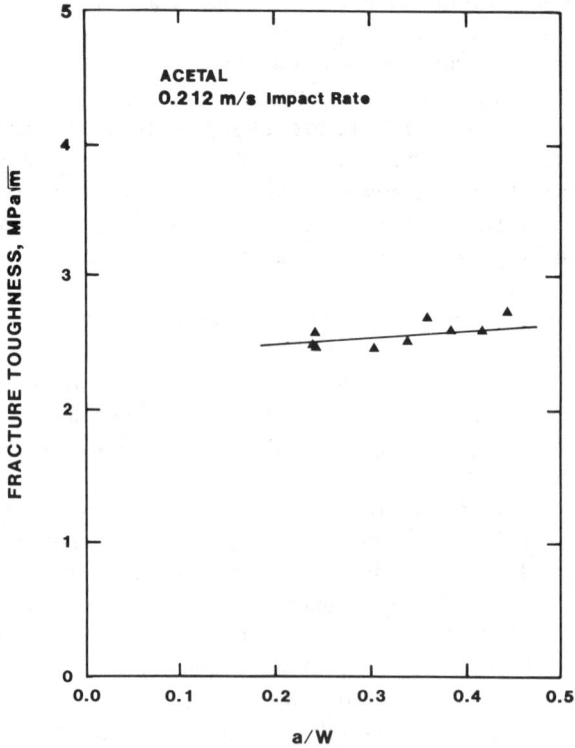

FIG. 7—Fracture toughness versus specimen crack-length-to-width ratio, a/W, for acetal specimens tested at 0.212 m/s.

a/W. With the exception of the test data obtained at the highest test rate, all acetal data followed this trend. For the 1.06 m/s test the increase of K_{Ic} with a/W was more pronounced. Figure 8 shows the curve fit lines for all test rates conducted on acetal. Data for PMMA, on the other hand, showed that K_{Ic} was constant with increasing a/W, except for data obtained at the 1.06 m/s rate. Figure 9 shows typical K_{Ic} versus a/W data for PMMA. As for the case of acetal, K_{Ic} was observed to increase with a/W at the highest test rate. Figure 10 shows the curve fit lines for all the test rates conducted on PMMA.

In evaluating the slight sensitivity of K_{Ic} with a/W observed for acetal at the lower loading rates, it is likely that this small increase in K_{Ic} with a/W is attributable to the local increase in strain rate at the crack tip accompanying higher a/W ratio tests and a strain rate sensitivity of acetal. As for the larger increase in K_{Ic} with increasing a/W observed for the 1.06-m/s tests for both PMMA and acetal, this is attributed to an increasing significance of specimen

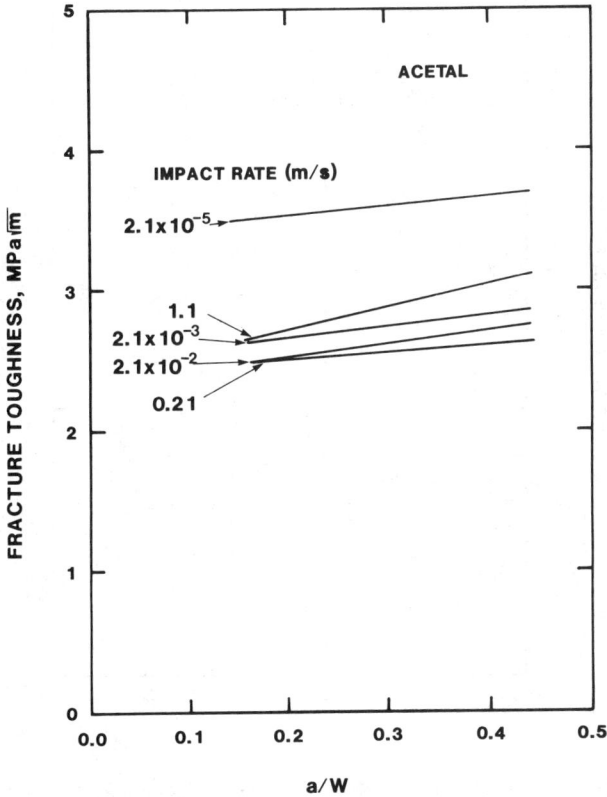

FIG. 8— *Fracture toughness versus specimen crack-length-to-width ratio,* a/W, *for acetal specimens tested at various loading rates.*

inertial effects. Others [2,5–7] have noted a dependence of fracture toughness on crack size for high-rate tests, unless corrections are made to the data. However, test procedures associated with these studies involve energy measurements rather than load measurements. Investigators who have used instrumented impact techniques where loads are measured [21–26] have discussed the difficulties in interpreting load records where specimen inertial effects are appreciable. Corrections to data and indirect derivations to ascertain failure loads are suggested [21–23], along with recommendations [24–26] for testing at lower rates. In the present study, no special analytical or experimental techniques were applied to the 1.06-m/s data. However, the variance of K_{Ic} with a/W shows that further investigation of this phenomenon is warranted.

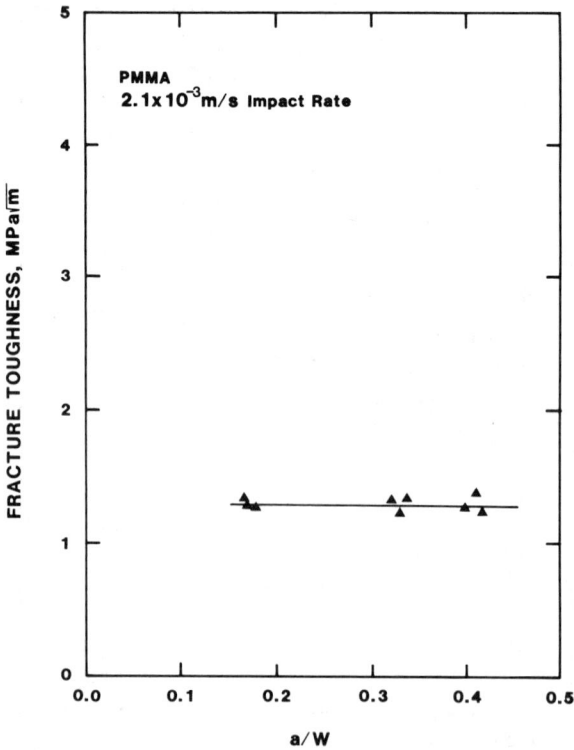

FIG. 9—*Fracture toughness versus specimen crack-length-to-width ratio,* a/W, *for poly-methyl methacrylate specimens tested at 2.12 × 10⁻³ m/s.*

Fracture Morphology

All specimens tested in this program exhibited a flat fracture surface typical of a plane-strain failure mode. A transspherulitic failure mode was observed for all acetal specimens. A typical SEM fractograph of acetal is shown in Fig. 11. It was noted that acetal specimens tested at the 2.12×10^{-5}-m/s rate exhibited a somewhat more uneven fracture surface than specimens tested at higher rates. This was evidenced by the transspherulitic fracture being distributed at slightly different levels on the crack propagation plane. The PMMA specimens exhibited an extremely flat surface, with characteristic parabolic features from crazes intersecting the main fracture plane [27–29]. A typical SEM fractograph illustrating this feature is shown in Fig. 12. These parabolic markings were not observed on specimens tested at the lowest test rate.

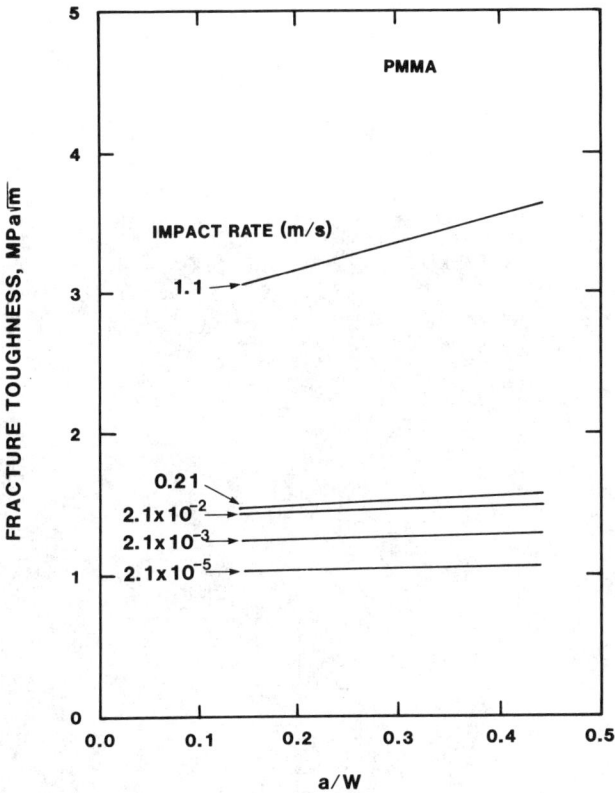

FIG. 10—*Fracture toughness versus specimen crack-length-to-width ratio*, a/W, *for poly-methyl methacrylate specimens tested at various loading rates.*

Summary and Conclusions

Instrumented impact fracture toughness measurements have been conducted on acetal and polymethyl methacrylate as a function of loading rate and specimen crack-length-to-width ratio. Based on these measurements, the following observations have been made:

1. Plane-strain failure was observed for all the specimens tested. Linear load/clip gage displacement and load/time records were observed for all the tests, except those at the highest test rate, in which oscillatory behavior was observed. This oscillatory behavior was attributed to specimen inertia effects.

2. Transitions in fracture toughness versus loading rate were observed for both polymers. The authors postulate that these transitions were related to

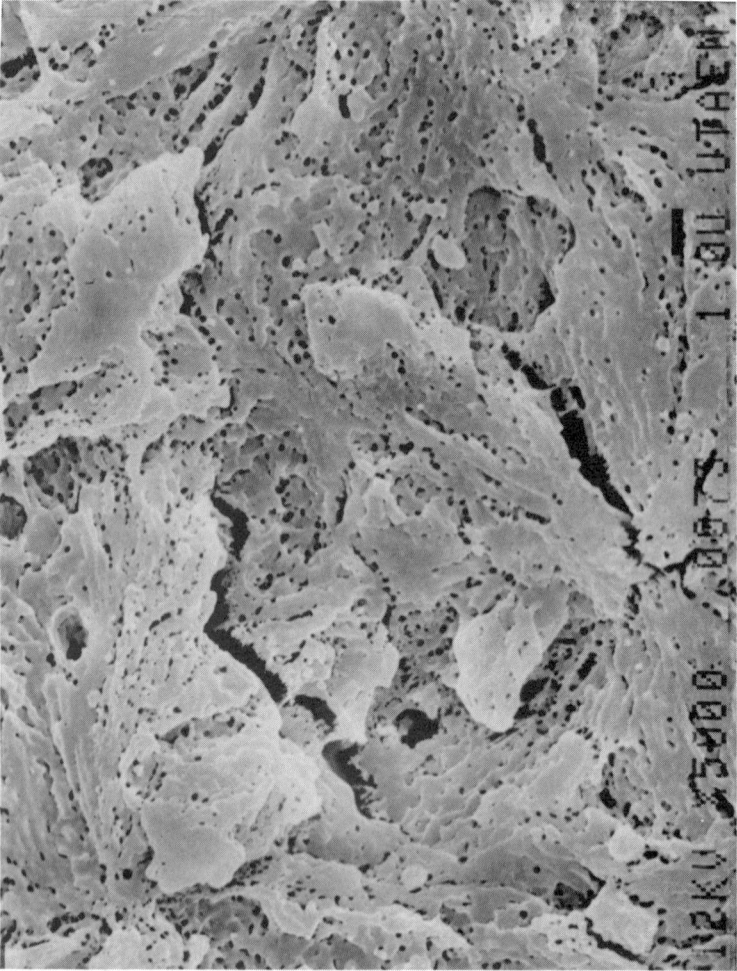

FIG. 11—Scanning electron fractograph of an acetal specimen impacted at the 2.12×10^{-2} m/s test rate, revealing transspherulitic fracture (approximate magnification, $\times 5000$).

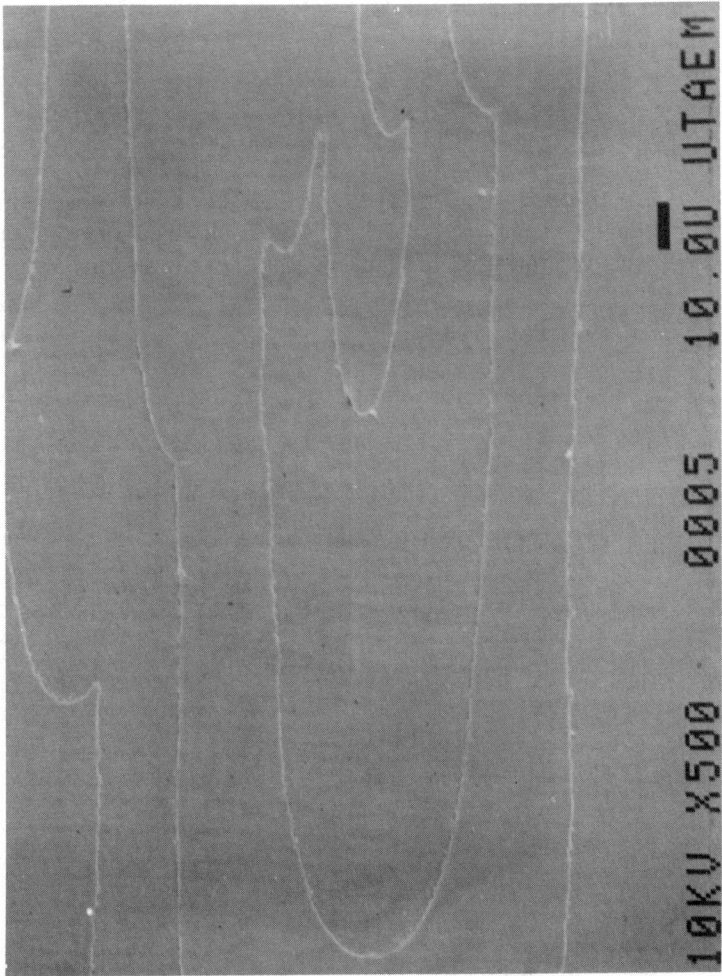

FIG. 12—Scanning electron fractograph of a polymethyl methacrylate specimen impacted at the 0.212-m/s test rate, revealing parabolic craze markings (approximate magnification. ×500).

molecular relaxation mechanisms (for PMMA) and a change in the thermal state at the crack tip (for both polymers).

3. It was found that fracture toughness was approximately constant with variation in specimen crack-length-to-width ratio, except for specimens tested at the highest loading rate, where K_{Ic} was observed to increase with a/W. This trend at the high rate was attributed to increased specimen inertia effects.

4. Scanning electron fractography showed acetal failed by a transspherulitic mode, while PMMA exhibited parabolic markings typical of PMMA impact failures.

Acknowledgments

All the tests performed in this program were conducted on a special high-rate servohydraulic system donated to the University of Texas at Arlington by Instron Corp. The authors wish to thank Instron for this essential support.

References

[1] Nikpur, K. and Williams, J. G., *Plastics and Rubber: Materials and Applications*, Vol. 3, No. 4, 1978, p. 163.

[2] Marshall, G. P., Williams, J. G., and Turner, C. E., *Journal of Materials Science*, Vol. 8, 1973, p. 949.

[3] Newmann, L. V. and Williams, J. G., *Polymer Engineering and Science*, Vol. 11, 1978, p. 893.

[4] Kisbenyi, M., Birch, M. W., Hodgkinson, J. M., and Williams, J. G., *Polymer*, Vol. 20, 1979, p. 1289.

[5] Birch, M. W. and Williams, J. G., *International Journal of Fracture*, Vol. 14, 1978, p. 69.

[6] Plati, E. and Williams, J. G., *Polymer Engineering and Science*, Vol. 15, 1975, p. 470.

[7] Williams, J. G. and Birch, M. W. in *Fracture 1977, Proceedings of the Fourth International Conference on Fracture*, Vol. 1, D. M. R. Taplin, Ed., University of Waterloo Press, Ontario, Canada, 1977, p. 501.

[8] Glover, A. P., Mucci, P. D. R., and Radon, J. C., *Engineering Fracture Mechanics*, Vol. 2, 1970, p. 165.

[9] Kobayashi, A. S. and Chan, C. F., *Experimental Mechanics*, Vol. 16, 1976, p. 176.

[10] Williams, J. G. and Hodgkinson, *Proceedings of the Royal Society (London)*, Series A, Vol. 375, 1981, p. 231.

[11] Goolsby, R. D. and Miller, J. M., *Proceedings, 42nd Annual Technical Conference*, Society of Plastics Engineers, Brookfield, CT, 1984, p. 561.

[12] Holt, D. L., *Journal of Applied Polymer Science*, Vol. 12, 1968, p. 1653.

[13] Broutman, L. J. and Kobayashi, J. in *Proceedings, International Conference on Dynamic Crack Propagation*, G. C. Sih, Ed., Noordhoff, Leyden, 1973, p. 215.

[14] Maxwell, B. and Harrington, J. P., *Transactions of the American Society of Mechanical Engineers*, Vol. 74, 1952, p. 579.

[15] Boyer, R. F., *Polymer Engineering and Science*, Vol. 8, 1968, p. 161.

[16] McCrum, N. G., Read, B. E., and Williams, G., *Anelastic and Dielectric Effects in Polymeric Solids*, Wiley, New York, 1967.

[17] Heijboer, J., *Journal of Polymer Science*, Vol. C16, 1968, p. 3755.

[18] Kambour, R. P. and Barker, R. E., *Journal of Polymer Science*, Vol. A-2, No. 4, 1966, p. 359.

[19] Radon, J. C. and Fitzpatrick, N. P., *Journal of Engineering Materials, Components, and Design*, Vol. 13, No. 9, 1970, p. 1125.

[20] Doll, W., *Engineering Fracture Mechanics*, Vol. 5, 1973, p. 259.
[21] Radon, J. C. and Turner, C. E., *Engineering Fracture Mechanics*, Vol. 1, 1969, p. 411.
[22] Turner, C. E. in *Impact Testing of Metals, ASTM STP 466*, American Society for Testing and Materials, Philadelphia, 1970, p. 93.
[23] Radon, J. C., *Journal of Applied Polymer Science*, Vol. 22, 1978, p. 1569.
[24] Ireland, D. R. in *Instrumented Impact Testing, ASTM STP 563*, American Society for Testing and Materials, Philadelphia, 1974, p. 3.
[25] Saxton, H. J., Ireland, D. J., and Server, W. L. in *Instrumented Impact Testing, ASTM STP 563*, American Society for Testing and Materials, Philadelphia, 1974, p. 30.
[26] Ireland, D. and Aleszka, J., "Dynamic Fracture Toughness Measurements from Instrumented Impact Testing of Precracked Charpy Specimens," Report TR 79-60, Effects Technology, Inc., Santa Barbara, CA, 1979.
[27] Kausch, H. H., *Polymer Fracture*, Springer-Verlag, Berlin, 1978.
[28] Berry, J. P., *Fracture Processes in Polymeric Solids*, Wiley, New York, 1964.
[29] Berry, J. P., *Journal of Applied Physics*, Vol. 33, No. 5, 1962, p. 1741.

Indexes

Author Index

Subject Index

A

Acceleration, 120, 126, 132
Accelerometer, 130, 284–285
Acetal, 351–367
Acrylonitrile-butadiene-styrene
  (ABS), 236–237
  bottles, 19
  ductile-brittle transition, 240–242,
    245–246
  fatigue testing, 282
  impact behavior of, illustration,
    166
  incipient crack formation testing,
    327–333
  terpolymers, 176
Aircraft materials and structure,
  224, 227
Aluminum alloys, 166
  energy absorption, 234
  impact data, 171
  Type 6061 T6, 221
Analog/digital (A/D) converter, 12
Analog/digital (A/D) resolution, 14
Antialiasing filter, 84
Aramid-reinforced composites, 219–
  235
ASTM Committee D-20 on plastics,
  45, 174
ASTM standards
  D 256-81: 19, 147, 162, 284, 285
  D 256-84: 237
  D 638-82a: 262
  D 790-81: 149, 266
  D 1709-15(1980): 148
  D 1822-83: 162
  D 1822-84: 147

D 2585-68(1980): 221
D 3029-82: 164, 165
D 3763-79: 19, 178, 183
E 23-82: 180, 182
E 399-83: 349, 353
Automotive applications
  dart impact evaluation of materi-
    als, 44–57
  impact data for steel and alumi-
    num in, 171
  impact stress states in, 28–30, 31–
    32
  impact testing materials, 24–43
  plastics in, 58, 162

B

Bending
  biaxial, 27, 39, 43, 160
    illustration, 31–32
    table, 28
  normal shear, 39
  uniaxial, 27, 43
    illustration, 31–32
    table, 28
Bismaleimide systems, 253, 259
Boat laminates, 192–197
Bottles, burst pressure, 221, 223,
  226–227, 233, 234
Boxcar function, 84–86, 94
  illustration, 88
Break
  elongation at, 148, 239
  energies, 282
  point, 34–35
  response, 42